D1196788

WETLAND and ENVIRONMENTAL APPLICATIONS of

GIS

Mapping Sciences Series

Editor-in-chief John G. Lyon

Titles include:

Aerial Mapping
Edgar Faulkner

Satellite Remote Sensing of Natural Resources
David L. Verbyla

Wetland and Environmental Applications of GIS
Edited by John G. Lyon and Jack McCarthy

WETLAND and ENVIRONMENTAL APPLICATIONS of GIS

Edited by
John G. Lyon and Jack McCarthy

LEWIS PUBLISHERS

Boca Raton   New York   London   Tokyo

DB# 1279436

Library of Congress Cataloging-in-Publication Data

Wetland and environmental applications of GIS / edited by John G. Lyon, Jack McCarthy.
    p.   cm. — (The mapping sciences series)
  Includes bibliographical references and index.
  ISBN 0-87371-897-6 (alk. paper)
  1. Wetlands—Remote sensing. 2. Geographic information systems. I. Lyon, J. G.
(John G.) II. McCarthy, Jack, 1956– III. Series.
GB622.W47   1995
551.4′57—dc20
                                        95-10772
                                          CIP

© 1995 by CRC Press, Inc.
Lewis Publishers is an imprint of CRC Press

No claim to original U.S. Government works
International Standard Book Number 0-87371-897-6
Library of Congress Card Number 95-10772
Printed in the United States of America 1 2 3 4 5 6 7 8 9 0
Printed on acid-free paper

GB
622
.W47
1995

Series Preface

The Mapping Sciences Series of books was conceived to serve important needs in the profession. Readers require contemporary information of a practical value, with high-quality theoretical detail. They need to know about methods and approaches for their given application. Authors have longed for a coherent book series that addresses the variety of disciplines and applications that come under the umbrella of mapping sciences. They require an outlet for their expertise and a series that can be readily identified by their audience. The combination of these needs and their solution has yielded the Mapping Sciences Series.

John Grimson Lyon, Ph.D.
Editor-in-Chief

Acknowledgment and Dedication

The editors and authors wish to acknowledge the fantastic assistance provided by Ms. Lynne Sterling of Lewis/CRC Publishers. A volume of this complexity could not have been completed without the high level of dedication, coordination, and thoughtful communication Lynne Sterling has supplied. We greatly appreciate all she has done, and dedicate this product to her in recognition of achievement beyond the call of duty.

The Editors

John Grimson Lyon is an Associate Professor of Civil Engineering and Natural Resources and has been a member of the Ohio State University faculty since 1981. His interests in remote sensor and GIS technologies began in 1976 while serving as a guest researcher at the Stanford University Remote Sensing Laboratory and as a Research Associate at the National Aeronautics and Space Administration (NASA) Ames Research Center at Moffett Field, California. These interests were further developed in his undergraduate thesis at Reed College in Portland, Oregon, and during Master's Thesis and Doctoral Dissertation efforts at the School of Natural Resources at the University of Michigan. These three documents focused on remote sensor and GIS technologies for evaluation of riverine wetlands on the lower Columbia River, and Great Lakes coastal wetlands on Lake St. Clair and the Straits of Mackinac.

Lyon's work at OSU has involved teaching, service, and research. This has included a number of efforts at characterizing wetlands remotely, and evaluation and inventory of wetlands using GIS. These and other related work efforts have been conducted for the U.S. Army Corps of Engineers, the National Oceanic and Atmospheric Administration Ohio and Michigan Sea Grant Programs, NASA, and others. Lyon has completed a 3-year term as a visiting scientist working on developing continental-scale data sets of Landsat Multispectral Scanner Scenes for the North American Landscape Characterization (NALC) program of the U.S. Environmental Protection Agency at the Environmental Monitoring and Systems Laboratory in Las Vegas, Nevada. Lyon is author of the Lewis bestseller *Practical Handbook for Wetland Identification and Delineation*, and is Editor-in-Chief for the Lewis book series on the Mapping Sciences.

Jack McCarthy has been the Manager of the ESRI's Regional Office in Boulder, Colorado since 1988. McCarthy acquired his early interests in GIS and remote sensor technologies as an undergraduate at the School of Natural Resources at the University of Michigan. He continued as a graduate student and Research Assistant at University of Michigan, writing a Master's Thesis on uses of small-format aerial photographs for quantifying insect damage to forests. His professional work has included appointments as a research associate concerned with development of GIS and remote sensor technologies at the School of Natural Resources at the University of Vermont and the School of Forestry, Northern Arizona University, Flagstaff, Arizona. He also worked as an applications specialist at the Measuronics Corporation in Great Falls, Montana. He most recently has pursued these interests with ESRI at the Olympia, Washington and Boulder offices with a focus on educating the user on GIS technologies, and on implementing systems to meet the growing needs of the GIS market.

Contributors

Kirt F. Adkins
Department of Civil
 Engineering
Ohio State University
Columbus, Ohio

Steve Bernath
Forest Practices Division
Washington State Department of
 Natural Resources
Olympia, Washington

Paul V. Bolstad
Department of Forestry
Virginia Polytechnic Institute and
 State University
Blacksburg, Virginia

Matthew Brunengo
Pacific Meridian Resources
Emeryville, California

Russell G. Congalton
Department of Natural
 Resources
University of New Hampshire
Durham, New Hampshire

Michael Garrett
Genasys II
Fort Collins, Colorado

Kass Green
Pacific Meridian Resources
Emeryville, California

James Hamlett
Environmental Resources Research
 Institute
College of Agricultural Sciences
Pennsylvania State University
University Park, Pennsylvania

Mason J. Hewitt
U.S. Environmental Protection Agency
Environmental Monitoring Research
 Laboratories
Las Vegas, Nevada

David E. James
Lockheed/Martin Environmental
 Systems & Technologies Company
Las Vegas, Nevada

Gary A. Jeffress
Conrad Blucher Institute for
 Surveying and Science
Texas A&M University
Corpus Christi, Texas

John R. Jensen
Department of Geography
University of South Carolina
Columbia, South Carolina

Wei Ji
U.S. National Biological Service
Southern Science Center
Lafayette, Louisiana

Bruce Kessler
Environmental Systems Research
 Institute (ESRI)
Colville, Washington

Lisa Lackey
Pacific Meridian Resources
Emeryville, California

Christopher Lee
Department of Earth Sciences
California State University–
 Dominguez Hills
Carson, California

Karen H. Lee
Lockheed Environmental Systems &
  Technologies Company
Las Vegas, Nevada

Ross S. Lunetta
U.S. Environmental Protection Agency
Seattle, Washington

John G. Lyon
Department of Civil Engineering
Ohio State University
Columbus, Ohio

Stuart E. Marsh
Arizona Remote Sensing Center
University of Arizona
Tucson, Arizona

Jack McCarthy
Environmental Systems Research
  Institute (ESRI)
Boulder, Colorado

Tawna Mertz
Environmental Resources Research
  Institute
College of Agricultural Sciences
The Pennsylvania State University
University Park, Pennsylvania

Loyd C. Mitchell
U.S. Fish and Wildlife Service
Ecological Services
Bloomington, Indiana

Gary Petersen
Environmental Resources Research
  Institute
College of Agricultural Sciences
The Pennsylvania State University
University Park, Pennsylvania

Elijah W. Ramsey III
National Biological Service
Southern Science Center
Lafayette, Louisiana

Uzair M. Shamsi
Chester Environmental
Pittsburgh, Pennsylvania

James L. Smith
Canal Forest Resources, Inc.
Charlotte, North Carolina

Stuart Smith
Forest Practices Division
Washington State Department of
  Natural Resources
Olympia, Washington

Baxter E. Vieux
School of Civil Engineering and
  Environmental Science
University of Oklahoma
Norman, Oklahoma

Donald A. Waechter
Conrad Blucher Institute for
  Surveying and Science
Texas A&M University
Corpus Christi, Texas

Donald C. Williams
Policy and Long Range
  Planning
North Central Division
U.S. Army Corps of
  Engineers
Chicago, Illinois

Section I: Introduction

Section II: Wetland Applications of GIS and Remote Sensing

Section III: Environmental Engineering Applications of GIS and Remote Sensing

Section IV: Additional Applications and Background

WETLAND and ENVIRONMENTAL APPLICATIONS of GIS

Section I

Introduction

CHAPTER **1**

Introduction to Wetland and Environmental Applications of GIS

John G. Lyon and Jack McCarthy

BACKGROUND

The goals and objectives of this book are to present current capabilities of spatial analysis and modeling using GIS and related technologies. These capabilities will be described and demonstrated using applications from the water resource science and engineering arenas. A combination of papers of general applications and specific applications provides the reader with an understanding of "state of the art" technologies. It also provides examples of methodologies that readers can apply to their own areas of endeavor.

The advances in computers, modeling, and planning techniques in the water resource engineering arena have been impressive. The use of GIS and other technologies has allowed more complex and larger projects to be attempted. Results of these efforts have been important in advancing knowledge within these disciplines. The lessons have also been valuable in developing experience with these tools, and for allowing researchers to assess spatial and temporal characteristics of water resources.

The following sections and chapters provide background material and examples of "state of the art" applications. These chapters were selected because their applications are important to water resource scientists and engineers. In particular, surface hydrology and related applications are emphasized. This is due to a variety of reasons, including the landscape scale and spatial distribution of the variables studied, and due to the high quality of results achieved by scientists and engineers in this area.

The materials presented here will allow the study of these technologies and applications in hopes of sparking interest in the reader. The efforts will demonstrate the utility of applying these techniques to the reader's research efforts. The user need only identify the utility of a given approach, and then work to adopt it within the existing suite of protocols currently applied to such applications or studies of water resources.

0-87371-897-6/95/$0.00+$.50
© 1995 by CRC Press, Inc.

3

GIS AND RELATED TECHNOLOGIES

In science and engineering, there are a number of roles that can be served by GIS technologies. As with any class of technologies, there are a variety of ways to employ the tools. The goal, of course, is to be innovative with the application of the tools. Therein lies the challenge and the reward for successful work.

At the lowest level of effort, the GIS and GIS data can be used to supply inventory information. The presence or absence of given land cover or water classes, or change in these variables on a spatial basis, can be valuable in planning and management. The variety or quantity of certain land cover or water types can be summarized by a certain watershed to produce statistics of interest. The capability to store and quantify data on a spatial basis is an inherent characteristic of GIS technologies.

At a highest level of technology, GIS can provide a spatial database of information to support modeling of phenomena. The GIS supplies the spatial data in a form that can be input to deterministic or statistical models. The spatial power of the GIS database is used in full by the model, and more detailed and spatially averaged results are produced.

This represents a high level of integration and achievement, that is now seen in the industry. It has taken awhile for such applications to develop, however. This is due to the absence of spatially integrated model for water resource phenomena. Many models use spatial data but average or summarize these data by watershed and/or subwatershed, and thereby lose much of the detail of spatial variability that often influences phenomena. This is the same level detail necessary to provide high quality model simulations.

GIS technologies can also facilitate the processing of data. Many spatially distributed variables are represented at different scales or resolutions. Likewise, map data often come from different relative scale, and again the scaling or interpolation issue is important. GIS technologies can supply the framework to help broach and solve the scaling question.

This work can be done in such a manner that the scaling or interpolation is based on known characteristics in the GIS database, rather than some approach involving area-averaging or kriging or the like. One can apply the desired solution to scaling within the framework of the GIS, and can potentially make a successful transformation using a potentially accurate and precise spatial basis for the interpolation.

In general, a strength of GIS is that it is possible to process the data sets using any type of numerical analysis procedure. In particular, certain procedures are valuable for data visualization and analysis, including image processing techniques, virtual reality, and simulation modeling. The digital approach to storing and processing spatial or image data is a fantastic boon to these analyses of data, in itself, and the capabilities have yet to be fully realized.

Image processing techniques can assist in the display and presentation of GIS data and results. For example, the use of spatial processing algorithms such as moving average filters can make more apparent the underlying spatial characteristics of data sets. The image can also be magnified to examine detail, or collapsed to present large areas for viewing by researchers.

GIS technologies can facilitate input of data sets to simulation models. This may take the form of input of one variable or "layer" of a GIS, or multiple variable or layer inputs to models. However, we still need to develop or identify models in application areas that are already amenable to the entry of spatially distributed data.

There is great potential here. Models must be formulated in the future to receive spatial data directly from the GIS. Many existing models favored by scientists and engineers have no spatial component in the selection of variables, nor do they have spatially distributed determinations of model results.

Many current and traditional model procedures (e.g., universal soil loss equation and its revised form) use summary information in the form of model coefficients. These types of models can also be improved by GIS or remote sensor technologies by conducting tests so that model coefficients can be further defined or refined through improved measurements of their characteristics. A more "natural" coefficient better defines the behavior of model variables, and allows the modeler to achieve high fidelity between natural systems and their model simulations.

Models often take the deterministic form, where the phenomena being studied are mathematically modeled or simulated using numerical descriptions of the physical, chemical, or biological principles underlying the phenomena of interest. This can result in complex models composed of submodels addressing each phenomenon with weighted contributions to the model result.

Deterministic models take a great deal of effort to simulate natural phenomena, and to present good and realistic results. The success of the model is commonly evaluated using verification procedures on new, independent data sets. Remote sensor data sets are often used to fulfill the requirement for an independent data set to check the model results for verification purposes.

Other modeling efforts involve statistical analyses of phenomena. The statistical approaches evaluate variables or phenomena as to the variability of their behavior. The goals are to test hypotheses, and to develop relational models of empirical origin. If these models are robust, the relationships can often be applied to more systems or locations, and to different times of the year or day, or to another selected variable.

Statistical models can also be applied later to predict other conditions at other places and times if the model represents the behavior of the variable. When statistical models of physical, chemical, or biological systems truly predict variables, users can apply the model to new situations. GIS and remote sensor technologies are helpful for the development and application of statistical models.

In particular, statistical measurements and models of certain variables have been demonstrated on a repetitive basis with existing sensors and technologies. Good applications include: the evaluation of suspended sediments in freshwater and coastal ecosystems, temperature of water bodies, crop residue and tillage practices for evaluation of nonpoint pollution, concentrations of chlorophyll in water and in plants, and a variety of other detailed applications. The use of statistical analyses has proven of great value in water resources studies over large areas.

Another valuable use of GIS and related technologies is the calibration of model coefficients in statistical models. Both statistical and deterministic models often consist of a number of submodel units. Coefficients used in either approach reflect the characteristics of nature, and they will adjust the contribution of variables or submodels to the overall model results.

To optimize the model simulation of natural phenomenon, the coefficients need to reflect the reality of the situation. As a given model begins to approximate nature, its further development often takes the path of improving the quality of coefficients. Many times, a number of experiments will be executed to better measure the level of a coefficient and thus to better have it mimic nature and help supply better model predictions.

In GIS and remote sensor experiments using statistical models can be greatly facilitated by the analysis of these individual coefficients. These analyses are driven to find the "sensitivity" of the overall model result or simulation to a given variable. Sensitivity analyses are part of a good modeling strategy because it is very desirable to understand the contributions of model coefficients to the overall results, and to ensure that each variable and/or submodel contribution is appropriate or similar to that of nature.

As mentioned earlier, the verification of results from model simulations is a useful capability of GIS. On many occasions, deterministic or statistical models are "run" by scientists and engineers. They often have no model inputs from GIS or remote sensor technologies. However, the model results are usually some variable that can be measured or evaluated using GIS and remote sensors. The use of separate results from GIS and/or remote sensor technologies supplies an independent verification of modeling results. A number of verification checks between model results and those of GIS or remote sensors will allow a validation of the model.

GIS and related technologies appear to have several different uses. A particularly valuable use of these tools is in project presentations. The managers or decision makers on any project appreciate the quantitative information derived from GIS analyses. Naturally, they also appreciate the high quality graphical products that support the project, and demonstrate the results of project analyses. Another valuable use is the public presentation of the results of project analyses using GIS. This is appropriate for public meetings and audiences beyond those of the immediate group. Current land and water planning efforts involve a wealth of information. GIS products can help to display the site conditions, and to show some of the variables that must be weighed to make a decision. The GIS displays can also help lead the audience through the same decision-making process, and help lend insight into the results of the planning study.

Specific uses of GIS products for presentation of projects include briefings for management or boards, book or article illustrations, presentations at public meetings, environmental analyses or environmental impact statements, and other efforts at communicating the results of a project.

APPLICATIONS

We have selected a number of applications to illustrate the capabilities of GIS and related technologies. The work is presented in chapter form and is organized into four sections. These contributions focus on applications related to surface water resources. This particular discipline or arena was chosen due to level of topical interest displayed by professionals, the inherent spatial distribution of the resources or problem, and general quality of advances of the work.

The second section is devoted to GIS and remote sensor technological applications related to wetland resources. This is because wetland issues are of great interest, and maintenance of wetlands is of vital concern. Wetlands are distributed across the landscape, and may have different spatial distributions depending on the region of the country. Their identification, mapping, and monitoring are important, as is the study of environmental variables and their influence on wetlands. These studies can be facilitated by using GIS and related technologies. Hence, a number of interesting and valuable applications have been completed, which we all can learn from and hopefully apply in future work.

The third section addresses environmental engineering applications related to surface waters. These applications examine issues and project results related to various surface water resource disciplines or concerns. Applications include the use of GIS for evaluation of nonpoint source pollution problems and opportunities, the routing sewer lines, the use of GIS in oil pollutant management and response, and use of other technologies such as remote sensing to facilitate analyses of water resources and the development of GIS databases and their application to environmental studies.

The fourth section discusses the use of GIS and the supporting role of other technologies. In particular, remote sensor data can be a vital data source for GIS. It may also supply independent measurements of resources that can provide valuable information. The chapters by Lee and Lunetta and by Lyon, Williams, and Lunetta demonstrate some of the value of remote sensor inputs to GIS. The works of Kass Green and associates, and Mason Hewitt and associates both demonstrate how complex questions can be addressed by using a variety of protocols and sources of data. These applications provide a good example of the level of detail that can be supplied by a number of GIS and related technologies.

The majority of materials presented here are from original manuscripts or sources. These chapters provide a structure for the book and evaluate a variety of techniques and applications. A few of these contributions have been published by the government, but have been poorly circulated or have appeared in publications where less than 300 copies were published and circulated. A small number of articles have appeared elsewhere, and are produced here because they serve an important educational role and are not generally available to the water resource professional. They are supplied via permission of the authors and publishers, and presented with acknowledgment of sources.

These sections and chapters will hopefully educate the new user. For the practicing professional or veteran, they will provide additional ideas to solve project-related problems. All of these methods have great potential to assist in the important work of evaluating surface water resources over large landscapes using quantitative analyses.

CHAPTER 2

The ABCs of GIS: An Introduction to Geographic Information Systems

Russell G. Congalton and Kass Green

ABSTRACT

This paper introduces the concepts of a geographic information system (GIS). A discussion of the sources of data, analysis techniques, software and hardware, and some example applications are presented. This introduction will help acquaint the reader with the very useful tool of GIS.

INTRODUCTION

A geographic information system (GIS) can be defined as a system for entering, storing, manipulating, analyzing, and displaying geographic or spatial data. These data are represented by points, lines, and polygons (Figure 1) along with their associated attributes (i.e., characteristics of the features which the points, lines, and polygons represent). For example, the points may represent hazardous waste site locations and their associated attributes may be the specific chemical dumped at the site, the owner, and the date the site was last used. Lines may represent roads, streams, and other linear features while polygons may represent vegetation types or land use.

In addition, geographic data can be represented in two data formats. The first of these formats is called raster or grid structure and is the older of the two data formats. Raster data are stored in a grid or pixel format which is referenced to some coordinate system (i.e., latitude and longitude). The size of the grid can vary and therefore the spatial resolution of the data is determined by the size of the grid. Raster data are computationally easier to manipulate but typically require larger amounts of storage. Digital remotely sensed data (i.e., satellite imagery) are a good example of raster or grid data. Vector or polygon data are the

Reprinted from the *Journal of Forestry* 90(11):13–20, published by the Society of American Foresters, 5400 Grosvenor Lane, Bethesda, MD 20814-2198. Not for further reproduction.

POINTS   LINES   POLYGONS

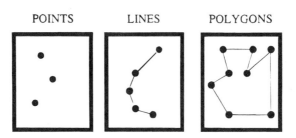

Figure 1   The three elements of geographic data.

second way geographic data may be represented. Vector data uses a series of points (i.e., x,y coordinates) to define the boundary of the object of interest. These types of data may require less storage and are preferred for display purposes because a truer rendition of the shape of the object of interest is maintained. However, some computations are especially difficult and time consuming if performed on vector data. The example below shows what an object would look like in both raster and vector formats (Figure 2). Note that the raster image has lost some of its true shape due to the gridding process. In this example, a large grid cell size was used to emphasize the differences between the two formats.

Recent technological developments and refinements in GIS computer hardware and software, and data acquisition techniques have revolutionized land management and land planning. Today, using GIS, land managers, planners, resource managers, engineers, and many others can use geographic data more efficiently than ever before to analyze management and policy issues. Geographic information systems link computerized maps (location data) to computerized databases which describe the attributes of a particular location. This linkage makes it possible for decision makers to access location and attribute data simultaneously to simulate the effects of management and policy alternatives. GIS is a powerful tool because a single user can quickly search, display, analyze, and model spatial information. In addition, maps and other data can be updated quicker and more accurately using GIS than with conventional methods.

This paper is intended to introduce the reader to the concepts of GIS. Others, notably deSteiguer and Giles (1981), have helped resource managers understand the usefulness of GIS. The goal of this paper is to highlight the basic concepts and fundamentals that must be understood in order to use GIS. The paper is divided into three main sections which correspond to the three aspects of a GIS: (1) data, (2) hardware and software, and (3) people. Each section contains the basic information needed to understand the concept. In some cases, more advanced information is also provided. The objective of this paper is not to provide extensive information about each subject, but rather to expose the reader to the many aspects of GIS and to encourage them to pursue these individual topics. At the end of this paper there are some sample applications and a list of publications containing more information about GIS.

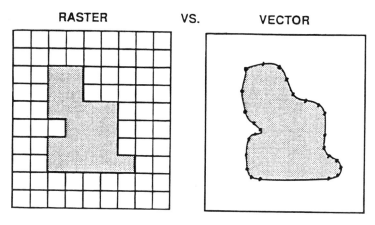

Figure 2 Depiction of the same shape in raster and vector format.

DATA

Spatial data are the "life blood" of any GIS. Somewhere between 80 and 90% of the effort and money required to run a GIS is used to acquire, input, update, and manipulate data. Therefore, it is critical that the GIS user have a thorough understanding of the many aspects of data acquisition and manipulation even before the system is turned on.

Map Projections and Coordinate Systems

A very important aspect of spatial data is ground registration. Failure to consider registering the spatial database to the ground can cause serious problems in the analysis and assessment stages. Surveyors understand well this registration process. In fact, they call a registered spatial database a multipurpose cadastre. In other disciplines, such as geography and the resource sciences, less attention is generally given to precise ground registration. This does not mean, however, that it is not a critical factor. In fact, understanding registration is so important that it is the first topic to be discussed in this paper. A failure to consider registration before beginning any GIS project can lead to serious problems during the later stages of GIS analysis.

Registering spatial data to the ground requires transforming the original data which is recorded in digitizer inches to some ground-based coordinate system. (A digitizer is an instrument that permits the electronic transfer of x,y coordinates into a computer by manually tracing a map and pressing a button to send an electronic pulse of the coordinate location to the computer.) A coordinate system is simply the two-dimensional (x,y) values that designate the position of a given point on the ground. Common coordinate systems include latitude-longitude, Universal Transverse Mercator (UTM), and state plane coordinates.

The procedure for registering spatial data to the ground is quite simple and would be even simpler if the earth was flat. However, because the earth is round, a projection system must be used to make a map flat. A map projection is an orderly, mathematical system of parallels and meridians used in creating a map. One of the commonly used map projections is the Mercator projection developed by Gerhardus Mercator in 1569. Projections allow us to flatten the earth only at the cost of one or more spatial attributes. Depending on the projection used, the attributes affected could be distance, shape, and/or direction. A good example of this projection problem is the large size of Greenland (much bigger than it really is) on world maps displayed using the Mercator projection.

Knowledge of map projections and coordinate systems is critical because overlay analysis is such an important tool in GIS. Obviously, maps of the same area generated with different projections will not overlay. One must consider not only the transformation of the map to ground coordinates, but also must know the map projection if data layers are to be overlaid. Although this may seem like a small problem, it is frequently overlooked when using spatial information. In addition, the ability to locate yourself precisely on the ground is very important. An assessment of the accuracy of any spatial data layer depends on being able to locate the necessary points on the ground (called ground control points). Without the ability to precisely locate ground positions, it is impossible to assess the accuracy of the data layer. This inability to assess the accuracy of a spatial database renders it useless for any analysis or decision making.

It is critical that one understands the importance of considering both the map projection and the coordinate system associated with the spatial information (Muehrcke, 1986; Snyder, 1987; Thompson, 1988). Registering the data to ground coordinates such as UTM or state plane is a relatively simple process. However, just because all the data has been registered does not mean that it will overlay. Careful attention must be given to the original map projection as well. Failure to consider either factor will greatly complicate the analysis using the spatial information.

Sources of Spatial Data

In a GIS, spatial data are expressed as points, lines, or polygons. The spatial relationship of the points, lines, and polygons to one another is called topology. All landscape features can be reduced to one of these three data types using x,y coordinate pairs. The data can then be entered into a computer where it is stored as topology for future analysis.

As previously discussed, collecting, importing, verifying, and updating spatial data is the most expensive component of any GIS. Knowledge of how each data layer is obtained is critical to the ultimate economic success of the GIS. Before any new data are collected an exhaustive search should be performed to verify that no substitute data exists. A good place to begin this search is the Earth Science Information Centers (ESIC) of the U.S. Geological Survey. ESIC provides information about the nation's mineral, land, and water resources through books, maps, aerial photographs, and other sources. The following is a partial description of some existing sources of spatial data.

Existing Sources

Obviously, collecting new information when satisfactory old information exists is a waste of resources. One of the most recognized sources of existing spatial data in the United States is the U.S. Geological Survey (USGS). The USGS provides spatial information not only in the form of hardcopy maps but also in digital format on computer tapes. In this way, the information can be readily input into a computerized GIS.

The USGS produces and distributes two kinds of spatial information: digital line graph (DLG) data and digital elevation models (DEM). The DLG data consists of vector format information about certain characteristics found on the maps including land use, land cover, transportation, ownership boundaries, and hydrography. These data are stored such that the spatial integrity of the map is retained. In other words, everything is in its proper place in relationship to everything else (i.e., topologically structured). The DLG is also referenced to a geographic coordinate system and can be readily tied to other new or existing spatial information. DLG data is available for most but not all of the United States.

Digital elevation models (DEM), also called digital terrain models (DTM), are digital files that contain a grid pattern of point elevations that can be used to simulate the topography of an area. These data are useful for generating three-dimensional information about a land area. Examples of this type of information include slope, aspect, volume, and surface profiles. The USGS provides digital elevation model data at two resolutions. A complete data base for the entire United States was derived from Defense Mapping Agency data for 1 by 2° (1:250,000 scale) maps. The map is split in half and a digital file created for each half. The grid interval for these data is 3 arc-seconds or approximately every 100 m. Because these data were produced by the Defense Mapping Agency, the data are often referred to as DMA data. In addition to the DMA data, USGS also has an incomplete coverage of the United States using a grid interval of 30 m. These data are available in 7.5-min blocks and are of much higher resolution than DMA data. These higher resolution data are frequently called DEM data.

Another excellent source of existing spatial information is the U.S. Census Bureau Topologically Integrated Geographic Encoding and Referencing (TIGER) system. TIGER contains all the digital data for the 1990 census map features including roads, railroads, rivers, census tracts and blocks, political areas, latitude-longitude, feature names, and classification codes used for the census. Other U.S. government agencies including the Department of Agriculture Soil Conservation Service and others, state agencies, and even local governments can be good sources of existing information.

New Sources

If no existing source of data can be found that is satisfactory for your purposes, then new data must be collected. Here are some excellent sources of new data.

Aerial Photography

It is very easy in this age of satellite imagery, video cameras, and global positioning satellites to overlook the many advantages and uses of aerial photography as a source of spatial information. However, to do this would be to make a huge mistake. Let's briefly look at some of the reasons why aerial photography will continue to be an important source of spatial information for a long time to come.

From an historical perspective, aerial photography is irreplaceable. Any project undertaken to review changes over time must rely heavily on aerial photographs. The reasons for this reliance on photography are obvious. Satellite imagery and other remotely sensed data are too new to be of much historical significance even if you consider the satellite age beginning in 1957 with the launch of Sputnik. In addition, despite the capabilities of off-nadir viewing satellites, aerial photography is still the standard for all topographic mapping. These satellites may allow for maps to be generated of areas of the world where no topographic data were available before, but these data will not replace aerial photography. The accuracy, precision, and detail required for many mapping projects dictates the use of aerial photographs.

Finally, and most importantly, the ease of use and simplicity of aerial photography with respect to other remotely sensed data must be considered. Aerial photography can be as simple as using the photograph to record some historical event (i.e., a picture is worth a thousand words), or as complicated as digitizing the photo and entering it into a computerized image processing system. Anyone can pick up an aerial photograph and begin to glean some information from it. However, photographs do have some disadvantages with positional accuracy due to distortion and displacement. Techniques are available to correct these problems and should be used to take advantage of this very common source of spatial data.

Remotely Sensed Data

Perhaps the most exciting development in the area of sources of spatial data is the ability to geocode and/or terrain process digital satellite data (Fischel and Labovitz, 1987). Geocoding refers to a process by which the remotely sensed data is very accurately registered to a ground coordinate system. This process is possible because knowing the precise location of the satellite in space is now feasible. Terrain processing refers to the removal of topographic displacement in the satellite data through the use of DEM data. As a result, digital satellite data such as Landsat and SPOT are excellent sources of new spatial information.

In addition to these geocoded products, the SPOT digital data can be used in a very special way. As mentioned, terrain parallax effects can be compensated for during the geocoding process. However, because the SPOT satellite has an off-nadir (i.e., nonvertical) ability, stereo SPOT imagery is possible. Therefore, using a SPOT stereo pair and working backward through the terrain parallax effects, digital elevation models can be derived from the SPOT imagery. Now, not only can digital data be very easily added into a GIS, but it can also be a source of one of the most important GIS data layers—elevation. Both of these facts have far reaching implications in our ability to quickly and efficiently update and revise GIS databases.

ANALYSIS TECHNIQUES

Once all the necessary existing and new data have been collected then it can all be registered to a common base map. As mentioned, the collection and registering of all these data from various sources can be an expensive, time consuming, and frustrating process. Once completed, however, the analysis can begin. GIS analysis techniques include overlay analysis, modeling, buffering, and network analysis. A discussion of each follows.

Overlay Analysis

The ability to analyze spatial data separates GIS from mere spatial databases. In early computer-aided design/computer-aided mapping (CAD-CAM) packages, data were frequently input such that each layer contained only a single attribute or label. For example, instead of all stream types being in one data layer with various labels, each stream type would be in a separate layer. There would be one layer for major streams, another for intermittent streams, etc. The appropriate layers could then be chosen to derive any map one wanted to produce. It is easy to envision the problems with such an approach. Each unique data layer had to be derived manually and entered into the database. Some of these layers could be quite labor intensive to derive. In addition, the number of layers needed in the database could quickly get out of control. Clearly, another approach was needed. That approach is what we now call a GIS. It is the ability of the system to create a new layer of information/data from two or more existing layers (Figure 3).

The ability to extract specific information from a data layer and combine it with other information from that same or some other data layer depends on the use of Boolean algebra. This procedure involves the use of the operators AND, OR, XOR, and NOT to manipulate spatial data by testing to see if a given condition or statement is true or false. It is then possible to combine data layers to form a new layer. For example, to find all locations where vegetation type B exists with soil type C, one would simply use the statement B AND C. A map or data layer could then be formed indicating all locations where this statement was true (see Map E in Figure 4). Similar examples can be derived using the other Boolean operators as shown in Figure 4. The statement "Map A **or** Map B" results in Map D while the statement "Map D **not** Map E" results in Map F.

The concept of overlaying data layers to obtain certain information is not a new concept to GIS. Many of us have used tracing paper and colored pencils to

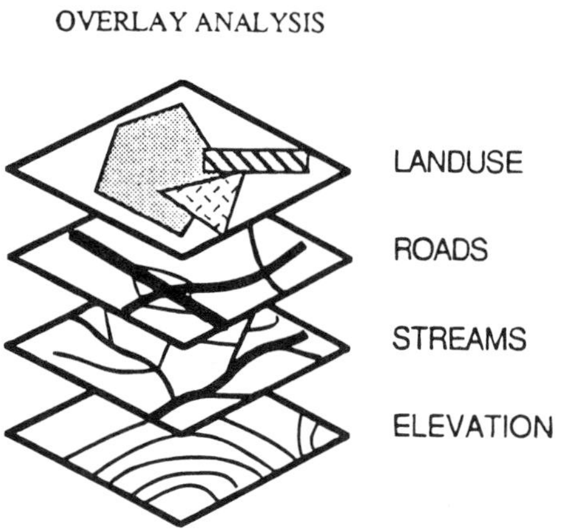

OVERLAY ANALYSIS

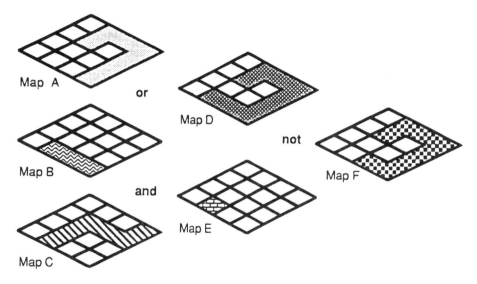

Figure 3 An example of overlaying various layers in a GIS.

LANDUSE

ROADS

STREAMS

ELEVATION

Figure 4 Some examples of Boolean operators used in overlay analysis.

produce transparent maps that could lay over the top of each other in order to derive some information. It only takes one try at this approach to quickly realize its limitations. It is not only time consuming and inaccurate, but can also result in a large number of polygons depending on the complexity of the original data layers. In many cases, some of these polygons will be what are called "slivers," which are caused when areas on two layers almost coincide, but not quite. These slivers must be "cleaned up" in order to have a useful new map/layer. Fortunately, GIS and its associated technology have overcome many of these overlay problems and therefore, overlay analysis has become a powerful GIS tool.

Overlay analysis can be divided into two general categories: point operations and neighborhood or region operations. Point operations can be anything from the Boolean operators we already discussed to simple weighting functions such as multiplication by some factor. In addition, point operations can involve more complex functions such as clustering, discriminant analysis, principal components analysis, and other such statistical techniques. Many published articles, books, and especially software user's manuals contain additional details about specific point operations (see list of references at the end of this paper).

Neighborhood or region operations differ from point operations in that they relate a point to its neighbors or to a specified region. This process is much more complicated than point operations and involves the use of the spatial component of the data in order to operate. It is in these neighborhood operations that the true utility of a GIS shines through. Measures of correlation and diversity as well as slope and aspect are common examples of neighborhood operations. Like point operators, much more information about neighborhood or region operators can be found in the literature.

Modeling

The word modeling can mean different things to different people. Three aspects of modeling spatial information will be discussed here: cartographic modeling, simulation approach, and predictive modeling.

The first aspect of modeling has been called "cartographic modeling" (Burrough, 1986). In this approach, when the user of spatial information is presented with a problem, the response should be a careful plan of what should be done. A more common response is to rush to the computer and start to work. Instead, "cartographic modeling" suggests detailed flow charts and careful planning to decide what data are important and how they should be used.

Another aspect of modeling is the simulation approach. In this case, the user tries to simulate some complex phenomenon using a combination of spatial and nonspatial information. This approach typically requires an expert who is knowledgeable enough to build such a simulation or model. It should be noted that rarely in these cases do any two experts agree on exactly how the model should be built. A good example of this type of modeling is evaluating wildlife habitat suitability. In this example, one might use the following spatial information layers: vegetation, elevation, aspect, slope, ownership, roads, and streams. This information could then be combined in some model with weights used to prioritize important layers. In addition, calculations of distances (i.e., distance from roads, distance to streams) and measures of diversity may be included in the model. This model is then used to evaluate areas of good habitat and determine where the habitat can be improved.

The final aspect of modeling to be discussed is predictive modeling. In this approach, statistical techniques are used to build a model that will be able to predict using the spatial information. The statistical tool used for building such models is most commonly regression analysis. The first step in this process is llect information about the phenomena one wishes to model. A subset of information is then used to statistically build the model. This model build-, is performed by looking at each layer of spatial information and each component of nonspatial information to see which are correlated to the phenomena one wishes to predict. Once the model is built, the model is tested using the remaining information.

An example will elucidate this explanation. Suppose one wants to predict the amount of snowmelt runoff from a forested watershed. These predictions currently are being made by point samples taken throughout the watershed. The predictions can be compared to the actual runoff statistics collected by stream gauges. One might hypothesize that using spatial data that completely covers the area should lead to better predictions than point samples. Therefore, one would put together a GIS with the necessary layers to predict runoff. These layers might include: vegetation, slope, aspect, snow extent, elevation, and soil type. In addition, the point sample data which includes snow depth and the amount of water in the snow may also be included. Some of this information remains constant over time while some changes daily. Therefore, a variety of conditions (i.e., years) should be represented in the information collected. In this example, one would collect runoff data for dry years, wet years, and average years. A subset of this runoff data would then be used to develop the model and select the necessary spatial and point sample data to be included in it. The remainder of the data would then be used to test the model. Once the model has been proved effective, it can be used to predict snowmelt runoff in future years.

One can easily envision applications of this modeling approach to many fields including planning, public health, energy services, transportation, and so on. This paper only briefly discusses modeling as an analysis tool in GIS. However, examples are commonly found in the literature, and the interested reader should have no trouble finding numerous papers describing this powerful tool.

Buffering

Buffering is a technique by which a boundary of known width is drawn around a point or linear feature (Figure 5). Some examples of point buffers may be a zone around a hazardous waste site or around a tree that is a nest for a particular endangered bird. Examples of linear buffers may be an area around a stream to prevent logging or an area around a utility pipeline to prevent digging.

Network Analysis

Network or corridor analysis is a technique by which a linear path is identified that represents the flow of some object through the area. Network analysis is especially useful in hydrology, transportation, and other disciplines that study the flow of an object. This flow is not limited to water but can also be used for vehicles, utility and communication lines, and animals. The following example shows the optimum pathway for a central sewage line in a small town (Figure 6).

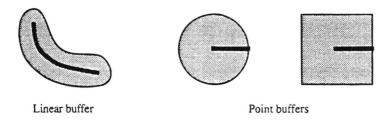

Linear buffer   Point buffers

Figure 5   Examples of various types of buffers.

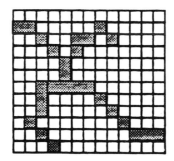

Figure 6   Example of network analysis.

ERROR ANALYSIS

In order to effectively use any GIS, one must understand the errors associated with spatial information. This knowledge is critical whether you are a user of a GIS or whether you are one of the suppliers of spatial information (i.e., data layers) used in the GIS.

Errors associated with spatial information can be divided into three groups: user errors, measurement/data errors, and processing errors (Burrough, 1986). User errors are those errors which are probably most obvious and are more directly controlled by the user. Measurement/data error deals with the variability in the spatial information and the corresponding accuracy with which it was acquired. Finally, processing error involves errors inherent to the techniques used to input, access, and manipulate the spatial information.

Included within the user errors are the age, scale, coverage, and relevance of the data. Errors result when outdated data are used because of a lack of current information. A good example of this problem is the use of old aerial photography because new photography is not available. Error also results when data of the wrong scale are applied to meet some objective. This situation is especially dangerous when small scale data are used to meet the objectives of some large scale project. An example of this problem might be using a statewide soils map to obtain soils information about a particular county or using digital elevation data from a 1:250,000 USGS map sheet for mapping on a 7.5-min quadrangle. In both cases, the source data are of insufficient detail to meet the required objectives. In addition, error results from using data sources that do not completely cover the area of interest. Many times only partial areas have been covered by the latest data and one is forced into the dilemma of choosing between full coverage using outdated data or using partial coverage new data. Finally, error can be caused by using indirect data layers as input into the GIS. An indirect layer is one that has been derived from some primary information. Probably the best example of this is a vegetation classification generated from satellite data or aerial photography.

Included within measurement/data error are those errors that are associated with variation within the data. Among the causes of variation are instrument error, field error, and natural variation. Instrument error is simply a measure of the limitations and quality of the instrument being used to collect the information. Similarly, field error is a measure of the limitations and quality of the person collecting the data. Natural variation, on the other hand, is a fact of nature. As much as one may hope, nature will not be pigeon-holed into nice, neat compartments. Therefore, although one must live with this error, every effort can be made to account for it in our use of spatial information. Measurement/data error is usually quantified in terms of positional accuracy or accuracy of content. In other words, am I where I think I am, and if so, are my surroundings what they should be?

Finally, processing error includes such factors as precision, interpolation, generalization, data conversion, digitization, and other methodologies performed on the data. A fact that is often overlooked is that a computer is designed for only a certain level of precision, and anything beyond that point results in round-off error. Almost every introductory course in computers discusses precision and the use of significant digits, yet often these factors are not considered when processing information. Techniques such as interpolation, extrapolation, and generalization are certainly subject to error. In these cases, information is derived about an area from a series of sample points. The best example of this technique for spatial information is the generation of a digital terrain model from a series of elevation points.

In addition, the way the data are entered into the GIS is subject to error. The data can either be scanned or digitized but errors can occur in both methods. Also, the way in which the data are stored and used in a GIS may include error. This fact becomes especially obvious when converting between vector and raster data. A good example of this situation is incorporating digital remotely sensed data into a GIS. Digital satellite data is recorded in pixel format which is also called raster or grid format. Other information in the GIS may be recorded in vector or polygon format. Therefore, in order to perform any analysis, the vector data may need to be converted to raster or vice versa. It becomes apparent how error could result from making smooth polygons into grid cells or making grid cells into smooth polygons. In addition to these processing errors, other problems arise when data layers are combined to perform some type of analysis. One can imagine problems associated with overlaying data layers, with boundaries, and with registration. All these factors and more result in error in the processing stage of using spatial information.

HARDWARE AND SOFTWARE

Now that the reader has a basic understanding of what a GIS is and how it works, it is important to discuss the factors that must be considered when selecting a system. The GIS software links the attribute data to the geographic features (represented by points, lines, and polygons and their topology), using a database management system. It also provides for the input, editing, and analysis functions previously discussed. The capabilities and costs of GIS software vary greatly from vendor to vendor and the selection of such software depends on the needs of the user. The following factors should be considered when choosing GIS software:

1. *Data input and editing functions.* Some software packages may offer rapid data query and retrieval and powerful analysis functions but have difficulty to operate data input and editing functions. Novice users can become quickly discouraged because these basic functions are so difficult; therefore, they never discover the excellent, more advanced functions. It is important to find software that handles both the simple and the difficult functions in a highly interactive yet easy-to-use manner. One should look for pull down menus, help screens, system prompts, and single character command strokes as indicators of well-developed software.

2. *Analysis functions.* Cartographic analysis tools such as polygon overlays, buffer generation, line and area measurement, and map production are essential to any software package. In addition, modeling capabilities such as network analysis and project simulation may also be important. Some packages perform certain functions better than others. The user should identify the functions most needed for his/her application and choose the system that is best suited for those functions.

3. *Flexibility of the system.* Flexibility refers to a software package's ability to interface with different computer systems and high level programming languages. Obviously, data from a variety of sources should be able to be quickly and easily entered into the system. Data conversion is time consuming, inefficient, and frustrating and should be avoided, if possible.

4. *Risk associated with the vendor.* The risk associated with any vendor depends on several factors: (1) number of years and type of experience with GIS, (2) number of users, (3) customer satisfaction record, (4) training offered, and (5) research and development.

5. *Cost to purchase, operate, and maintain the system.* The cost to purchase, operate, and maintain a GIS is most likely the major consideration of potential GIS customers. The initial start-up costs for a GIS have dropped dramatically in the last few years, however, it is still a considerable investment. It is important to note that the cost of good employees is as important as the cost of software and hardware.

6. *Type of data base management system.* The two database management systems most used in GIS are hierarchical and relational. In a hierarchical system, data are related by level, in much the same way as a family tree is structured; there is a "parent-child" relationship between layers. Access to data is rigid and restricted to a series of hierarchical pathways between data layers.

Most GIS software packages use a hierarchical data base management system. A relational database allows the user to relate or compare attributes in different data layers regardless of the structure of the data. This approach is more complex and expensive and is found only in a few GIS systems at the present time.

There are virtually as many possible hardware configurations for GIS as there are users. Again, as with software, the hardware should be matched to the user's needs. With the continuing development of hardware, GIS is possible in almost every work setting. Computer sizes range from microcomputers and advanced workstations to mini- and mainframe computers. Size is mainly a function of speed, disk space, random access memory (RAM), number of users, types of input/output devices (i.e., tape drives, streaming tape, scanners, digitizers, plotters, and printers), and cost. There always seems to be a trade-off between buying the right computer and then being able to afford the necessary peripherals. This dilemma is similar to the one we face when buying a car. Do we buy a basic model with a big engine or do we settle for a smaller engine and get the stereo, air conditioner and sun roof? Careful consideration of your application is critical before buying the right software and hardware. Additional information about selecting the proper hardware and software can be found in Guptill (1988) and Parker (1989).

PEOPLE

An integral and yet largely forgotten and unnoticed component of a GIS is people. Geographic information systems need people in order to operate. Without well-trained people and an adequate staff, it is likely that an investment of thousands of dollars on state-of-the-art equipment and data will be wasted. GIS definitely requires a financial and philosophical commitment of human resources. Continued funding is critical to provide the necessary training and database maintenance. Despite this brief discussion of people compared with the longer discussions on data and hardware/software, people are the most important resource. Failure to make a strong commitment to the people operating the GIS, dooms it to failure.

APPLICATIONS

No discussion of GIS would be complete without mentioning some of the applications of spatial data. Although there are many examples, only a few will be discussed here. A very important application of spatial data is for wildlife habitat assessment. One can easily imagine the usefulness of spatial data for many calculations performed in habitat analysis. These data can be used to calculate the home range or territory of a particular species. They can also be used to eliminate areas where a certain animal would rarely or never be found. For example, slope data could be used to eliminate extremely steep slopes or elevation data could be used to eliminate elevations above a certain level. In addition, by simply moving a 3×3 pixel window over the vegetation map, calculations of juxtaposition and interspersion can be made. Juxtaposition measures the number of different vegetation types in a certain area while interspersion measures the value of the edges. They are both very important to wildlife since food and cover must be in close proximity in order for the habitat to be useful to the animal. Obviously, other proximity analysis such as distances to water and roads is also possible. In wildlife habitat assessment the spatial data and its derivatives (i.e., juxtaposition, etc.), are input to some model to predict habitat quality. This model is typically based on knowledge compiled by one or more wildlife biologists for the species of interest. The model is a combination of the spatial data, its derivatives, and other information necessary to predict habitat quality. Frequently, the more important variables in the model are given higher weights. From this analysis, one can identify areas of good, marginal, and poor habitat. This identification is the first step toward intelligently improving the habitat.

Another application of spatial data involves a different kind of modeling approach. Predicting snowmelt runoff from a watershed involves the use of a statistical model generated using regression analysis. As in the wildlife example, the same spatial data are useful. In this case, however, the regression analysis chooses the important variables instead of the wildlife biologist. In predicting snowmelt runoff the important parameters are the snow covered area (SCA) and the snow water equivalent (SWE), or the amount of water in the snow. The spatial data considered for this regression model include elevation, slope, aspect, vegetation, thermal emittance, near infrared reflectance, and soils. In addition, data collected at specific sample locations on the ground are also important. All these data are tested by the regression analysis to see which are significant in predicting runoff. The best model is the one that has the highest correlation coefficient. The model is then used to predict snowmelt runoff over the entire watershed and not just a few ground sample points.

These two examples demonstrate different types of modeling approaches using spatial data. It should not be inferred from these examples that GIS only applies to natural resources. As discussed in the introduction to this paper, GIS has many applications including everything from aiding the census to evaluating the habitat of the spotted owl. Instead of describing more examples, we recommend that the interested reader pick up any of the major journals in their discipline and they will be sure to find examples of GIS. In addition, we encourage anyone starting to use spatial data as a tool for decision making to review the literature and learn as much as possible from past successes and failures.

Finally, we would just like to offer this warning. The use of GIS is a very powerful decision-making tool. However, it is just a tool. There is a great temptation to rely too heavily on the computer and to stop thinking. There is also a tendency to collect too much spatial data. One more layer of data is not always the answer. Therefore, be warned and GIS will serve you well.

REFERENCES

Burrough, P. A., *Principles of Geographical Information Systems for Land Resources Assessment*, Oxford University Press, 1986, 193 pp.

de Steiguer, J. and Giles, R., Jr., Introduction to computerized land-information systems, *J. Forestry*, 79(11), 734, 1981.

Fischel, D. and Labovitz, M., Questions and answers about Geocoded Products, EOSAT Corporation, Lanham, MD, 1987, 7 pp.

Guptill, S. C. (Ed.), A process for evaluating geographic information systems, US Geological Survey Open File Report 88–105, Reston, VA, 1988.

Muehrcke, P. C., *Map Use: Reading, Analysis, Interpretation*, JP Publications, Madison, WI, 1986.

Parker, H. D., (Ed.), *The GIS Sourcebook*, GIS World, Inc, Fort Collins, CO, 1989.

Snyder, J. P., Map projections—A working manual, USGS Professional Paper 1395, U.S. Government Printing Office, Washington, D.C., 1987.

Thompson, M. M., Maps for America, U.S. Government Printing Office, Washington, D.C., 1988.

BIBLIOGRAPHY

Here is a partial list of publications helpful in learning more about GIS:

Aronoff, S., *Geographic Information Systems: A Management Perspective*, WDL Publications, Ottawa, Canada. 1989, 294 p.

Burrough, P. A., *Principles of Geographical Information Systems for Land Resources Assessment*, Oxford University Press, 1986, 193 pp.

GIS. Special Issues of *Photogrammetric Engineering and Remote Sensing* published by ASPRS annually in November.

Goodchild, M. (Ed.), *The Accuracy of Spatial Databases*, Taylor & Francis Ltd., 1989, 350 pp.

Proceedings of the Annual GIS/LIS Conference co-sponsored by the Am. Soc. for Photo. & Remote Sensing and the Am. Congress on Surveying and Mapping, available from 1986–present from ASPRS.

Proceedings of the NCGA Mapping and Geographic Information Systems Annual Conference, available from the National Computer Graphics Association

Proceedings of the URISA Annual Conference, available from the Urban and Regional Information Systems Association.

Rhind, D. and Mounsey, H. (Eds.), *Understanding GIS*, Taylor & Francis Ltd., 1990, 240 pp.

Ripple, W. (Ed.), GIS for Resource Management: A Compendium, ASPRS, Bethesda, MD, 1987, 288 pp.

Ripple, W. (Ed.), Fundamentals of GIS: A Compendium, ASPRS, Bethesda, MD, 1989, 248 pp.

Star, J. and Estes, J., *Geographic Information Systems: An Introduction*, Prentice Hall, Englewood Cliffs, NJ, 1990, 303 pp.

Section II

Wetland Applications of GIS and Remote Sensing

CHAPTER 3

Introduction to Wetland Applications

John G. Lyon and Jack McCarthy

Wetland issues have become a major source of interest to the professional and to the public. Unlike other environmental issues that are localized or found only in certain areas, wetlands are found almost everywhere. From deserts to alpine and arctic environments one finds wetlands in either a permanent or temporal status. Because wetlands are to be regulated and preserved, and because they are ubiquitous, there is the high probability that wetlands will in some way influence a given engineered project or other activity.

The history of wetland issues has been one of rapid activity. From the earliest settlement of the United States, people have been changing wetland areas. Now there is a whole litany of projects for preserving or improving wetlands, just as there were incentives to drain the same areas in past years. Time has witnessed a variety of priorities related to wetlands. The present priority is preservation, avoidance of disturbances, and a management scenario including a long-term perspective for maintenance and enhancement of wetland functions.

There is now a great need to inventory the extent of wetland resources and to do so on a regular basis. Currently, local inventories are available but most areas have little or nothing. The U.S. Fish and Wildlife Service has the National Wetlands Inventory (NWI), which provides a nice product for the general public and for general governmental programs. However, local and regional governmental or conservation groups may require more detail, or may require additional inventories over time rather than one evaluation alone that NWI supplies. This need is a result of maturation in the interests of the public, and of state and local governments.

After basic abundance data are collected and become available in inventory form, it is important to move onto other protocols and procedures. The objective is to derive additional value from wetland inventory data sets. Thus, it becomes important to manage the resource for future benefits. From the need comes the requirement to collect information on wetland type, local abundance, permitting requirements, project plans or location of project or local of project corridor, and the capability to run spatially based models to evaluate scenarios and various plans.

To keep track of a changing resource such as wetlands creates a need for information about the resource. The information includes the historical and current extent of wetlands, or inventories of wetlands. It is desirable to locate wet-

0-87371-897-6/95/$0.00+$.50

lands within planning projects, to avoid any impact, and to design the site plan around and away from the resource. Large area inventories of wetlands are necessary to plan and permit large projects or corridors. These just represent a few of the information needs that must be met in the future.

There are any number of solutions to the need for information on wetlands. Due to the spatial and temporal distribution of wetlands over large areas, technologies which capture the synoptic view of the earth are favored. In particular, GIS and remote sensor technologies have proven particularly useful in evaluating the distributed wetland resources across the landscape.

There is a great need for quantitative methods to address basic questions. The questions include a variety related to the resource. GIS and related technologies such as remote sensing can be used to supply a number of these requirements for quantitative information. Remote sensor technologies can supply a good deal of the input variable information. They can: (1) supply inventory data on the extent of wetlands, (2) identify the wetland resource as to type, (3) characterize the general wetland land cover type, (4) identify submergent and emergent wetlands, and (5) supply details about the resource using multiple spectral analysis of remote sensor data.

GIS and remote sensor technologies supply information of a more general nature. With remote sensor or aerial photography data as input, the wetlands can be located and characterized as to type. These data include inventory and change information for management on a local or individual site basis. The information can help the landowner best comply with prevailing regulations, and help design a site plan to avoid wetland resources.

In a regional wetland inventory, satellite and high altitude image data sets can provide a valuable resource or focal point for data analyses. They represent the "base map" coverage, and a start for research efforts. Naturally, it is desirable to use a variety of other data sources in the analysis of remote sensor data. In particular, other remote sensor data of similar scale or detail can supply important information, and sources should be identified and used in analyses as researchers see fit.

In the following chapters and sections, a number of remote sensor sources are mentioned along with one or two characteristics that make them valuable to the evaluation of wetland resources on a regional basis. The types of sources have been identified as being beneficial in analyses of wetland areas, and have been shown to be compatible with other products. Further, current and historical aerial photographs are available from archive, as are the NWI maps. These sources can be employed over large areas of the continental United States.

In formulating the approach to inventorying wetlands, it is important to remember to derive quantitative estimates of wetlands. A major part of such an effort is to decide on the definition of wetlands important to the results of the study. In many projects, it is common to have a number of objectives related to identification of wetlands. The definition of wetlands for the given application becomes a vital issue. Hence, a major concern of any wetland study using GIS and remote sensor technologies is the matter of definition. Wetlands mean different things to different people. A basic definition for use by regional and local studies has been proposed. The definition (Lyon, 1993) is that of the Potential Jurisdictional Wetland (PJW), where land and water areas exhibiting one or more of the three federal criteria: plants, soil, and hydrology. The PJW (Lyon, 1993) can often be identified from aerial photographs or remote sensor data. This definition allows the user to develop a remote sensor or aerial photography based identification of wetlands (PJW). The actual determination of whether an area is a true jurisdictional wetland (JW) must be made from the ground due to the detailed requirements (USACE, 1987).

Often, the general hydrological conditions of the site can be estimated by examining a number of aerial photograph or remote sensor images over time. This approach is an extremely valuable capability, which can supply general information on the hydrology. Average conditions are determined from a number of photographs over time. A given set of photographs will often contain one or two examples of extreme conditions, such as a drought year or a year with above average rainfall conditions. A major advantage of these technologies is the capability to store and process input data. The creation of a database allows the later manipulation and evolution of a number of conditions of the database. The evaluation of scenarios allows the user to examine the sensitivity of variable contributions to the submodel or model results. This capability is one of the most valuable contributions of GIS and related technologies for data analyses.

The capability to evaluate scenarios is one of the most valuable contributions of GIS for inventory and modeling. GIS will advance the modeling and planning industry just as computer aided drafting and design (CADD) has advanced the engineering industry. CADD made available the ability to make small changes and produce new map products. Gone is the problem of altering plans and drafting new plans.

CADD provides the opportunity for the engineer to engage in various scenarios. It allows them to redraft the subdivision to avoid wetlands, and to maximize the number of lots thus fulfilling two expectations. GIS can conduct the same sort of scenarios, and process data based on attributes of the data. Both systems allow the user to conduct much more ambitious efforts with less people power and cost as compared to past methods.

A big question involved in GIS and remote sensor evalutions of wetlands is the difference between jurisdictional wetlands and potential jurisdictional wetlands. Previously, the term potential jurisdictional wetlands (PJW) was discussed. In each wetland project the issue of definition must be addressed. It is straightforward for GIS and remote sensing to inventory and map PJW or wetlands in a general sense.

To go toward jurisdictional wetland identification implies that ground truth data should be collected. This requires field evaluations on the ground, with sampling of vegetation, hydric soils, and wetland hydrology. This level of detail is very difficult to supply remotely. Clearly, these sort of evaluations must be done at the field level and the resulting data can be used with remote sensor and GIS technologies to supply results or supply products for public meetings.

The issue of generating maps of jurisdictional wetland is moot. These analyses are best conducted locally and in the field. However, there is great interest in regional inventories of wetlands. This stems from municipalities and their desire to catologue the resource that they must avoid and manage.

The need for a regional product can be met with the use of GIS and remote sensor technologies. A combination of field records, historical maps or evaluations, current and historical aerial photographs, available remote sensor data, and other sources can be mobilized to deal with the issue. The GIS can be used at once to organize the information, supply an inventory and inventory products such as maps, and can be a "living source" through updates to the GIS database.

The dreams of scientists and engineers are not easily confined. Numerous people thirst for the opportunity to conduct a continental and/or a global inventory of wetlands. Such an inventory would be of great interest to the scientific community and the public (NRC, 1995).

Currently, the U.S. Fish and Wildlife Service (1987) has a publication estimating the change in historic quantities of wetland and the contemporary condition. It still remains to supply a true estimate of wetland quantities for the nation, continent, and for the world. The inventory needs to be conducted with modern remote sensing and aerial photography methods which were previously unavailable. The inventory must be conducted with an assessment of accuracy and a documentation of quality control and quality assurance procedures.

Capabilities of these technologies is great when they are applied to applications that exploit their strengths. In particular, remote sensor capabilities allow the user to evaluate wetlands spatially and temporally. Several chapters in this book will illustrate the capabilities of remote sensor data to identify and inventory wetlands. These papers will show how historical aerial photographs can help document the extent and condition of wetlands.

REFERENCES

Lyon, J., *Practical Handbook for Wetland Identification and Delineation*, Lewis Publishers, Boca Raton, FL, 1993, 157 pp.

National Research Council (NRC), *Wetlands: Characteristics and Boundaries*, National Academy Press, Washington, DC, 1995, 268 pp.

U.S. Army Corps of Engineers (USACE), Corps of Engineers Wetland Delineation manual, Technical Report Y-87-1, Department of the Army, Washington, D.C., 1987.

U.S. Fish and Wildlife Service (USFWS), Draft National Wetlands Priority Conservation Plan. U.S. Department of Interior, Washington, D.C., 1987.

CHAPTER 4

Analytical Model-Based Decision Support GIS for Wetland Resource Management

Wei Ji and Loyd C. Mitchell

ABSTRACT

This chapter describes the development of an analytical model-based spatial decision support system focusing on wetland restoration planning. Theoretical and technical issues concerning the integration of environmental models into the GIS environment are discussed. This study indicates that analytical environmental models generally lack a fully developed interface to a GIS, and additional system integration is often required to develop model-embedded automated spatial analysis capabilities. Several key techniques are developed through this model integration. The resultant wetland value assessment decision support system proves to be a powerful, operational tool for wetland management decision-making.

INTRODUCTION

Wetlands, generally defined as transitional areas between permanently flooded deepwater environments and well-drained uplands (Watzin and Gosselink, 1992), are important ecological systems that contribute a wide array of biological, social, and economic benefits. Wetlands provide habitat for rare, endangered, and commercially or recreationally important fish and wildlife species. They also serve as focal points for outdoor recreation, and provide an important function in water quality improvement, floodwater storage, storm surge reduction, and groundwater recharge (Greeson et al., 1979; Mitsch and Gosselink, 1986; Tiner, 1984). However, approximately 53% of the wetlands estimated to have originally occurred in the conterminous United States were lost by the mid-1980s, primarily due to human-induced land-use conversions (Dahl, 1990; Dahl and Johnson, 1991). Although wetland losses since that time appear to be slowing, they are still occurring, and society is placing greater demand on the remaining wetlands.

Wetland managers are facing increasing challenges to slow or reverse wetland loss trends by protecting or enhancing existing wetlands, restoring former wetlands, or creating new wetlands. Wetland managers usually have limited funding with which to perform those activities. To maximize the return from these funds, managers may be required to quantitatively estimate the benefits and priority of various management projects to wetlands and associated fish and wildlife resources. Also, increasing demand on a shrinking wetland base may require wetland managers to make quantitative impact assessments to guide future policy decisions regarding the use of limited wetland resources by competing users.

Solving such complex resource management problems often involves a multidisciplinary approach. It requires computerized analytical modeling techniques to manipulate large quantities of spatial-temporal data according to a defined set of objectives or constraints, and a mechanism to provide quick responses and a graphical visual means for supporting dynamic management decision-making. For example, an ongoing coastal wetland conservation program in Louisiana, funded under the Coastal Wetland Planning, Protection, and Restoration Act (CWPPRA; Public Law 101-646, Title III), mandates that a prioritized list of conservation projects be compiled annually. Assessing individual projects and prioritizing a group of projects requires input from wetland ecologists, engineers, economists, soil scientists, planners, and other professions.

Environmental modeling can be particularly useful as a planning tool to assist in wetland management project assessment and implementation alternatives, and in impact assessment and resource allocation problem solving. Environmental models, linked with GIS capabilities, can allow managers to gather and display large amounts of spatially and temporally related data, to analyze those data within the framework of resource response strategies and economic constraints, and to make prioritization decisions based on those analyses.

The Southern Science Center (SSC) of the National Biological Service* has developed digital GIS wetland habitat databases for several states bordering the Gulf of Mexico and has conducted wetland trend analyses using GIS technology since the early 1980s. In 1992, SSC initiated an advanced spatial analysis research project on the development of a GIS-based multifunctional spatial decision support system for environmental resource management (Ji et al., 1993). As a pilot development, a decision support function designed to support activities required by the CWPPRA has been completed. This system provides not only an operational wetland management tool, but also a basis for the development of other decision support capabilities under the multifunctional development concept.

This chapter will focus on wetland restoration planning while addressing key issues related to the integration of environmental models into the GIS environment for resource management decision support. These technical discussions are preceded by an analysis of the relationship between environmental modeling and GIS, particularly with respect to the integration of modeling with GIS, to provide a theoretical framework and technical guideline for this special aspect of GIS-based resource management technologies.

ISSUES OF ENVIRONMENTAL MODELING WITH GIS

The environmental model is a simplified presentation of environmental reality. There are generally three types of modeling approaches that can be incorporated into environmental resource management: *physical modeling, concep-*

*Formerly National Wetlands Research Center of U.S. Fish and Wildlife Service.

tual modeling, and *analytical modeling*. These three approaches represent not only different levels of understanding of environmental processes, but also different stages of resource management planning, and therefore they determine the degree and characteristics of the model integration with the GIS.

The physical model is a visualized presentation of the underlying environmental problems and reflects primary physical features of environments. As a preliminary description of reality, the physical model may give only illustrations of relationships among addressed spatial features to provide a reference in resource management decision-making. Examples of such models in spatial analysis are vector-based thematic maps and raster-based classified satellite imagery. The physical model builds a very preliminary connection with spatial analysis by requiring only a "loose integration" with GIS techniques, mainly through digital database development and map production. This is defined as *first degree integration*. Many GIS activities that support environmental modeling are currently at this stage.

The conceptual model is widely used in environmental resource management and research and usually is not in a form which can be visualized. Instead, such models are commonly translated by resource researchers or managers into functional diagrams embracing environmental state variables and forcing functions, as well as management policy elements in some cases. These components are interrelated based on the understanding of the processes of the underlying problem. The conceptual model requires a higher level integration (*second degree integration*) with GIS. Typically, the conceptual model could (1) serve as a guideline for GIS database generation in defining specific environmental variable layers, (2) provide approaches for simple map analysis in determining spatial boundaries and temporal components of the database, and (3) assist in specifying procedures for GIS database manipulation to query spatial information of environmental parameters (e.g., data attribute identification), verifying the interrelationship of model variables (e.g., map overlay), or improving the understanding of modeled environmental processes (e.g., coastal wetland trend analysis). Most map analysis work in environmental resource management with GIS can be related to this type of modeling procedure.

The analytical model refers to a model based on explicit rules either in a quantitative form (e.g., a mathematical equation), or in a qualitative heuristic reasoning format (e.g., "if a condition is true, then a result or an action will happen"). Analytical models require profound knowledge of the modeled environmental system, and usually are domain specific. To conduct such modeling in the GIS environment, one often encounters two problems: (1) current commercial GIS products are developed for generic map production and analysis based on relatively simple relational data models that lack spatial analytical or expressive modeling power for specific, complex decision-making tasks (Ji, 1993); and (2) analytical models usually have been developed without reference to GIS and therefore are difficult to directly interface with the GIS software (Goodchild, 1993). To overcome these technical difficulties, a tighter integration is needed, which is defined as *third degree integration*. This type of integration technique is generally characterized by using domain rules or model bases tightly coupled with the GIS database, and therefore it is considered as rule based or knowledge based. The resultant integrated system is often called a GIS-based spatial decision support system (SDSS) (Densham, 1991). Other names are used interchangeably in the GIS literature, such as "environmental decision support system" (Frysinger et al., 1993; Abel et al., 1992), "resource decision support system" (James and Hewitt, 1992), "modeling support system" (DePinto et al., 1993), "geographical modeling system" (Bennett et al., 1993), and "expert GIS" (Ji et al., 1992).

In practice, the boundaries of the three types of environmental modeling and corresponding integrations are often indistinct. For instance, an SDSS for integrated analytical modeling usually includes functions of the first and second degree integrations for database management and information query. Therefore we refer to the third degree integration as a *complete model integration* and the first and second degree integrations as *incomplete model integrations*. A complete modeling integration typically requires (1) system design and programming functions such as spatial database and rule-base development and management, (2) embedding analytical models into the GIS environment, (3) designing automated modeling strategies, and (4) creating a customized interface to organize system operations. These technical issues are discussed in the following sections as they relate to the pilot system developed for coastal wetland restoration planning under the CWPPRA.

THE MODEL, DATABASE, AND DECISION RULES

The Model: Functions, Structure, and Variables

In designing an SDSS, the nature, structure, and input and output formats of the model significantly affect the architecture and functionality of the SDSS. For example, if different models are needed for different and dynamic environmental scenarios, then a system function needs to be developed to direct automated model selection or switching; and, if policy decisions will be based not only on the deterministic model output but also on visualized spatial relationships of studied environmental components, then a graphical display capability should be part of the system.

Model Function

In this study, the models being integrated into the GIS environment are part of the wetland value assessment (WVA) methodology (Mitchell, 1993). The WVA contains analytical, habitat-based wetland resource assessment models for use in prioritizing wetland restoration project proposals submitted for funding under the CWPPRA. The WVA models are used to predict changes in wetland quality and quantity that are expected as a result of a proposed project. The results of the WVA models can be used to measure the effectiveness of a proposed project in planning wetland restoration funding priority.

Model Structure

A WVA model has been developed for each of four coastal Louisiana habitat types: fresh/intermediate marsh, brackish marsh, saline marsh, and cypress-tupelo swamp. Each model consists of a hierarchy of quantitative expressions. On the first level, a set of functional relations defines the quantitative relationship between wetland variables, which are considered important in characterizing the particular wetland type, and *suitability indices*, which define wetland quality. The second level contains a mathematical formula for each specific wetland type that combines the suitability indices for all model variables into a single, overall value for wetland quality, termed the *habitat suitability index*. The third level calculates the *average annual habitat unit* based on suitability indices and habitat suitability indices for specified time periods (target years) in model analysis, and the size of the project site. This modeling procedure is also repeated to calculate the average annual habitat unit for the same wetland area without the restoration project. Comparing habitat units with and without restoration projects provides the net gain in wetland quality due to the restoration project. The value of the net gain is used to normalize the engineering cost (average annualized cost) so as to calculate the cost efficiency of an evaluated project, a criterion for prioritizing the project proposals being evaluated.

Model Variables

Environmental state variables for each WVA model were selected to characterize wetland habitats (Mitchell, 1993). Six variables are used for marsh models, including vegetation components (Variables 1 and 2), marsh edge and interspersion (Variable 3), water depth (Variable 4), water quality (Variable 5), and aquatic organism access (Variable 6). Three variables, limited to hydrologic factors, are employed for the swamp model: water depth and duration (Variable 1), water flow and exchange (Variable 2), and average high salinity (Variable 3). The values of these variables are estimated or predicted based on the domain expert's interpretation of existing data (e.g., aerial photography, satellite imagery, GIS data, and water quality data obtained from field monitoring stations), as well as interviews with knowledgeable individuals.

The aforementioned functional and structural features of the WVA models serve as a framework for the system in database manipulation, rule base design, interface programming, data query and display, and modeling result reporting.

The Model Input and Output: GIS Database

A GIS database usually provides three functions for integrated modeling: (1) it supplies the model input data; (2) it provides supporting information resources to decision-makers; and (3) it handles modeling outputs. The GIS database for WVA modeling consists of GIS coverages for each proposed project site, model input files, the project record file, satellite imagery, digital line graph (DLG) data, field hydrologic data, and the model output files, the last of which consists of ranking lists of selected and/or rejected project proposals.

GIS Coverages and Model Input Files

GIS coverages for evaluated project sites are prepared from the existing National Wetland Inventory (NWI) database developed for the Gulf of Mexico coast at SSC. Specific project coverages are segmented from those NWI maps in ARC/INFO[1] (ARC/INFO is the registered trademark of Environmental Systems Research Institute, Inc., Redlands, CA) format with digitized project boundaries delineated by project sponsors. NWI digital maps describe wetland habitats based on the U.S. Fish and Wildlife Service (USFWS) classification system, which arranges ecological taxa in a system for resource managers, furnishes units for mapping, and displays spatial distributions of environmental features (Cowardin et al., 1979). However, the USFWS system was not designed for specific analytical resource modeling and, thus, few attributes of the map can be directly entered into WVA model calculation. To make a general purpose-based GIS coverage like the NWI map suitable for specific analytical modeling like WVA, the coverage needs to be supplemented with new layers to hold model variable items and their values (see Section 4).

Project Record File

This file contains the information associated with project proposals under evaluation. For each competing project, the project name, the project description, the hydrologic basin in which the proposed project is located, the project sponsor, and the project type are coded and stored in a special ASCII format. This file not only provides records of all projects for review by the decision makers, but, more importantly, it also serves as a database accessible to automated modeling process to associate project record information with final project evaluation ranking.

Satellite Imagery

Satellite imagery data sets have been generated with Landsat Thematic Mapper data for the entire Louisiana coastal area. That raster-based imagery is used as a backdrop for the displayed project site coverage to provide decision-makers with spatial connections of target habitat classes in the context of wetland restoration. For efficient data search and information query, as well as reducing GIS database storage requirements, satellite imagery was transformed from a 24-bit RGB format to a single band of 8-bit image with a look up table (LUT) color (see RGB function in ERDAS user's menu, 1991; ERDAS is the registered trademark of ERDAS, Inc., Atlanta, GA).

---

[1] The mention of trade names of commercial products in this chapter does not constitute endorsement or recommendation for use by National Biological Service, U.S. Department of the Interior.

DLG Data

U.S. Geological Survey DLG data were acquired and converted to ARC/INFO format. Individual coverages were merged to form a large data set for the Louisiana coastal region. These data provide information on hydrologic and artificial structures such as highways or canal systems in or near the proposed restoration site.

Field Hydrologic Data

Field hydrologic data provide supporting information on hydrologic conditions in or adjacent to the proposed project sites. These data were obtained from field hydrologic stations and reformatted and integrated into the system database for review by decision-makers through the system's spatial query function.

Model Output Files

Model outputs are considered a dynamic part of the SDSS database in that they do not exist before modeling and that they vary depending on specific modeling scenarios. In the WVA application, the model output consists of a list of ranked selected or rejected projects, with associated project record information. The output file is integrated into the SDSS through a customized interface for reviewing, formatting, and printing.

Analytical Modeling: Assumption, Procedure, and Rule Expression

Decision Assumptions

The WVA model operates under the assumptions that optimal conditions for general fish and wildlife habitats within a given wetland type can be characterized, and that existing or predicted conditions can be compared to that optimum to provide an index of habitat quality; that index is further assumed to have a linear relationship with the suitability of a coastal wetland system to provide fish and wildlife habitats. With these assumptions, and relevant engineering costs, the model output can provide a measure of the cost efficiency of a specific wetland restoration project in comparison with other competing projects.

Decision Procedures

WVA modeling is conducted through the following procedure (Figure 1): first, the decision-maker provides initial inputs including the project name and anticipated engineering costs; second, the system, based on pre-embedded information about the specific project, determines modeling strategies on which model (wetland type) will be used, how many target years will be applied to the model analysis, and whether a model switching (wetland type change due to the restoration project) will occur and under what condition; then the system re-

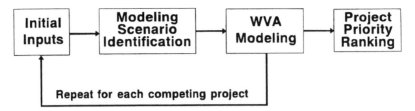

Figure 1   The wetland value assessment modeling procedure.

trieves model variable values from the database and predicts the environmental suitability with the suitability index and the habitat suitability index, and combines the project area size factor and the time component into the wetland change process, ending up with the average annual habitat unit. The system repeats this modeling procedure for each competing project. Finally, projects are ranked in order of priority by comparing cost efficiencies (see the section on model structure).

Rule Expressions

The rules for predicting the change of wetland habitat quality or "suitability" for a given wetland type as values of given variables change are based on the knowledge of the domain experts, and are expressed as deterministic relations in IF/THEN format. For example, to predict a suitability index (SI) for a brackish marsh due to average annual salinity [expressed in parts per thousand (ppt)] change caused by a proposed project, we have rules as follows:

$$\text{If } 3 \leq \text{salinity} < 6, \text{ then } SI = (0.233 \times \text{salinity}) - 0.4$$

$$\text{If } 6 \leq \text{salinity} < 10, \text{ then } SI = 1.0$$

$$\text{If salinity} \geq 10, \text{ then } SI = (-0.15 \times \text{salinity}) + 2.5$$

MODEL INTEGRATION

A complete, third degree model integration with GIS is essentially a software development task. As indicated earlier, to integrate an analytical model into a GIS, one would encounter GIS-architecture and model interface-related technical difficulties. The architecture of most commercial GIS software is unsuitable for complex analytical modeling. This is due in part to the simplicity of the relational data model used for many current GIS packages, which is generally inadequate to handle the complexity of analytical model inputs and modeling processes (Crosbie, 1993). Analytical models usually lack direct interface to the GIS in three areas: (1) the existing GIS database lacks attribute layers that can be directly used as model inputs; (2) the specific model structure leads to a situation where there are no existing GIS functions able to convert those attribute values into adequate model input formats and efficiently carry intermediate and final modeling results; and (3) the analytical modeling procedures are usually too complex to be implemented in the ways with which the existing GIS handles simple spatial queries and organizes the system program and data resources. Solving these problems has been the main task in the WVA decision support system development effort which has resulted in some unique approaches to model integration.

The ARC/INFO GIS was chosen as the host GIS for the WVA model integration and implementation project. WVA models were embedded into this environment by programming the models into executable system files with the ARC Macro Language (AML). Advantages of this model embedding approach are: (1) it allows model access to the ARC/INFO relational database (attribute files), and INFO variables, such as wetland variables and the acreage of the study site, can be directly converted into AML variables as model inputs, facilitating a tight coupling of the GIS database and model; (2) as a programming language, AML possesses the necessary capabilities, such as decision-making (e.g., IF-THEN-ELSE logical branching), looping, calculation, string operations, and external ASCII file manipulation, to program rule-based procedures such as the WVA; and (3) AML can also be used to customize the model-implementing environment by creating a menu-driven interface. The following discussion summarizes the basic features of the host GIS environment and the major technical approaches used in this model integration.

Host GIS Environment

This project utilized ARC/INFO Version 7.0.2 software running on a SUN SPARCstation 10 workstation under the SunOS 4.1.3 UNIX operating system (UNIX is the registered trademark of AT&T, New York, NY). The workstation was equipped with a 19 in., high resolution color monitor, 32 MB of RAM, a 2-GB disk, a 644-MB CD-ROM drive, and a 1/4-in tape drive. As a leading commercial GIS package, ARC/INFO's large user community provides opportunities for extensive sharing of technical information and data. There is potential that this system will be applicable to the requirements of other users.

Model Embedding Approach

Model Input Formatting

A typical analytical model is composed of two parts: facts (inputs) and decision rules (functional expressions). The model embedded in a GIS environment acquires "facts" from the GIS database, which provides one of the primary functional linkages between the model and the host GIS. To ensure the effectiveness of this linkage, the model input format should be compatible with the way in which the host GIS database represents spatial attributes of data in its attribute file. For this purpose, the WVA model inputs are represented with a multilevel format as *Classes* (wetland types) containing *Objects* (wetland variables) that possess *Properties* (variable values). The *Class-Object-Property* structure of the WVA model inputs can be made equal to the *File-Item-Value* structure of the ARC/INFO relational database.

Variable Layer Creation

With this format matching, WVA model variable layers can be created by adding a model input file represented by a *Class*-attributed GIS coverage with appropriate *Objects* (variable items) associated by their *Properties* (variable values) determined by domain experts. For example, *fresh marsh variables* and their *values* for the conditions with and without the restoration project can be stored in a *fresh marsh coverage's* polygon attribute file (PAT file) as special layers. Under the system operation, this can be completed by preparing a formatted wetland variable (ASCII) file and integrating it into the project coverage's attribute file (PAT) with a specially designed program operated in the ARCEDIT environment. Model variable layer creation will make a general purpose-based GIS coverage suitable for a specific modeling.

System Variable Manipulation

Appropriately defining and manipulating system variables is one of the key factors in ensuring the efficiency of integrated analytical modeling. System variable issues in this development included variable classification, variable set-up, and variable-naming strategies.

There are four types of system variables used in developing an SDSS for wetland value assessment:

1. *Control variables*, such as a counter used to keep track of the number of the competing projects entering for ranking
2. *Associate variables*, such as variables for wetland types and engineering costs, which are associated with given projects
3. *Operational variables*, such as wetland variables, time variables (project target years), the area of the project site, and model output variables, which are directly involved in WVA model operations
4. *Intermediate variables*, which carry the intermediate results of model calculation.

The first three types of variables are defined as AML *global variables* with unique names that can be passed among different system programs or modeling stages, while intermediate variables are defined as AML *local variables* which are used only in the same model to avoid intermodel confusion. Control variables are considered top level variables and are set up (initialized) from the system-starting program. Associate variables are characterized by their changeable values, and are interactively set up by selecting appropriate menu choices. Operational variables are set up by running particular programs or carrying on the results of the current model, and special wetland variables, including the area of the study site, are retrieved from corresponding ARC/INFO attribute files by converting these INFO variables into AML variables. Intermediate variables, considered bottom level variables, are set up within the current modeling program.

There are two variable-naming methods used in this modeling system development: (1) variable names associated with *numerical subscripts*, and (2) variable names composed of more than one other variable name, a *compound variable name*. With the first method, model variables can be directly manipulated in AML loop operations for modeling. The compound variable naming method takes advantages of AML capabilities in variable and string manipulations and is used to keep track of competing projects and the associated model outputs in decision making. In this way, a variable name may consist of two, three, or more combined variable names to carry multipart messages (Ji, 1993).

Model Interpreter: Modeling Scenario Identifier

Environmental resource management processes usually involve varied decision-making scenarios. The challenge in handling different modeling scenarios through GIS-based SDSS operations is to automate the modeling process rather than to rely excessively on the interactive interface, which may reduce the system's efficiency and user-friendliness. The WVA modeling process varies from model types (wetland types) and model switching due to planned wetland quality changes. To identify the modeling scenario for a specific project under evaluation, predetermined project conditions need to be input into the SDSS to set up automated modeling strategies. In this system development effort, two special techniques are employed to address this technical challenge: (1) a special "modeling scenario head" of the model input file is designed to embed information of the predetermined project conditions, and (2) a modeling interpreter, a controlling part of the modeling, is programmed to read the modeling scenario head and determine modeling scenarios. This scenario identification process is completed with the following steps as shown in Figure 2. First, the user selects the coverage of the project for evaluation and enters initial inputs and anticipated project costs. Second, the modeling interpreter locates the corresponding model input file according to the coverage, reads the modeling scenario head, and then stores the information as system variables. Finally, those variables are verified to program the modeling procedure for the given wetland restoration project.

Modeling Organizer: Customized Interface

A customized system interface, a controlling component of an SDSS, provides the user with a way to organize system resources in automated modeling.

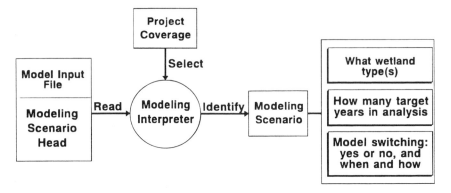

Figure 2   The wetland value assessment modeling scenario identification.

A customized interface of a GIS-based SDSS needs to represent features of both the GIS and SDSS; the former mainly depicts cartographic representations of the spatial data as well as manipulates databases, while the latter performs functions such as model input retrieval, modeling scenario identification, and modeling operations. In this effort, a customized, menu-driven interface has been designed to fulfil the following goals:

1. Increase ease of use of the system. The interface with specially designed layouts consists of an AML form menu sequence that accommodates all WVA modeling-related operations. The interface also provides the user with on-line technical assistance, and allows exiting the system or returning to previous operations from any phase of the modeling.
2. Provide a means to associate and manipulate related system resources for analytical modeling. ARC/INFO coverages for project sites, the project record file, rule bases, decision output files, and supporting environmental information can be searched, subset, or displayed through interface manipulation.
3. Organize the multistep modeling procedures into a simple, integrated form.

SYSTEM FEATURES AND FUNCTIONS

As shown in Figure 3, the resultant system includes four components: a WVA model base, a GIS database, a modeling interpreter, and a customized interface. The model base is tightly coupled with the GIS database with the techniques described earlier. This integrated unit serves as the information and knowledge source of the system. The modeling interpreter provides an inference mechanism in identifying modeling scenarios for automated modeling. The system can be invoked in the ARC environment and is operated through the customized interface, a sequence of AML menus with specially designed layouts. Figure 4 shows top-down operations of the system, including the WVA modeling, the GIS database manipulation, and the model base manipulation. The following discussions focus on major application functions of the system.

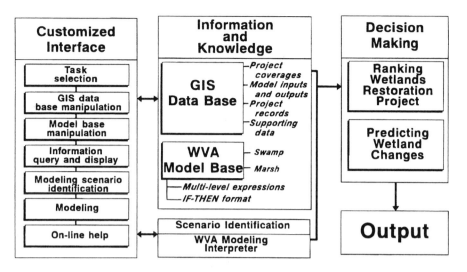

Figure 3   The wetland value assessment decision support system.

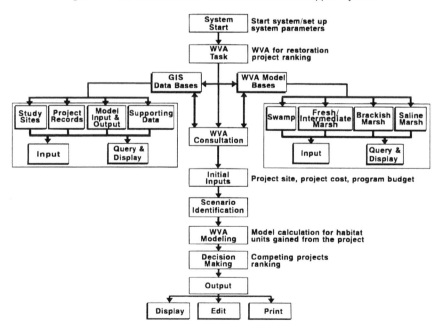

Figure 4   The wetland value assessment decision support manipulation.

Wetland Restoration Priority Analysis

This is the fundamental capability of the system which was alluded to in previous sections. The priority analysis, or WVA modeling, follows a four-step procedure: (1) the project site ARC/INFO coverages are prepared. This is done by segmenting an existing large GIS database covering the study areas, and corresponding model input files, and then integrating the input file into the coverage's relational database to create the model layers; (2) interactively select the project name and enter initial model inputs in the modeling environment (a customized menu) including the project's average annual cost, fully funded cost, as well as total restoration project budget; (3) identify the modeling scenario for the specific project evaluation and model the cost efficiency of the proposed project; and (4) after modeling all competing projects, the system makes decisions on each project's priority ranking with predicted wetland changes. Final output will consist of a list of accepted projects and associated project information. The output file can be either displayed on the screen, reformatted, or printed.

Data Browsing and Updating

In environmental modeling, resource managers often need to visually verify study areas, review project-related data, and update models and information resources. This system can provide a convenient means to satisfy these requirements. In the model base environment, the user can interactively select WVA models to review model variable definitions and verify decision rules. The system also allows the user to update embedded models by modifying program source code. In the GIS database environment, the user can interactively select and display project coverages in the graphic window with user-defined colors and access the project record file, specific model input files, and project coverage's attribute files. These data also can be updated through the customized interface.

Supporting Information Query and Feature Identification

Analytical modeling may provide only part of the solutions that support the resource manager's decision-making. Managers generally rely on additional information to complement their understanding of complex environmental resource processes. In the case of WVA, the resource manager's decisions may be based not only on modeling results, but also on analysis of historical environmental parameter changes, the distribution pattern of the wetland habitats in the project area and surrounding hydrologic basin, and the functional connectivity of the project area with surrounding environments. For this purpose, system capabilities have been developed to conduct spatial query functions on supporting environmental information including satellite imagery, DLG data, and field hydrologic data as previously described. In addition, the system is able to identify wetland habitat features, display the features' spatial distributions, and analyze statistical characteristics such as the acreage of a specific habitat type and the maximum and minimum acreage of geographic units (polygons) within the habitat.

In the GIS database-manipulation environment, the spatial query function employs a technique developed for spatial data searching, segmenting, and overlaying based on the areal extent of a displayed project boundary coverage. The system is able to automatically search large satellite imagery or DLG databases, "grab" the matched part, and display it on the graphic window simultaneously with the overlaid project site coverage. A similar technique is being developed to query the field hydrologic information.

To identify habitat features automatically through SDSS, seven habitat classes are defined and coded, based on the USFWS classification, as "emergents," "shrub/scrub," "forested wetlands," "flats," "aquatic bed," "open water," and "uplands." An interface menu is created to set up identification selectors for these habitat classes. The habitat class can be superimposed over the displayed project site coverage with specially designed color shading patterns after the corresponding identification selector (a menu button) is invoked.

CONCLUSIONS

This study indicates that analytical models generally lack direct interface to commercial GIS packages because: (1) existing GIS databases lack adequate attribute layers that can be directly used as model inputs; (2) there are no existing GIS functions able to convert GIS data attributes into adequate model input formats and efficiently carry modeling results; and (3) modeling procedures are usually too complex to be efficiently implemented using existing commercial GIS functional capabilities.

This SDSS development effort suggests that the following steps are required to effectively and efficiently integrate an analytical model into the GIS environment: (1) design model input formats in manner compatible with the host GIS, (2) create model variable layers in the existing GIS attribute file, (3) adequately define and set up system variables to carry modeling inputs, outputs, and intermediate results, and (4) customize the system interface to improve interactive control of system resources required for automated modeling.

The WVA methodology, an analytical modeling procedure, is a powerful tool for wetland restoration planning. Integrating this model with a GIS has proved to be an efficient approach for developing a GIS-based spatial decision support system. The system is capable of efficiently handling large quantities of spatial-temporal data for specialized environmental modeling and providing a means for visualizing dynamic environmental variables associated with wetland resource management decision-making.

ACKNOWLEDGMENTS

Dr. James B. Johnston, the Chief of Spatial Analysis Branch of SSC, planned and supervised this research cooperation between SSC and Ecological Services.

The authors wish to thank Drs. Robert Stewart and Gerald Grau, Director and Assistant Director of SSC, for their encouragement, support, and suggestions to foster this project; and David Frugé and Russell Watson, Field Supervisor and Assistant Field Supervisor, U.S. Fish and Wildlife Service, Ecological Services, Lafayette, for their consistent support to this cooperative project between SSC and that office. Thanks are also extended to Vince Sclafani, JCWS, for his support in satellite imagery and DLG database generation and management, as well as GIS network operations; and Sue Lauritzen, SSC, for her work in preparing high quality graphical illustrations.

An earlier version of Figure 3 appeared in the previous publications (Ji et al., 1992, 1993; Ji, 1993). Figure 4 and a part of the contents of Section 4 are adapted from Ji (1993). The authors acknowledge the copyright permission granted by American Society of Civil Engineers, American Society for Photogrammetry and Remote Sensing, and Aster Publishing Corporation.

REFERENCES

Abel, D. J., Yap, S. K., Ackland, R., Cameron, M. A., Smith, D. F., and Walker, G., Environmental decision support system project: an exploration of alternative architectures for geographical information systems, *Int. J. Geographical Information Syst.*, 6, 193, 1992.

Bennett, D. A., Armstrong, M. P., and Weirich, F., An object-oriented modelbase management system for environmental simulation, in Proceedings of Second International Conference/Workshop on Integrating Geographic Information Systems and Environmental Models, Breckenridge, CO, 1993.

Cowardin, L. M., Carter, V., Golet, F. C., and LaRoe, E. T., *Classification of Wetlands and Deepwater Habitats of the United States*, U.S. Department of the Interior, Fish and Wildlife Service, FWS/OBS-79/31, 1979 (reprinted in 1992), 131.

Crosbie, P., A new approach to environmental modeling, in Proceedings of Second International Conference/Workshop on Integrating Geographic Information Systems and Environmental Modeling, Breckenridge, CO, 1993.

Dahl, T. E., *Wetland Losses in the United States: 1780's to 1980's*, U.S. Department of the Interior, Fish and Wildlife Service, Washington, D.C., 1990, 21 pp.

———— and Johnson, C. E., *Status and Trends of Wetlands in the Conterminous United States, Mid 1970's to Mid 1980's*, U.S. Department of the Interior, Fish and Wildlife Service, Washington, D.C., 1991, 28 pp.

Densham, P. J., Spatial decision support systems, in *Geographical Information Systems: Principles and Applications*, Vol. 1, Maguire, D. J., Goodchild, M. F., and Rhind, D. W., Eds., Longman Scientific & Technical, Harlow, 1991, Chap. 26.

DePinto, J. V., Atkinson, J. F., Calkins, H. W., Densham, P. J., Guan, W., Lin, H., Xia, F., Rodgers, P. W., Slawecki, T., and Richardson, W. L., Development of GEO-WAMS: a modeling support system for integrating GIS with watershed analysis models, in Proceedings of Second International Conference/Workshop on Integrating Geographic Information Systems and Environmental Models Breckenridge, CO, 1993.

ERDAS user's menu, Core Vol. 2, Ver.7.5., ERDAS, Inc., 1991.

Frysinger, S. P., Copperman, D. A., and Levantino, J. P., Environmental decision support systems: an open architecture integrating modeling and GIS, in Proceedings of Second International Conference/Workshop on Integrating Geographic Information Systems and Environmental Models, Breckenridge, CO, 1993.

Goodchild, M. F., The state of GIS for environmental problem-solving, in *Environmental Modeling with GIS*, Goodchild, M. F., Parks, B. O. and Steyaert, L. T., Eds., Oxford University Press, Oxford, 1993, Chap. 2.

Greeson, P. B., Clark, J. R., and Clark, J. E., Wetland functions and values: the state of our understanding, in Proceedings of National Symposium on Wetlands, American Water Resource Association, Minneapolis, MN 1979, 674.

James, D. E. and Hewitt, M. J., III, To save a river: building a resource decision support system for the Blackfoot River drainage, *Geo Info Syst.*, 2, 37, 1992.

Ji, W., Integrating a resource assessment model into ARC/INFO GIS: a spatial decision support system development, in Proceedings of ACSM/ASPRS Annual Conference, Lewis, A. J., Ed., American Society for Photogrammetry and Remote Sensing, New Orleans, LA, 1993, 159.

Ji, W., Johnston, J. B., McNiff, M. E., and Mitchell, L. C., Knowledge-based GIS: a expert system approach for managing wetlands, *Geo Info Syst.*, 2, 60, 1992.

Ji, W., Mitchell, L. C., McNiff, M. E., and Johnston, J. B., A multifunctional decision support GIS for coastal management, in Proceedings of the Eighth Symposium on Coastal and Ocean Management, Vol. 1, Magoon, O. T., Wilson, W. S., Converse, H., and Tobin, L. T., Eds., American Society of Civil Engineers, New Orleans, LA, 1993, 94.

Mitchell, L. C., *Wetland Value Assessment Methodology and Community Models*, File Report, U.S. Fish and Wildlife Service, Ecological Services, Lafayette, LA, 1993.

Mitsch, W. J. and Gosselink, J. G., *Wetlands*, Van Nostrand Reinhold, New York, 1986, 539 pp.

Tiner, R. W., Jr., *Wetlands of the United States: Current Status and Recent Trends*, U.S. Dept. Inter., Fish and Wildl. Serv., Washington, D.C., 1984, 59.

Watzin, M. C. and Gosselink, J. G., *The Fragile Fringe: Coastal Wetlands of the Continental United States*, Louisiana Sea Grant College Program, Louisiana State University, Baton Rouge, LA; U.S. Fish and Wildlife Service, Washington, D.C.; and National Oceanic and Atmospheric Administration, Rockville, MD, 1992.

CHAPTER 5

Use of a GIS for Wetland Identification, The St. Clair Flats, Michigan

John G. Lyon and Kirt F. Adkins

ABSTRACT

A recent trend in science and engineering is the greater awareness of the resource value of coastal wetlands. Geographic information systems (GIS) provide a powerful tool for storing, analyzing, and displaying both spatial and nonspatial data associated with wetlands. A study area on the St. Clair Flats in eastern Michigan was chosen for a wetland prediction model that incorporated GIS and remote sensor data. If soils, previous wetland classifications, and water depth/land elevation are known, wetland communities can be predicted, and the progressive change of the wetland can be potentially monitored. This procedure was performed by merging land cover classifications with soils, and elevation/bathymetry data to form a model for predicting wetland communities. The predictions were then checked by aircraft MSS data taken in 1985.

INTRODUCTION

In recent years a greater awareness of the resource value of coastal wetlands has brought about a need to protect and conserve these ecosystems. Wetlands are defined as "areas which are periodically or permanently inundated with water and which are typically characterized by vegetation that requires saturated soil for growth and reproduction." This definition includes areas commonly referred to as bogs, fens, marshes, sloughs, swamps, and wet meadows (Herdendorf et al. 1986; USACE, 1987).

Traditionally wetlands were conserved as waterfowl breeding and feeding areas, or to a lesser extent as spawning and nursery habitat for fish. Today it is realized that wetlands have many benefits, including flood control, water purification, shore erosion protection, control of nutrient cycles, accumulation of sediment, and supply of detritus for the aquatic food web (Herdendorf et al., 1986). A valuable resource such as wetlands should be monitored so that temporal variations in the environment can be observed. These changes can then be used in up-to-date land suitability analyses for human activities (development, roads, cottages, etc.), and for intelligent resource management and assessment.

GIS can provide a powerful tool for storing, analyzing, and displaying both spatial and nonspatial data associated with wetland areas. Once land and resource use alternatives are formulated by a natural resource agency through modeling, GIS technologies can be used to evaluate each alternative in terms of environmental impacts, economic implications, land areas involved, and potential use conflicts (Parker, 1988).

While GIS are very effective in these types of analyses, they will not be of much value without any data, or with very low quality data. Overall this tends to make good quality spatial and biological data for wetlands very scarce and often hard to acquire. While maps often contain the marsh areas on them, the positional and feature data contained in these areas is often of questionable quality.

Only recently have detailed studies such as Lyon (1979) and Herdendorf et al. (1986) started to provide details concerning the diversity of plant and animal species contained in wetlands, as well as the operations of these dynamic ecosystems and their interrelationships with surrounding land cover communities.

Remote sensing technologies such as satellite images and aerial photos have begun to help resource managers get a good idea of the species composition and areal extent of the vegetation communities, as well as neighboring land use which could affect wetland activities (Lyon, 1979). Aerial photos and satellite images provide a way of collecting data and monitoring change without disturbing the fragile wetland areas.

BACKGROUND

The study area for this project is located on the northeast shore of Lake St. Clair, which lies on the east side of central Michigan. Lake St. Clair is bordered on the southwest and west by Detroit and its northern suburbs, respectively. The lake is noted as a midway point between Lakes Huron and Erie; the water flowing from Lake Huron runs through the St. Clair River to Lake St. Clair, which empties into the Detroit River, which in turn flows south to Lake Erie. The lake and rivers form the boundary separating the United States and Canada (Ontario), and also provide the Great Lakes navigation route between Lake Huron and Lake Erie.

The St. Clair River empties into the northeast corner of Lake St. Clair and forms the St. Clair Flats (Figure 1), a massive, deltaic deposit consisting of a number of active and inactive distributaries at the mouth of the river. This delta region is the largest in the Great Lakes system and is characterized by shallow water areas and low ridges that are covered by marshes and wetlands. These marshes are comprised of vegetation such as willows, cattails, and sedges, as well as submergent vegetation.

In the delta area the main factor affecting marsh type is the water level; which fluctuates over long periods due to variable rainfall in the Great Lakes Basin, and over short periods of time due to wind generated seiches on the lake.

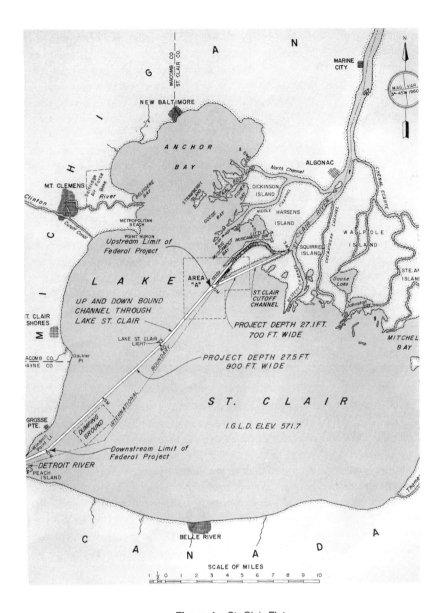

Figure 1   St. Clair Flats.

The wetlands of the St. Clair Flats are dynamic ecosystems which slowly change to uplands due to the trapping of sediment and the build-up of organic matter. This conversion can be slowed by variable water levels, but it seems to be a natural process on this coast. Monitoring this land cover change is very important in the assessment of resources and can help in the understanding of the processes associated with a delta type wetland.

One main process associated with wetland areas is the relationship of vegetation and environment. The vegetation making up a wetland is a very strong indicator of environmental condition, because aquatic plants respond to many environmental factors. These factors include water depth, topography, soils, pH, turbidity, and flooding. Since these vegetation types are easily observed, and because there are distinct differences in the dominant species at different water depths and land elevations, the differences in plant composition can be used to differentiate wetland areas into communities.

Many of the wetland communities exhibit particular reflectance characteristics in the near-IR and red sections of the spectrum, as well as distinct morphologies (Lyon, 1979). Since vegetation growth patterns are related to area environmental factors such as water depth, inundation periods, elevation, and soils, the changes in plant species composition indicate the wetland boundary (Lyon, 1979).

Thus, a GIS can be used to predict wetlands types if three main levels are developed: soils from soils maps, bathymetry/elevation from topographic and bathymetric maps, and previous wetland classifications from aerial photos. These predictions can then be checked by satellite or airborne MSS data. From this, a better understanding of certain wetland communities can be gained, and an evaluation of the accuracy of the predictions can be made.

OBJECTIVES

The objective of this study was to analyze and attempt to predict wetland change, and to gain a better understanding of vegetation class relationship to soils and elevation/bathymetry data. The wetland communities in the St. Clair Flats are dynamic ecosystems which are defined in terms of vegetation, soils, and hydrology, but are difficult to map by traditional methods due to water level fluctuations and their effect on species composition (Lyon, 1979). Field work on species composition and lake level give only a single record at a sample, and the observer's view is often restricted. Presumably, if soils, previous wetland classifications, and water depth and land elevation are known, wetland communities can be predicted, and the natural progressive change of wetlands to uplands can be estimated and then monitored. In turn, recent data such as aircraft MSS data can be used to check the predictions and evaluate the model.

METHODS

The basic outline for this study was as follows: land cover classifications from the Michigan Land Cover/Use System (Appendix A), which were derived from aerial photos taken in 1974, were merged with soils, and elevation/bathymetry data to predict wetland communities for 1985. These predictions were then compared to a computer classification performed on multispectral scanner data which was taken in 1985. Given the knowledge about wetland characteristics, the predictions of the natural change in wetland locations and community types, as well as predictions of new upland area, presence of wetlands could be evaluated.

The image processing, digitizing, and GIS work for this project was performed on an ERDAS system, housed at the Ohio Department of Natural Resources. To predict wetland change, past land cover data had to be acquired. For this a 1:24,000 scale map was used which contained land cover classifications delineated from color and color-IR aerial photos taken in 1974 (Figure 2). The wetland communities were classified in the Michigan Land Cover/Use Classification System (Appendix A). The smallest area that can be classified from this scale photograph is about 2 acres in size (Lyon, 1979).

Figure 2   Aerial photos delineating land cover classifications (1974).

The Michigan system is an expansion of the USGS Anderson system. Levels 1 and 2 of the Michigan system are identical to the Anderson System, but levels three and four are optimized for use with wetlands found in Michigan.

A multispectral scanner image of the St. Clair Flats taken in October of 1985 was also used (Figure 3). This imagery was taken by the United States Environmental Protection Agency's Daedalus 1260 scanner, which records 11 channels of reflected and radiant energy. The data were taken at an altitude of 20,000 ft above ground level, which provides an IFOV of 50 × 50 ft (0.057 acre). The bands used for this study were 0.45 to 0.50 μm (blue), 0.50 to 0.55 μm (green), and 0.65 to 0.69 μm (red). A small area of the St. Clair Flats was selected for the analysis to make a more compact and clear data group. The southwest side of Harsens Island and the east side of Little Muscamoot Bay were chosen because of the variety of land cover and the existence of submerged vegetation.

The subset of the MSS scene was first entered into a clustering algorithm to identify the land cover classes. The computer was told to select 20 different clusters for this area (Figure 3; see color section). After the computer classified the pixels, more recent color-IR photos were used to identify the clusters and aggregate them to the 8 classes presented in Figure 3.

Next, the Michigan Land Cover/Use classification from Lyon (1979) (Figure 4) was digitized into a geo-referenced system (State Plane Coordinates). The seven land cover classes for the study area were then digitized from the 7.5-min

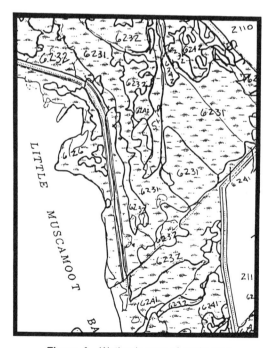

Figure 4   Wetlands map of study area.

quad and subsequently labeled (Figure 5). Notice that the Michigan System classification does not include submerged aquatics nor deep marsh areas in Little
Muscamoot Bay.

A soils layer was then digitized from the maps provided in the Soil Survey
of St. Clair County, MI (Figure 6). This figure presents the two predominant soil
types in the study area: the Bach Series and the Sanilac Series. The Bach Series

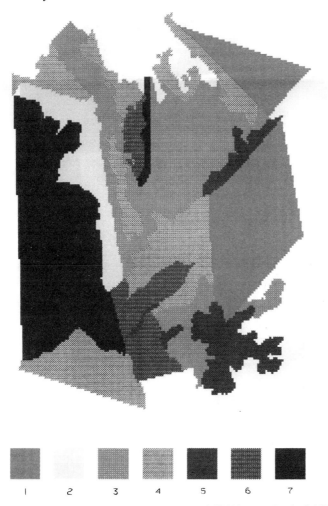

Figure 5   The variable name is: Landcover of SW Harsens by Aerial Photo

Value	Class Name	No. of Points	%
1	Cropland	3651.	17.48
2	Buttonbush (>6 in. water)	1335.	6.39
3	Shallow Marsh (Cattails)	4450.	21.31
4	Shallow Marsh (Sedges)	4069.	19.48
5	Deep Marsh (Cattails)	1474.	7.06
6	Deep Marsh (Bur reed)	1387.	6.64
7	Water	?	?

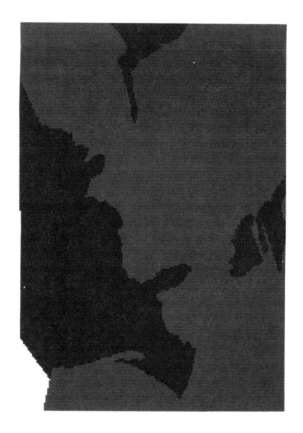

The variable name is : SW Harsens Soils

Figure 6   GIS image of soil types.

Value	Class Name	No. of Points	%
1	Water	11216.	24.50
2	Bach Series (Bc) Very Fine Sand	32162.	70.27
3	Sanilac Series (SaA) Loamy Very	2394.	5.23

consists of nearly level, very poorly drained soils that formed in limey lacustrine sediments of very fine sandy loam. Most areas of Bach soils were covered with wetland grasses, sedges, and reeds. The Sanilac Series was made up of somewhat poorly drained soils formed in limy, waterlaid sediments of loamy very fine sand. These soils occurred on the higher areas of the delta and can support crops and deciduous trees.

The next layer that was to be entered was the elevation/bathymetry data. While this type of data was available for areas with more relief, good accurate data for the nearly level wetland areas was very scarce. Yet water depth was the most important factor in determining wetland vegetation communities, with a change of 0.5 to 1.0 ft (0.15 to 0.30 m) being significant for aquatic plants.

From the 7.5-min quad it was apparent that the high point on Harsens Island near the study area is 577 ft above sea level, only 2 ft (0.60 m) above the mean lake level of 575 ft. Elevation was also very important for differentiating wetland and upland communities, where 1 ft (0.3 m) of elevation change is thought to strongly affect the make-up of the plant cover. An elevation/bathymetry map was finally completed, and this information was incorporated.

A simple wetland prediction model was developed to identify wetlands. The basic wetland prediction model formulated for this study combines soils, elevation/bathymetry, and the 1974 land cover classification in a linear fashion to produce a predicted wetland type. The formula included these parameters:

$$y = ax_1 + bx_2 + cx_3 + e \qquad (1)$$

where: y = Predicted wetland community
x_1 = Soils data
x_2 = Elevation/Bathymetry data
x_3 = 1974 Land cover classification from photos
a,b,c = Weighting factors for each variable
e = "Random noise" or error term in model

The model shown was first used with the weighting factors (a,b,c) equal to one. In this fashion the true weighting for the independent variables could be evaluated and appropriate adjustments could be made. Sample calculation for two land cover classes can be found in Appendix B.

RESULTS

The relationship of wetland plant species and environmental factors such as water depth and soils was found to be fairly accurate as evaluated from aircraft data. It is also seen from this project that potentially a wetland prediction model can be developed and applied to real-life situations through the use of a GIS.

Another result is the realization of the suitability of aircraft MSS data for this type of study over aerial photos. The pixel size of 0.057 acres from this scanner data will give a much more accurate assessment of the acreage covered by the different wetland vegetation communities. These areas often cover areas much less that the 2 acre minimum required by the aerial photo interpretation method. In addition, the MSS data can identify submerged vegetation communities such as Eel grass, which can be important foraging areas for dabbling ducks.

A better understanding of the use and function of GIS systems was developed, especially polygon digitizing routines. The errors associated with this type of data input were observed first hand and will serve as a reminder when more work of this type is being performed. The compilation of many varied data sources with different scales and pixel sizes was found to be fairly complicated. It was also seen that rectification of scenes with large areas of water can be very tough to achieve.

REFERENCES

Burrough, P. A., *Principles of Geographic Information Systems for Land Resources Assessment*, Oxford University Press, New York, 1986, 193 pp.

Campbell, J. B., *Introduction to Remote Sensing*, Guildford Press, New York, 1987.

Herdendorf, C. E., Raphael, C. N., and Jaworski, E., *The Ecology of Lake St. Clair Wetlands: A Community Profile*. U.S. Fish and Wildlife Service. Biological Report 85(7.7), 1986, 187 pp.

Johnston, K. M., Natural resource modeling in the geographic information system environment, *Photogramm. Eng. Remote Sensing*, 53, 1411–1415, 1987.

Lillesand, T. M. and Kiefer, R. W., *Remote Sensing and Image Interpretation*, John Wiley & Sons, New York, 1987.

Lyon, J., *Analyses of Coastal Wetland Communities: The St. Clair Flats, Michigan*, M.S. Thesis, School of Natural Resources, Univ. of Michigan, Ann Arbor, MI, 1979, 80 pp.

Parker, H. D., The unique qualities of a geographic information system: a commentary, *Photogramm. Eng. Remote Sensing*, 54, 1547–1549, 1988.

U.S. Army Corps of Engineers, Wetlands Delineation Manual. Technical Report Y-87-1, Department of the Army, Washington, D.C., 1987.

U.S. Department of Agriculture, Soil Conservation Service. *Soil Survey of St. Clair County, Michigan*. U.S. Government Printing Office, Washington, D.C., 1974.

APPENDIX A

Michigan Land Cover/Use Classification System for
Wetland Areas Found in This Study

21   Cropland, Rotation and Permanent Pasture

54   Great Lakes

61   Forested (wooded) Wetlands

6126   Willow-buttonbush associations (less than 50% cover—more than 6 in. of water)

62   Nonforested (non-wooded) Wetlands
   623   Shallow Marshes
      6232   Bur-reed, bulrushes, sedges, blue-joint grass
   624   Deep Marshes
      6241   Cattail predominates
      6242   Bur-reed, rushes, and sedges

APPENDIX B

A sample wetland prediction is presented for two distinct classes seen on Harsens Island. Each attribute in each GIS layer is given a number, so that composite overlays can be developed. The addition of overlying samples from each layer gives an idea of the strength of the prediction for that particular category: each category has an optimum composite number.

Soils: Bach = 1
                              Sanilac = 2

Bathymetry/elevation: > 1.0 water depth = 8
                            1.0 – 0.5 water depth = 1
                            0.5 – 0.0 water depth = 2
                            575 – 576 land elevation = 3
                            576 – 577 land elevation = 4

Land cover from 1974:   6242 = 1
                            6241 = 2
                            6126 = 3
                            6231 = 5
                            2110 = 8

Bur-Reed, Deep Marsh

Bur-reed prefers Bach Series soil and water depths from 0.5 to 1.0 ft. A pixel in the south center of the study area, in the mouth of the small cove was chosen for the calculation.

Soils (Bach) = 1
Bathymetry (0.5 to 1.0 ft depth) = 1
1974 Classification (6242-Bur-reed, sedges) = 1

$$y = 1 + 1 + 1 = 3 \qquad (2)$$

Thus all factors were optimum for Bur-reed and its corresponding optimum composite number is 3. This prediction can then be checked with the scanner classification—it is seen that the mouth of this cove is a yellow-green color on the original (5), showing Bur-reed in deep marsh (Figure 7; see color section).

Crops are more easily grown on Sanilac series soils and elevations above 576 ft. A pixel in the far northeast corner of the study area was chosen for the calculation.

Soils (Sanilac) = 2
Elevation (577 ft) = 4
1974 Classification (2110-cropland) = 8

$$y = 2 + 4 + 8 = 14 \tag{3}$$

So the factors were again optimum and the corresponding optimum composite number for cropland is 14. The scanner data confirms this prediction.

CHAPTER **6**

Modeling Mangrove Canopy Reflectance Using a Light Interaction Model and an Optimization Technique

Elijah W. Ramsey III and John R. Jensen

ABSTRACT

At 20 sites, incorporating mixtures of black, red, and white mangroves, canopy reflectance spectra were derived from high resolution spectral data taken from a helicopter platform. Canopy characteristics were predicted from the canopy reflectance spectra by using measured and estimated data as inputs into a light-canopy interaction model within a optimization routine. Pertinent to average conditions typifying the area and time of the study, the light-canopy interaction model accomplished two goals. Using the model as a predictor, a sensitivity analysis suggested that little error in modeling the near-nadir view canopy reflectance (R_{cv}) would result from assuming an average soil reflectance of about 0.1, at leaf area index (LAI) values above 2, at near-infrared (NIR) leaf reflectances higher than about 0.45, and at sun elevation angles >40°. Moderate errors could result from assuming a spherical leaf angle distribution (LAD). Relatively high errors could result from errors in estimating visible leaf reflectances (and NIR reflectances <0.45) and percent skylight. Differences between canopy hemispherical reflectance (R_c) and R_{cv} were dominated by percent skylight variation, while differences between R_c and R_{cv} were moderate to slight at a sun elevation above 20 to 30°, a near spherical LAD, a soil reflectance near 0.1, a LAI up to 4, and a NIR leaf reflectance less than 0.7.

Simulated canopy reflectance spectra were close predictors of obtained spectra, with R^2 values >0.97. Mean predicted LAI values were 2.6 ± 0.86 (mean ± 1 standard deviation) and were highly related to LAI values derived from field measurements. Seventy-eight percent of the modeled LAI variance was predicted by a normalized difference vegetation index transform of the field canopy spectra data. Predicted LAD values had a near spherical mean value, while the mean difference between input (estimated from laboratory measurements) and predicted leaf reflectances was nearly zero.

INTRODUCTION

Worldwide, the prospect for coastal wetlands is bleak in light of existing conditions and projected changes (NRC, 1987). Detrimental human-induced influences have increased to the point that a large proportion of the global mangrove wetlands is threatened with destruction. In southwest Florida, tremendous growth and development pressure has resulted in appreciable losses in mangrove wetlands (Lugo and Snedaker, 1974). Further compounding these human-induced stresses, the eustatic sea-level rise has serious implications for mangrove wetlands. Predominantly, the sea level is rising globally from 1 to 5 mm yearly; however, increasing rates are predicted to cause a 50 to 150 cm rise in sea level by the year 2100 (NRC, 1987). If the "greenhouse effect" is confirmed, even these devastating levels may be outpaced (NRC, 1987).

Management practices that consider both the function and value (amenities provided from within the wetland system and beyond its boundaries) of these wetlands must be formulated. However, plant types, the distribution and extent of plant communities (Tomlinson, 1986), and the uses and structural aspects (Odum et al., 1982; Saenger et al., 1983) of a majority of mangrove wetlands are poorly known. Regionally, this information is needed to understand the structure and function of mangrove ecosystems in a holistic context. Provided on a repetitive basis, improvements in the understanding of basic processes such as succession and primary production would result in gaining critical information used in environmental surveys and the resources to make management decisions and develop sustainable management practices.

Even though statistical methods were useful in relating canopy reflectance to canopy structure (e.g., leaf area index and height) and species composition (Ramsey and Jensen, 1995), these statistical techniques were inferential only. They could not provide a direct link between the incident light, canopy describers, and light detected at the sensor. Further, canopy properties such as leaf reflectances and transmittances could not be directly related to canopy reflectance. To fully exploit remote sensing of mangrove wetlands from aircraft or satellite platforms, relationships must be understood between the canopy reflectance and the canopy structure (e.g., leaf angle distribution, leaf area index) and species (leaf spectral properties) (Badhwar et al., 1985; Goel and Grier, 1986). Canopy modeling can provide the link between the incident light, canopy characteristics, and the light detected at the sensor (Kimes, 1985; Steven, 1985). Potentially, from measurements of reflected radiation from the canopy, an accurate canopy model will predict not only structural information but also will estimate species composition (Franklin and Strahler, 1988), and thereby identify the community type within the mangrove wetland ecosystem. The prediction ability must be independent of illumination conditions; therefore, the canopy model either must reduce the influence of variable incident light character or incorporate this influence in the model (Horler et al., 1983; Kimes and Sellers, 1985).

The problem is one of simplification. Plant environments are not understood well enough to be described in exact physical terms (Norman, 1979). The researcher must select those aspects of the canopy and light climate that are likely to be particularly significant and simplify the canopy model to terms justified by knowledge of canopy components (e.g., leaf azimuthal symmetry, generalized leaf inclination distribution) (Evans, 1965). If necessary, complexities can be introduced and their relative performance judged without initially confronting the problem in all its intricacy (Evans, 1965). We believed a reasonable beginning was to implement a physical model; simplify the model based on the data available and knowledge of the system; conduct sensitivity tests to evaluate the appropriateness of the model formulation, inherent with simplifications; and compare the model predictions to observations.

Thus, the study objective was to begin to answer the question of whether the spectral definition and discrimination of mangrove communities can be improved by using high spectral resolution data (e.g., airborne visible and infrared imaging spectrometer) as inputs into a simplified radiative transfer model. To accomplish this objective, we applied site-specific high resolution spectral data to the study of mangrove communities. A light-canopy interaction model within a optimization routine was used to predict canopy reflectance, leaf area index (LAI), leaf angle distribution (LAD), and a leaf reflectance factor (LRF) based on the difference between measured and predicted leaf reflectances.

DATA COLLECTION

Data were collected in southwest Florida from October 1988 to April 1989 (see Ramsey and Jensen [1995] for a more detailed discussion). Twenty field sites were chosen that encompassed five mangrove communities defined by Lugo and Snedaker (1974) (Figure 1), while the higher number of field sites offered replication of each category. Seventeen of the 20 sites were occupied on the ground. The field data collected within a 30.5 by 30.5 m area centered at each site were canopy closure, litter samples at selected sites, leaf samples, and subsequent leaf reflectance. Observations recorded at each site included: canopy and understory species type, percent, and height. Location markers, visible from areal and ground surveys, were installed at 17 sites. In October, near-nadir (about 5 to 15° zenith angle) canopy radiance upwelling spectra were obtained from a helicopter platform. Helicopter spectral measurements were centered at the flag markers.

CANOPY REFLECTANCE SIMULATIONS IN SOUTHWEST FLORIDA

Several investigators have developed mathematical models to predict spectral reflectance of a vegetative canopy (a summary is contained in Vygodskaya and Gorshokova, 1989). Most models show no fundamental differences, but differ primarily in the required input parameters and in the way the canopy architecture is modeled (Bunnik, 1977). The model chosen for this analysis was first

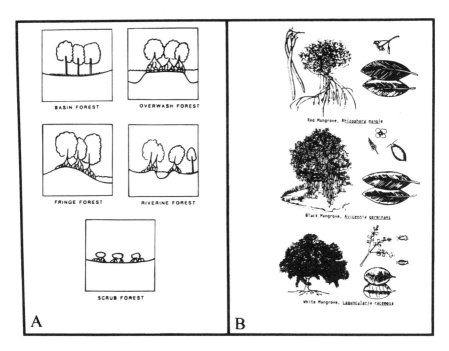

Figure 1   Mangrove community types (A) and species (B) present within the study area. (Adapted from Odum et al., 1982. With permission)

developed by de Wit (1965) and later extended and improved by Goudriaan (1977). The original computer algorithm that simulated light and canopy interactions was taken from Goudriaan (1977), and altered to accommodate the inputs and outputs of spectra. This model was chosen because: (1) the model has been widely used (Bunnik, 1977; Vygodskaya and Gorshokova, 1989); (2) the canopy reflectances generated by the model are comparable to many other models; (3) the model uses a nine class (each 10°) leaf inclination angle distribution; (4) the model has the ability to incorporate both the sun and view zenith positions (nine classes); (5) a range of leaf inclination angle distributions and skylight distributions can be implemented; and (6) the level of the model (e.g., integration over both view inclination and azimuth) can be altered to suit the task and quality of the data available.

Canopy Reflectance Model Assumptions and Inputs

Because of the complexities in defining all aspects of the light-canopy interactions, some fairly accepted simplifications were incorporated at the onset. First, a one-layer uniform canopy covering a Lambertian soil was assumed (Kimes et al., 1986). Little difference exists between a dense canopy and an equivalent homogeneous scene (Kimes et al., 1986); and, in fully formed canopies, the canopy model is insensitive to soil reflectance (Goel et al., 1984; Kimes, 1984;

Badhwar et al., 1985). Second, leaves were assumed to be symmetric azimuthally (Gates, 1970; Norman, 1979), and leaf inclinations were assumed to be adequately described by a generalized distribution function (Evans, 1965; Gates, 1970; Norman, 1979; Sellers, 1987). Finally, the upper and underside of leaves were not distinguished and the leaves were assumed lambertian in both transmittance and reflectance (de Wit, 1965).

Even without these complexities, asymmetries can exist in the canopy reflectance, with respect to illumination and view geometries, that reduce the appropriateness of approximating the canopy hemispherical reflectance from recordings at one view angle (commonly near-nadir; Horler et al., 1983). The causes of these asymmetries are functions of complex relationships that may be very different for different canopy reflectance data sets (Kimes, 1984). However, when modeling the light climate in canopies, two important problems can occur. First, at a fixed solar zenith angle and canopy depth, the probability of gap (view depth into canopy) will decrease as the off-nadir view angle increases (Kimes, 1985). Thus, at increasing view angles, there is an increasing return potential of the incident light, and therefore, an increasing relative reflectance. Second, the magnitude of the off-nadir increase in canopy reflectance is modified by the solar zenith angle; the increase is highest at large solar zenith angles and smallest when the sun is at zenith (Kimes, 1984).

In mangrove canopies, at least one cause of canopy reflectance asymmetry (i.e., off-nadir increases in canopy reflectance) may be somewhat minimized by two factors. First, in tropical forests the probability of gap is generally more uniform from nadir to the canopy horizon than in northern temporal forests (Evans, 1965). Second, the canopy depth is fairly narrow (about 4 m), while the light interception is high (about 95%) (Odum et al., 1982). Further, even though soil reflectance can significantly influence the reflectance distribution in incomplete canopies (Kimes, 1984), this aspect may be minimized by the combined high light interception of the canopy with the low reflectance at all wavelengths of the moist litter, or standing water, that characterized the background at most sites.

Inputs to the modified canopy reflectance model matched the conditions pertinent to the time of the helicopter radiometer data acquisition at each site. These inputs are as follows.

1. The sun zenith position was entered in one of nine sky sections.
2. The ratio of skylight and direct sunlight to total downwelling light was input at each of the 252 wavelengths defining the radiometer spectral characteristics (Ramsey and Jensen, 1995). The character and magnitude of downwelling light were generated by using a single scattering approximation (Turner and Spencer, 1972). The atmospheric optical state was generated from the meteorological visibility (Naples Airport, FL) by using an empirical relationship derived by Elterman (1970).
3. The type of soil reflectance spectra depended on the mangrove community type. Five background reflectance spectra types were available from field sample spectral analysis (Ramsey and Jensen, 1995).

4. Leaf reflectance spectra were entered. These spectra were available from flat-plate reflectance analysis (Ramsey and Jensen, 1995). The leaf transmittance was assumed to be equal to the reflectance if the reflectance was less than 45%, while conversely, if the reflectance was larger than 45%, the transmittance equaled 90%-leaf reflectance (assuming a 10% absorption of light).

5. A parameter, XL, was introduced as an estimate of LAD deviation from a spherical distribution. A value of XL of 0 characterizes a spherical LAD, while values from 0 to -0.4 show an increasing tendency for a more vertical LAD and 0 to 0.6 for a more horizontal LAD (Goudriaan, 1977). By using equations outlined in Goudriaan and a LU decomposition routine (Press et al., 1990), each XL value can then be used to estimate the leaf inclination distribution used in the model as a weighting factor for the scattered fluxes.

6. The skylight distribution was assumed uniform (Goudriaan, 1977).

The model capabilities were examined in two steps: (1) the sensitivity of canopy reflectance to different factors influencing the canopy and light interactions was performed by using the model. This was accomplished with a limited analysis of model/variable sensitivity; and (2) the performance of the model in predicting canopy reflectance for each site was examined by comparing these predicted values to those calculated from direct observation.

Sensitivity Analysis of the Canopy Model

Sensitivity analyses were performed to ascertain the canopy characteristics that significantly influenced the prediction of canopy reflectance. The sensitivity was restricted to average conditions prevalent during the helicopter radiometer data acquisition. For example, the sensitivity of the model to variations in LAI could be expected to vary substantially with sun inclination, soil reflectance, and other canopy factors. However, only the predominant sun inclination and soil reflectance were entered during the LAI/model sensitivity analysis. This restriction made the sensitivity analysis less general but still pertinent to this study. The standard inputs into the sensitivity analysis were as follows: (1) leaf inclination angle distribution—spherical ($X_L = 0.0$); (2) sun inclination—50°; (3) leaf reflectance—0.1 Visible (VIS), 0.6 near infrared (NIR); (4) leaf transmittance—0.1 VIS, 0.3 NIR; (5) mean canopy LAI—2.7; (6) background reflectance—0.04 VIS, 0.15 NIR; and (7) skylight irradiance ratio—0.4 VIS, 0.24 NIR.

The sensitivity of the canopy was examined at the two wavelength regions, VIS and NIR, by holding all input variables constant while incrementally changing the variable of interest and noting the model prediction at that increment value. Figure 2 depicts the model predictions related to changes in the six input parameters. Two curves are related to each wavelength range. One set is related to hemispherical canopy reflectance (R_c), and one set is related to canopy reflectance calculated at an average view angle from 5 to 15° (R_{cv}) (Goudriaan, 1977). The hemispherical reflectance is the ratio between upward emittance and downward irradiance (Bunnik, 1977), while the view-dependent canopy reflectance is calculated assuming a Lambertian reflector.

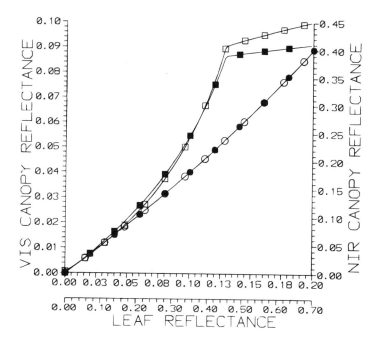

Figure 2a   Canopy reflectance sensitivity plots at visible (VIS; circles) and near-infrared (NIR; squares) wavelengths generated by using the light-canopy interaction model and alternatively incrementing one parameter while holding all other parameters constant. The solid symbols refer to canopy reflectance predicted at near nadir view and the open symbols refer to hemispherical reflectance. The plots are canopy reflectances at the visible (left) and near-infrared wavelengths vs.: (A) Leaf reflectances (0.0 to 0.20 VIS and 0.0 to 0.70 NIR).

Considering all sensitivity analyses at near nadir view (R_{cv}), the model sensitivity was dominated by changes in leaf reflectance (in the NIR region between 0.0 to 0.5 leaf reflectance), percent skylight, and sun elevation. Conversely, a slow increase in canopy reflectance was found at both wavelength regions for changes of LAD, background reflectance, and LAI (VIS region). More substantial but still moderate changes in canopy reflectance at NIR wavelengths were associated with changes in mean canopy LAI.

As shown in the plots of measured red and black mangrove leaf reflectance variance spectra (Ramsey and Jensen, 1995), the maximum deviation in black or red leaf reflectance is about 0.04 at VIS and about 0.06 at NIR wavelengths. Consulting the leaf reflectance sensitivity plots, below 0.5 this level of variation could result in about a 0.035 change in the visible and about a 0.15 change in the NIR wavelength regions of canopy reflectance. However, most measured NIR leaf reflectances were above 0.5, resulting in less than about a 0.02 increase in canopy reflectance. The LAI findings show a difference in the results associated with VIS and NIR wavelengths. At VIS wavelengths the canopy reflectance approaches an asymptote at a low LAI of <1, while at NIR wavelengths the curve

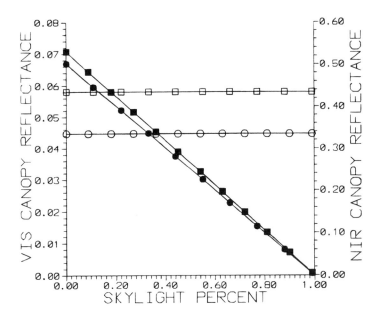

Figure 2b   The ratio of skylight to total downwelling light.

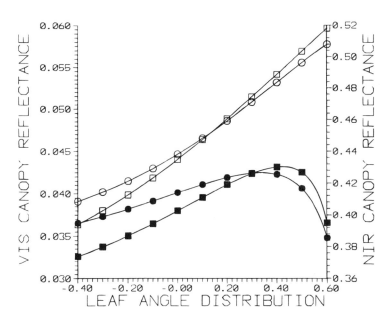

Figure 2c   Leaf inclination distribution angle from a nearly vertical (−0.4) to a nearly horizontal (0.6) distribution.

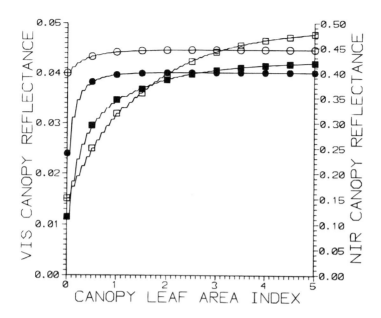

Figure 2d   Leaf area index (m² leaf area/m² ground area).

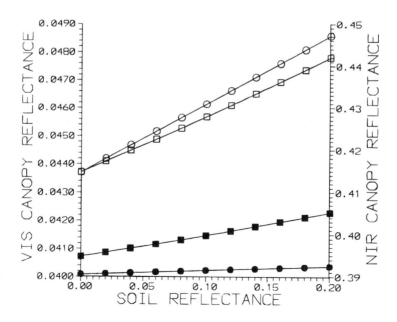

Figure 2e   Soil reflectance.

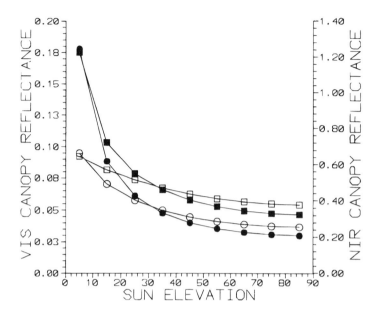

Figure 2f   Sun elevation angle (0° is the horizon).

is starting to level off at a LAI between 2 and 3. This implies that within full mangrove canopies with an average LAI of about 2.7, in the VIS region the canopy reflectance is nearly independent of changes in LAI, while in the NIR region a change of about one LAI unit is equal to a 0.05 change in canopy reflectance. Sun elevation curves indicate little change above about 40 to 50°, and similarly, soil reflectance changes caused little change in canopy reflectance (about 0.0003 VIS and about 0.02 NIR). LAD changes from −0.2 to 0.2 caused minor changes in canopy reflectance of about 0.004 in the VIS and about 0.04 in the NIR regions; however, a 20% change in skylight percent produced changes of about 0.02 and about 0.14 in the VIS and NIR regions of canopy reflectance, respectively.

Canopy reflectances calculated from near-nadir view (R_{cv}) fluxes closely followed hemispherical canopy reflectance (R_c) curves when related to changes in leaf reflectance and followed moderately close at sun elevations higher than 10 to 20°. R_c was constant with respect to skylight percent, while more rapid increases of R_c vs. R_{cv} were shown for variation in leaf angle distribution, soil reflectance, and LAI in the NIR region. Pertinent to this data collection are crossovers of R_{cv} and R_c curves in the graphs of skylight percent (at about 0.35 VIS and about 0.24 NIR) and LAI (at about 2 NIR). These results suggest that R_c was closely related to R_{cv} as a function of small changes in skylight and LAI (NIR) during the acquisitions. However, non-Lambertian effects could result in about 0.02 VIS and about 0.12 NIR, about 0.008 VIS and about 0.03 NIR, and about 0.005 VIS differences in canopy reflectance as a function of LAD, soil reflectance, and LAI, respectively.

Model Predicted and Observed Canopy Reflectances

This part of the research evaluated the light-canopy interaction model as a predictor of observed canopy reflectance, R_{cv}. The canopy model was used at a general level to predict canopy reflectance at 20 sites characterized by fully formed canopies. An optimization technique was used to minimize the weighted sum of squares between the observed canopy reflectances and the modeled canopy reflectances by successive iterations performed on sequential sets of parameter values. (A full description of the optimization technique is given in Himmelblau, 1972.)

Added to simplifications incorporated in the canopy model and those previously described in the input list, two more alterations were implemented in preparation for the optimization. First, the leaf reflectance was adjusted by using a constant factor (LRF) throughout the VIS and NIR wavelength ranges. Second, a soil reflectance weighting factor was introduced by incorporating water absorption. The forms of these adjustments were:

Predicted Leaf Reflectance = Observed + (Observed * LRF)
Predicted Soil Reflectance = Soil − (Water Absorption * Factor).

Four parameters were entered into the optimization: LAI, LAD (X_L), and the leaf and soil reflectance weighting factors. Initial values and parameter restrictions (in parentheses) were: LAI—from field measurements (1 to 5), LAD—spherical (−0.4 to 0.6), LRF—0.0 (−0.7 to 0.7), and SOIL—0.0 (0.0 to 0.05). The restrictions applied during all optimizations, except in the case of the LAI parameter when the optimization stalled at intermediate values at the upper or lower extremes. Leaf area index upper extremes were increased during sites GB1, KI2, and S3 optimizations to 5.5, and decreased to 0.0 during sites RM1 and KI1 optimizations (Table 1).

The optimization was initiated by using spectral inputs of mean canopy, leaf, and soil reflectances between 450 nm and 900 nm, defining 153 wavelength bandwidths. The wavelength range was restricted due to relatively high noise in the leaf spectra below about 450 nm and above about 900 nm. At each site, mean canopy leaf reflectance was estimated by weighting measured spectra by the percent species occurrence and summing. However, at sites where leaf reflectance spectra had not been obtained, leaf spectra were taken from a site of similar type. For example, a red leaf spectra taken from a basin site would be substituted for a missing red leaf spectra at another basin site. Soil reflectance spectra were taken from a pool of five typical spectra obtained from field and laboratory analyses (Ramsey and Jensen, 1995).

As indicated in Figure 3 and Table 1, predicted leaf area index (LAI) values have a mean and one standard deviation of 2.6 ± 0.86, while the soil reflectance

Table 1   Measured and Predicted Canopy Characteristics

Sites	Category[a]	Percent[b] (red/black/white)	LAI_o[c]	LAI_p[d]	LAD[e]	LRF[f]
HQ1	BASIN	0/100/ 0		2.40	0.26	−0.40
HQ2	BASIN	50/ 50/ 0		2.10	−0.13	−0.34
HQ4	BASIN	70/ 30/ 0		2.81	−0.06	−0.21
TH1	BASIN	40/ 30/ 30		1.50	0.01	−0.16
HK1	BASIN	20/ 80/ 0		3.92	0.24	0.24
GB1	BASIN	35/ 65/ 0	2.7	2.54	−0.05	0.02
DK1	BASIN	20/ 80/ 0	3.2	3.20	0.00	0.00
S1	BASIN	60/ 35/ 5	3.9	3.50	0.38	−0.13
ML1	BASIN	30/ 70/ 0		4.17	0.20	0.28
CS1	BASIN	30/ 70/ 0		3.35	−0.18	0.07
TH3	FRINGE	100/ 0/ 0		2.20	0.18	−0.56
KI1	SCRUB	0/100/ 0	0.9	1.41	−0.06	0.18
KI2	RIVERINE	75/ 25/ 0	3.6	3.29	0.57	0.29
AK1	OVERWASH	90/ 5/ 5		1.21	0.03	−0.70
RM1	OVERWASH	95/ 5/ 0		1.91	−0.03	0.06
RM2	OVERWASH	78/ 2/ 20		2.57	−0.12	0.10
SH1	OVERWASH	90/ 5/ 5	1.6	1.52	0.06	−0.11
MQ1	OVERWASH	90/ 5/ 5	3.5	2.90	−0.09	−0.20
S2	OVERWASH	90/ 0/ 10	2.2	2.30	−0.04	−0.33
S3	OVERWASH	80/ 20/ 0	3.4	3.48	0.06	0.26

[a] Refers to categories defined in Lugo and Snedaker (1974).

[b] Refers to percent of red, black, and white mangrove present.

[c] Refers to field measurements.

[d] Refers to predicted values.

[e] Values indicate difference from a spherical distribution (0.0); decreasing values indicate a more vertical distribution and increasing values indicate a more horizontal distribution.

[f] Values reflect the constant leaf correction factor added to the measured leaf spectra for model optimization.

correction factor was nearly 0 in all cases (<0.03). Leaf angle distribution (LAD) has a mean near spherical ($X_L = 0.0$) with a positively skewed distribution, and the leaf reflectance correction factor has a mean and one standard deviation of −0.06 ± 0.28.

Even though leaf correction factors and LAD values centered near the expected value of 0.0, there were high deviations between the entered and predicted leaf reflectance spectra (e.g., sites AK1 and TH3) and predicted LAD from spherical (e.g., sites HQ4, KI2, and S1). High leaf reflectance differences may have resulted from using a constant percent leaf correction factor in the visible and near-infrared regions, from the method of generating the mean leaf reflectance for each site, or from substituting nonrepresentative leaf spectra for missing spectra (e.g., sites TH3 and AK1). High LADs may have resulted from estimating the leaf inclination distribution by using a generalized expression (X_L). Further study is needed to ascertain the validity of the predicted values.

Simple correlation results indicated that predicted LAI was not significantly related to predicted LAD or percent species composition, but positively correlated to the predicted leaf factor (r = 0.54), estimated canopy height (r = 0.67), field measured LAI (r = 0.97), and normalize difference vegetative index NDVI

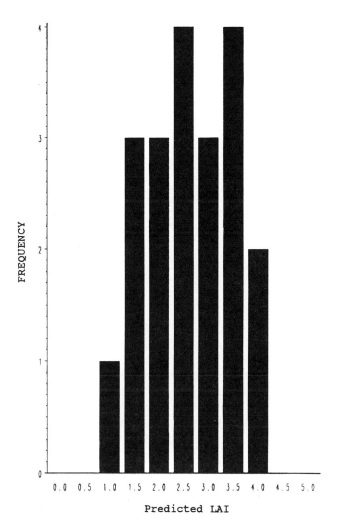

Figure 3a   Histograms depicting the distribution of predicted LAI (A), LAD (B) (as related to X_L), and the leaf correction factor (C).

created from the canopy spectra generated from field measurements ($r = 0.89$). Statistics describing the correlation between predicted LAI and NDVI forecast a R^2 of about 0.78, a regression slope coefficient related to NDVI variance of about 19.6, and a regression intercept of about −13.6. Figure 4 shows a plot of NDVI vs. predicted and measured LAI values. Predicted LAD variance was not significantly related to any measured canopy describer, such as LAI, height, predicted leaf factor, and percent species composition. The predicted leaf factor was correlated significantly only to predicted LAI, while percent species composition was not related significantly to any measured or predicted canopy parameter.

Overall, the model predictions were satisfactory. In the 20 site optimizations, R^2 was >0.97 and most often nearly 0.99. The highest differences between the

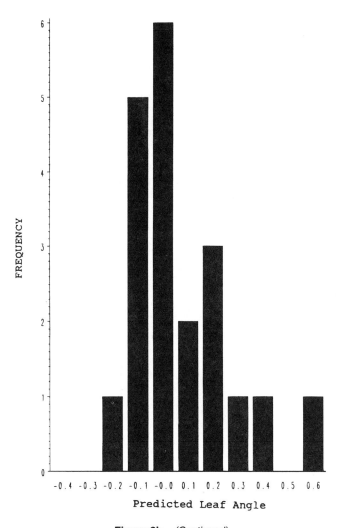

Figure 3b   *(Continued)*

observed and predicted canopy reflectances occurred in the visible wavelength region and in the transition region between low (about 680 to 690 nm) and high reflectances (about 710 to 750 nm). Figure 5 illustrates five sets of observed and predicted canopy reflectance spectra.

SUMMARY AND CONCLUSIONS

A preliminary study was performed to explore uses of remote sensing to discriminate among different mangrove community types. To accomplish this goal, site-specific, high resolution spectral data were applied to the study of man-

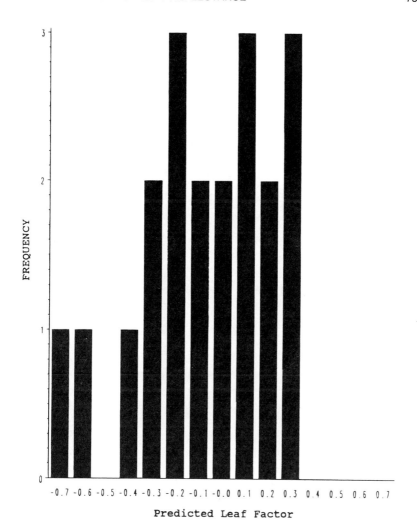

Figure 3c   *(Continued)*

groves. A light-canopy interaction model within a optimization routine was used to predict canopy reflectance, leaf area index (LAI), leaf angle distribution (LAD), and a leaf reflectance factor based on the difference between measured and predicted leaf reflectances. The model capabilities were examined in two steps. First, the sensitivity of canopy reflectance as a function of various input parameters was investigated at two wavelength regions, visible (VIS) and near-infrared (NIR), as restricted to average conditions prevalent during the acquisition. Second, the performance of the model was examined by comparing predicted canopy and leaf reflectances and LAI to those obtained from direct measurements, and by analyzing correlations between predicted canopy parameters (e.g., LAI, LAD, percent species composition, and leaf correction factor).

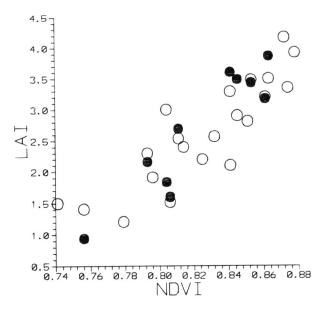

Figure 4 Normalized difference vegetation index vs. predicted LAI (open circles) and measured LAI (solid circles) obtained from field measurements. (From Ramsey and Jensen, 1995).

The sensitivity analysis suggested that little error in modeling the near nadir view canopy reflectance (R_{cv}) would result from assuming an average soil reflectance of about 0.1, at sun elevation angles higher than about 40°, and LAI values and NIR leaf reflectances higher than about 2 and 0.45, respectively. Moderate errors could result from assuming a single spherical LAD, and relatively high errors could result from errors in estimating VIS leaf reflectances (and NIR leaf reflectances <0.45) and percent skylight. Errors estimating canopy hemispherical reflectance (R_c) from R_{cv} are dominated by percent skylight variation. In a percent skylight range from about 20 to 40%, R_c differs from R_{cv} by about 0.015 in the VIS and about 0.1 in the NIR. Differences between R_c and R_{cv} are moderate to slight at a sun elevation above 20 to 30°, a near spherical LAD, soil reflectances near 0.1, a LAI up to 4, and a NIR leaf reflectance less than 0.7. Future work will use new instruments and technologies to define leaf spectral properties (instead of the stacked leaf method) and will more directly measure canopy LAI in order to decrease the possibility of error in model input data.

Model canopy reflectance spectra closely simulated obtained spectra; R^2 values were >0.97 and most often about 0.99. The highest differences between observed and predicted canopy reflectances occurred in the visible and at the beginning of the NIR wavelength regions. Predicted LAI values had a mean and one standard deviation of 2.6 ± 0.86 and were highly related to LAI values derived from field measurements. Predicted LAI values were not significantly related to predicted LAD or percent species composition but were positively cor-

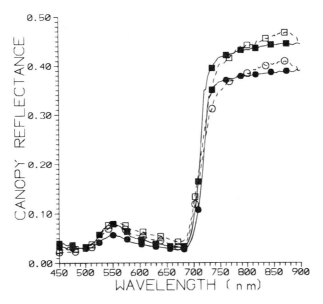

Figure 5a   Model derived (R$_{cv}$) reflectances (solid line-solid symbols) and canopy reflectances generated from site-specific light upwelling recordings [helicopter] (dashed line-open symbols). (A) (circles) MQ1—Overwash, about 90% Red (about 25′), about 5% Black about (30′), and about 5% White (30′), large episodic over-wash island, (squares) TH3—Fringe, 100% Red (~9.6 m).

related to the predicted leaf correction factor and field estimated canopy height. Normalized difference vegetation index values generated from the canopy spectra were highly correlated to predicted LAI values; 78% of the LAI variance was predicted by the NDVI variance. Predicted LAD values had a near spherical mean value and a positively skewed distribution, while the mean and one standard deviation difference between input (estimated from laboratory measurements) and predicted leaf reflectances was −0.06 ± 0.28. Singular high differences between input and predicted leaf spectra and LAD from spherical only occurred at two and three sites, respectively.

   Even though model simulations of canopy reflectance closely replicated canopy reflectances derived from observations, and predicted LAI values closely followed field-generated LAI values, model prediction of canopy LAD and mean leaf reflectance spectra could not be validated. However, LAD and leaf reflectance correction factor means were near the expected value of 0.0. Finally, results are encouraging considering the extreme range of environments and environmental conditions simulated by the model that was used in only the most general form. Refinement of the model to include view and solar azimuth should improve the model accuracy (Chen, 1983, 1985); however, before this complexity is introduced, further experiments are needed to establish the canopy reflectance models' prediction accuracy.

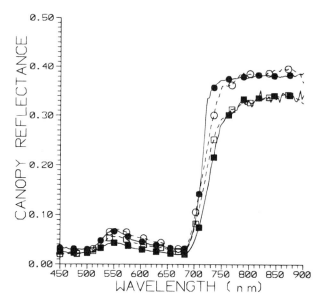

Figure 5b   (circles) HQ4—Basin, about 70% Red (about 25 to 30'), and about 30% Black (about 30 to 35'), (squares) S3—Overwash, about 80% Red (about 25 to 30'), about 20% White (about 30 to 35') mangroves with a sparse understory of Red mangroves (2 to 4') located on a peninsula.

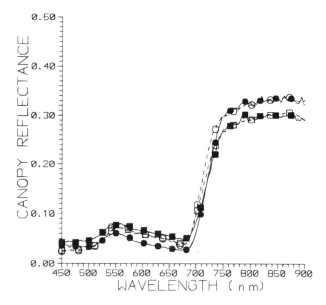

Figure 5c   (circles) S2—Overwash, about 90% Red (about 11') and about 10% White (about 15'), (squares) stunted interior reds on small island, KI1—Dwarf, 100% Black (from 2 to 4' near saltern to 5 to 7' on the outer edge surrounding saltern) with a light gray sandy soil and restricted tidal input due to presence of shell berm.

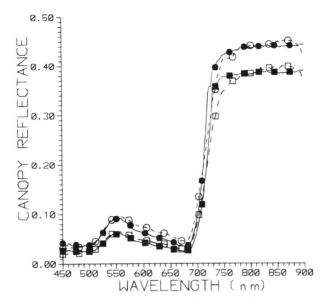

Figure 5d   (circles) HQ1—Basin, 100% Black (about 1 m), (squares) DK1—Basin, about 80% Black (about 45′), about 20% Red (about 25 to 35′), and some Whites, canopy mangroves, and a sparse Red mangrove understory (mostly 5 to 7′, some about 10′).

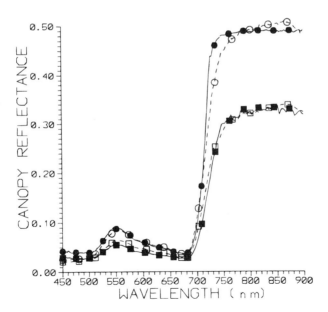

Figure 5e   (circles) Site Basin, about 60% Red (about 40 to 45′), about 35% Black (about 50′), and about 5% White (about 50′), clean understory, interior of island with fringe Reds, (squares) KI2—Tidal Creek, about 75% Red (30 to 40′), and about 25% Black (30 to 40′) mangroves a sparse understory, and large change in canopy closure with transect direction.

REFERENCES

Badhwar, G. D., Verhoef, W., and Bunnik, N. J. J., Comparative study of Suits and SAIL canopy reflectance models, *Remote Sensing Environ.*, 17, 179–195, 1985.

Bunnik, N. J. J., The multispectral reflectance of shortwave radiation by agricultural crops in relation with their morphological and optical properties, Wageningen, Netherlands, DEEL 78-I, t.m. 78-9, 1977, 175 pp.

Chen, J., Kubelka-Munk equations in vector-matrix forms and the solution for bidirectional vegetative canopy reflectance, *Appl. Optics*, 24(3), 376–382, 1985.

Chen, J., The reciprocity relation for reflection and transmission of radiocarbon by crops and other plane-parallel scattering media, *Remote Sensing Environ.*, 13, 475–486, 1983.

de Wit, C. T., Photosynthesis of leaf canopies. Agriculture Research Report No. 663, Centre for Agricultural Publications and Documentation, Wageningen (Agriculture Research Reports No. 663), 1965, 57 pp.

Elterman, L., *Vertical-attenuation model with eight surface meteorological ranges 2 to 13 kilometers*, Air Force Cambridge Research Laboratories, AFCRL-70-0200, 1970.

Evans, G. C., Model and measurements in the study of woodland light climates, in *Light as an Ecological Factor*, Bainbridge, R., Evans, G. C., and Rackham, O., Eds., John Wiley & Sons, New York, 1965, 53–76.

Franklin, J. and Strahler, A. H., Invertible canopy reflectance modeling of vegetation structure in semiarid woodland, *IEEE Trans. Geosci. Remote Sensing*, 26(6), 809–825, 1988.

Gates, D. M., Physical and physiological properties of plants, in *Remote Sensing With Special Reference to Agriculture and Forestry*, National Academy of Science, Washington, D.C., 1970, 224–252.

Goel, N. S., Stregel, D. E., and Thompson, R. L., Inversion of vegetation canopy reflectance models for estimating agronomic variables. II. Use of angle transforms and error analysis as illustrated by Suits' model, *Remote Sensing Environ.*, 14, 77–111, 1984.

Goel, N. S. and Grier, T., Estimation of canopy parameters for inhomogeneous vegetation canopies from reflectance data. I. Two-dimensional row canopy, *Int. J. Remote Sensing*, 7(5), 665–681, 1986.

Goudriaan, J., Crop micrometeorology: a simulation study, Wageningen, Netherlands Centre for Agricultural Publishing and Documentation, 1977, 249 pp.

Himmelblau, D. M., *Applied Nonlinear Programming*, McGraw-Hill, New York, 1972.

Horler, D. N. H., Dockray, M., and Barber, J., The red edge of plant leaf reflectance, *Int. J. Remote Sensing*, 4(2), 121–136, 1983.

Kimes, D. S., Modeling the directional reflectance from complete homogeneous vegetation canopies with various leaf-orientation distributions. *J. Opt. Soc. Am.*, 1(7), 725–737, 1984.

Kimes, D. S., Modelisation of the optical scattering behaviour of the vegetation canopies. Proceedings of the Third International Colloquium on Spectral Signatures of Objects in Remote Sensing, Les Arcs, France, December 16–20, 1985, 157–163.

Kimes, D. S. and Sellers, P. J., Inferring hemispherical reflectance of the earth's surface for global energy budgets from remotely sensed nadir or direction radiance values, *Remote Sensing Environ.*, 18, 205–223, 1985.

Kimes, D. S., Newcomb, W. W., Nelson, R. F., and Shutt, J. B., Directional reflectance distributions of a hardwood and pine forest canopy, *IEEE Trans. Geosci. Remote Sensing*, GE-24 (2), 281–293, 1986.

Lugo, A. E. and Snedaker, S. C., The ecology of mangroves, *Annu. Rev. Ecol. Syst.*, 5, 39–64, 1974.

National Research Council (NRC). *Responding to changes in sea level, engineering implications*. Committee on Engineering Implications of Changes in Relative Mean Sea Level, National Academy Press, Washington, D. C. 1987, 148 pp.

Norman, J. M., Modeling the complete crop canopy, in *Modification of the Aerial Environment of Plants*, Barfield, B. J. and Gerber, J. F., Eds., American Society of Engineers, St. Joseph, MI, 1979, 249–277.

Odum, W. E., McIvor, C. C., and Smith, T. J. III, *The ecology of the mangroves of south Florida: a community profile*, U.S. Fish and Wildlife Service, Office of Biological Services, Washington, D. C. FWS/OBS-81/24, 1982, 144 pp.

Press, W. H., Flannery, B. P., Teukolsky, S. A., and Vetterling, W. T., in *Numerical Recipes, The Art of Scientific Computing*, Cambridge University Press, New York, 1990, 31–38.

Ramsey, E. W. III and Jensen, J. R., Remote sensing of mangroves: relating canopy spectra to site-specific data, *Photogramm. Eng. Remote Sensing*, in press, 1995.

Reyna, E. and Badhwar, G. D., Inclusion of specular reflectance in vegetative canopy models, *IEEE Trans. Geosci. Remote Sensing*, GE-23(5), 731–736, 1985.

Saenger, P., Hegeri, E. J., and Davie, J. D. S., (Eds.), Global status of mangrove ecosystems, *The Environmentalist*, 3 (Suppl 3) 1983.

Sellers, P. J., Canopy reflectance, photosynthesis, and transpiration. II. The role of biophysics in the linearity of their interdependence, *Remote Sensing Environ.* 21, 143–183, 1987.

Steven, M. D., The physical and physiological interpretation of vegetation spectral signatures, Proceedings of the Third International Colloquium on Spectral Signatures of Objects in Remote Sensing, Les Arcs, France, December 16–20, 1985, 205–208.

Tomlinson, P. B., *The Botany of Mangroves*, Cambridge University Press, Cambridge, 1986, 413 pp.

Turner, R. E. and Spencer, M. M., Atmospheric model for correction of spacecraft data, Proceedings of the Eighth International Symposium on Remote Sensing of Environment, Ann Arbor, MI, October 2–6, Environmental Research Institute of Michigan, Ann Arbor, 1972, 783–793.

Vygodskaya, N. N. and Gorshokova, I. I., Calculations of canopy spectral reflectances using the Goudriaan reflectance model and their experimental evaluation, *Remote Sensing Environ.*, 27, 321–336, 1989.

The Use of Archival Landsat MSS and Ancillary Data in a GIS Environment to Map Historical Change in an Urban Riparian Habitat

Christopher Lee and Stuart Marsh

ABSTRACT

Recent changes in the condition of the riparian habitat of the Tanque Verde Creek in Tucson, Arizona prompted an investigation of the dynamics of the native plant communities, hydrology, and climate in the area. The project was an initial attempt to assess and monitor the changing condition of native vegetation along the Tanque Verde Creek during 1983 to 1989, and to discover if mapped changes in vegetation patterns could be correlated with available ground data.

Analysis techniques involved the use and evaluation of archival Landsat Multispectral Scanner (MSS) multitemporal satellite data coupled with pertinent hydrologic and meteorologic measurements. The MSS data were used to develop indices of vegetation condition for 1983 to 1989. A wide array of multitemporal comparisons were constructed between the vegetation indices and well water levels, temperature, and precipitation. No clear statistical relationships between changes in vegetation indices, well water levels, and climatic variables could be established, however. Lack of direct correlation is probably the result of the coarse spatial resolution of the satellite data coupled with local spatial and temporal variations in terrain and subsurface hydrology. Nevertheless, multitemporal maps of changing vegetation condition and well water levels both show a clear and marked decline after 1985. In addition, the multitemporal satellite data provide a viable and important historical view of changing vegetation condition within the Tanque Verde riparian area.

Reprinted with permission from *PE&RS*, copyright by the American Society for Photogrammetry and Remote Sensing: C. Lee and S. Marsh, "The Use of Archival Landsat MSS and Ancillary Data in a GIS Environment to Map Historical Change in Urban Riparian Habitat" (unpublished).

INTRODUCTION

Recent questions concerning the condition of the riparian habitat of Tanque Verde Creek in Tucson, Arizona indicated the need for an improved understanding of the dynamics of native plant communities and their relationship to hydrologic and climatic factors. This study was an initial attempt to assess and monitor the changing condition of native vegetation along Tanque Verde Creek and to determine if these changes could be related to variations in ground water levels or available meteorologic variables. The specific goals of the study were to determine if archival satellite data could document changes in vegetation in the study area and to ascertain if any of the available well or meteorological data correlated with vegetation change as monitored by satellite. Obviously, a great deal more information could be acquired and used for more detailed (tree-by-tree) analyses. However, this study was the logical first step in establishing if, when, and where vegetation changed during the past decade and how it might correlate with readily available field information.

A variety of analysis techniques were utilized in the study. Satellite data were examined using standard image processing software (ERDAS)[1] and the results of the digital processing were combined with geo-referenced well and hydrologic information using geographic information system (GIS) software (ARC/INFO).

Study Site

Tanque Verde Creek is an intermittent stream which flows for 25 km from its headwaters in the Rincon Mountains to its confluence with Pantano Wash in southern Arizona. The lower 10 km of the stream course and associated riparian habitat constituted the study area (Figure 1). Fremont cottonwood, Goodding willow, Arizona walnut, and other riparian trees are present at low densities along the stream channel which varies in width between 50 and 200 m along a relatively low gradient (0.2 m/km) at an average elevation of 800 m. Mesquite communities occupy the broad floodplain terraces adjacent to the stream channel. Mesquite is present in many sizes and it comprises three-quarters of all large woody plants in this community. Mexican elderberry, velvet ash, netleaf hackberry, and desert willow comprise the majority of the remaining tree stems. Wolfberry, graythorn, and hackberries are the dominant understory shrubs (Stromberg et al., 1992).

The hydrogeologic setting of the study area changes near the confluence of Tanque Verde Creek with Agua Caliente Wash. Below the confluence there exists a structural trough of thick unconsolidated, highly permeable sediments which overlie older consolidated rocks (Stromberg et al., 1992). About 2 km above the confluence, faulting and subsequent erosion have resulted in a relatively shallow depth to the older, lower permeability, consolidated sedimentary rocks.

[1] Trade names are included for the benefit of the reader and do not imply an endorsement of the product by California State University or The University of Arizona.

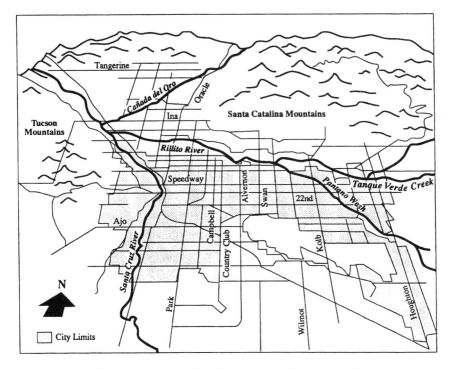

Figure 1   Location of the Tanque Verde Creek study site.

Because of the greater hydraulic conductance and thickness of the younger, unconsolidated sediments, the downstream section of the study area has been more intensely developed in recent years as a source of high quality, low cost groundwater. Rates of groundwater withdrawal from the underlying aquifer increased in 1988 primarily because of the activation of several large capacity wells by the city of Tucson.

METHODS

Data Description

Data utilized in this study consisted of digital satellite imagery, tabular information, and a variety of derived GIS data layers.

Landsat Multispectral Scanner (MSS) images (80-m spatial resolution) for 1983 to 1986, 1988, and 1989 were acquired for the purposes of the study. No cloud free 1987 MSS images were available. Images from the early part of June were chosen in an attempt to analyze vegetation condition at its most naturally stressed level as a result of soil moisture deficiencies just prior to the onset of the summer monsoon. It is during this part of the year that phreatophytic vegetation relies most heavily on direct withdrawals of available groundwater (Meinzer,

1927; Robinson, 1958). Thus, overdrafting of aquifer reserves would be most dramatic on riparian vegetation behaving as phreatophytes during this period. A retrospective analysis of daily meteorological records indicated that there were no significant rainfall events prior to the image acquisition dates. The Landsat images were used to derive normalized difference vegetation index (NDVI) (Tucker, 1979), and soil adjusted vegetation index (SAVI) (Huete, 1988) images for the 6 years.

A May 1988 SPOT image was also utilized to produce a land use/land cover classification map to aid in the analysis. However, it was not compared directly to the MSS images primarily because the differences in indices and classification introduced by its finer spatial resolution (20 m). Landsat Thematic Mapper (TM) data also provide a viable means of performing riparian assessments (Hewitt, 1990). However, consideration of cost and the availability of historical data make the assessment of the Landsat MSS data a cost effective and logistically valid first step in projects of this kind.

Water level data were available from both state and local agencies. Based on the consistency of measurement techniques, we elected to utilize the local data set in our analyses. Data are collected during the minimum use periods of December and January, and after wells have been shut down for 1 to 2 weeks allowing water levels to reach a static state. In addition to well data, meteorological records were acquired from the Tucson Magnetic Observatory station which is the only station in close proximity to the study area. Monthly temperature and precipitation records from the past 25 years were compiled and analyzed for the study.

Satellite Data Preprocessing

As a first step the original Landsat data were subjected to a radiance transformation using previously derived calibration coefficients and the header records of the individual scenes. Using image processing software (ERDAS), the digital numbers (DNs) recorded by the MSS were converted to physical values of radiance (Price, 1987; Robinove, 1982) so that a meaningful comparison could be made between images recorded on different dates by two different Multispectral Scanners (Landsat 4 and 5). Band 4 images, which are originally recorded in 6-bit format, were then scaled to 8-bit format to allow for the derivation of vegetation index values during the data extraction phase.

Analysis of multidate imagery also requires the matching of the spectral characteristics of the images by normalizing the scene histograms. This process acts as a first order correction for differences in atmospheric conditions between different acquisition dates. This is not to be confused with absolute corrections for atmospheric scattering and absorption. Normalization procedures were applied because historical optical depth measurements that could be utilized in a radiative transfer code (e.g., LOWTRAN) (Holm et al., 1989) were not available and because of the lack of appropriate targets in the study area to be able to apply a dark object correction for atmospheric scattering (Chavez, 1988).

The 1989 image was chosen as the standard for normalizing the other images. This choice was based on the fact that the image exhibited the least amount of offset of lowest DN values between bands and the greatest dynamic range of values throughout the image. The mean, standard deviation, and minimum and maximum values were in close agreement for all bands in all corrected images except for the green band (Band 1) of the 1986 image. This was due to an area of clouds in the image which distorted the image statistics. This was not considered a problem, because although all bands of the images were processed, only Bands 2 and 4 are used in the derivation of the vegetation indices.

Once the spectral preprocessing had been accomplished, the six images were geometrically rectified to a map base using 45 ground control points resulting in a root mean square (RMS) error of 0.74 pixels, meaning that points on the image corresponded to within 60 m of their ground location.

The May 1988 SPOT image was subjected to the same radiometric and geometric correction techniques but did not require the matching of histograms because it was not compared directly to the multitemporal Landsat data.

Generation of Vegetation Indices and Maps

Numerous previous studies involving the monitoring of vegetation condition and change have been based on the normalized difference vegetation index or NDVI, (Chilar et al., 1991; Tucker et al., 1983, 1985; Marsh et al., 1992). The differential reflection of green vegetation in the visible and near-infrared portions of the electromagnetic spectrum provides the theoretical basis for this method. The NDVI can be calculated from a variety of sensors carried on environmental satellites (i.e., NOAA-AVHRR, SPOT, and Landsat TM and MSS). The formula for the MSS data takes the following form: $NDVI = CH4 - CH2/CH4 + CH2$, where CH2 and CH4 are the reflectances in the visible red (0.6 to 0.7 μm) and near-infrared (0.8 to 1.1 μm) channels, respectively. Ultimately, the NDVI is determined by the degree of absorption by chlorophyll in the red wavelengths, which is proportional to leaf chlorophyll concentration and by the reflectance of near-infrared radiation which is proportional to green leaf condition and area.

A more refined version of the vegetation index that compensates for the reflectance of background soil, called the soil adjusted vegetation index (SAVI) (Huete, 1988), was also calculated from the MSS data employing the formula: $SAVI = CH4 - CH2/(CH4 + CH2 + 0.5) * (1.5)$. Interactive extraction and comparison of the SAVI and NDVI values for selected sites of known vegetation cover supported the concept that SAVI values exhibit a greater sensitivity to partial ground cover than the traditionally used NDVI. Based on these findings, the SAVI vegetation index imagery was chosen as the primary satellite product for analysis of vegetation status in this study. Image subsets corresponding to the study area were extracted from the SAVI master images for the six available dates.

At this point the vegetation subset images contained 8-bit data values (0 to 255). By inverting the formula originally used to create the single band index image from the red and near-infrared bands of the Landsat MSS data the true index values could be retrieved from the imagery. These index values ranged from 0.0 (absence of vegetation) to 0.7 (highest levels of vegetation). The 8-bit image data was retained on the image processing system for both display purposes and the extraction of pixel array values corresponding to the well level data for regression analysis.

For conversion of the image raster data to vector format and the subsequent use of the vector data for temporal analysis and map generation, the image datasets needed to be reduced in storage size as well as complexity while still retaining the highest possible level of information. To accomplish these goals the original range of values were re-coded to seven classes corresponding to 0.1 increments in the 0.0 to 0.7 range previously calculated for the vegetation index data. The raster files were then imported into ARC/INFO vector format. Using an empirical approach it was decided to group the lowest three classes (0.0 to 0.3) into a single class (Class 1) and renumber the remaining classes into a five class scheme. Classes 2, 3, 4, and 5 correspond to different vegetation conditions within the riparian zone (adjoining polygons with the same class code were also joined).

RESULTS AND DISCUSSION

Multitemporal Data Analysis

As the first step in the analysis, a series of maps depicting the spatial distribution of vegetation index classes for 1983 to 1989 for the image subsets were generated (see Figure 2 [in color section] for 1984 example).

Statistics generated during map compilation were displayed as a series of change matrices to facilitate analysis (Figure 3). The cells forming the diagonal in each matrix indicate the percentage of pixels in each vegetation class that remained unchanged between the 2 years. The cells above the diagonal are percentages of increase for various classes between the 2 years analyzed. The cells below the diagonal are the percentages of class decrease. These general trends are indicated by the plus symbol (+) in the upper right corner and the minus symbol (−) in the lower left corner. Class 1 is non-vegetated surfaces and Classes 2 to 5 include sequentially greater amounts of vegetation.

Analysis of the 5 year-to-year change matrices (3A to 3E) revealed the following patterns: (1) the percentages are always larger below the diagonal than above (there is more vegetation class decrease than vegetation class increase); and (2) the percentages above the diagonal drop sequentially from year to year and more cells are occupied by zero values through the years of the study (the amount of class increase drops in the latter years of the study).

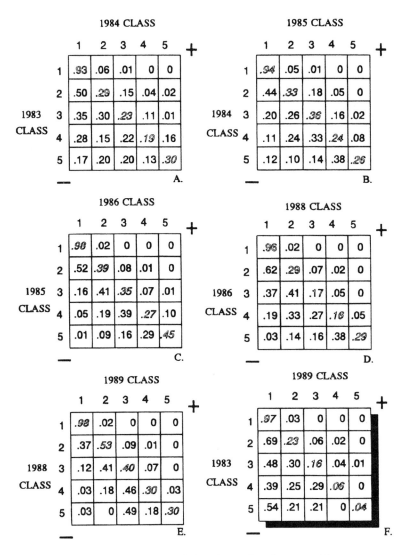

Figure 3 Vegetation index class value change matrix.

Analysis of the cumulative change from 1983 to 1989 (F) reenforces the yearly patterns of decline. Ninety-seven percent of the nonvegetated surface remained in that class for the entire duration of the study. Only 31% of the Class 2 pixels (the lowest vegetation class; the hinge between vegetated and nonvegetated) remained vegetated throughout the 6-year study period. Between 39 and 69% of the vegetation class pixels (Classes 2 to 5) changed to nonvegetated status while only minor increases (1 to 6%) took place in any vegetation class during the same period.

The change matrix analysis clearly documents the trend to decreasing vegetation index class values, but it does not provide insight into the spatial distribution of change. This is provided by visual analysis of the vegetation class maps. An increasing trend in vegetation cover and class exists for 1983, 1984, and 1985 with the peak for the study period being reached in 1985. The 1986 map exhibits a slight decrease from the previous year, with the trend accelerating for 1988 and 1989. The extent and density of the riparian zone shows a marked decline from levels exhibited in the 1983 to 1986 maps. These patterns are exhibited throughout the subset but are especially acute in the Sabino and Aqua Caliente tributaries.

SPOT Multispectral Data Analysis

An unsupervised (maximum likelihood) classification was run on the SPOT multispectral data for the Tanque Verde study area. The resulting classification produced 16 unsupervised classes. The results were displayed and compared to land cover/land use maps of the area. Based on this comparison the 16 classes were combined into six general land cover types: (1) bare soil; (2) urban cover; (3) residential cover; (4) desert scrub (mixed cover); (5) desert scrub (mesquite); and (6) dense cover vegetation. Figure 4 is the resulting classification map (Figure 4; see color section). Inspection of this map demonstrates the difficulty of discriminating classes that are mixtures of desert vegetation and urban and residential cover. The bare soil class can be easily recognized as open fields and washes. The urban cover class represents mixtures of bright and dark manmade materials. The residential cover class is generally a mixture of desert vegetation with bright structures. Dense cover vegetation are generally planted or maintained areas such as golf courses. The desert scrub mixed cover class is representative of Catalina Mountains foothills vegetation; the mesquite class is the most pertinent to this study and coincides with the riparian zones along the washes.

Vegetation Indices and Climatic Data

A 25-year data-set of monthly precipitation and temperature data was used to derive a series of plots illustrating the relationships between meteorological variables as well as their relationship to vegetation indices for the study period. The 25 year mean was calculated for both temperature and precipitation using monthly values from 1955 to 1979 (Figures 5 and 6). These figures show the departure from the 25 year mean for 1965 to 1989. Only 5 of the 25 years plotted show both increases in temperature and decreases in precipitation. Furthermore, 1988 and 1989 are the only 2 consecutive years exhibiting these conditions, and the magnitude of the departures (especially temperature) are the greatest of the 5 years exhibiting these conditions. Additional plots were generated to compare temperature and precipitation values to mean values of the vegetation indices (1983 to 1989) for the vegetation index subsets

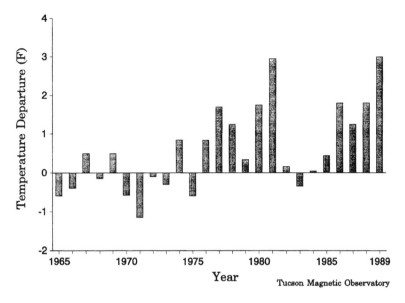

Figure 5 Twenty-five year record of annual mean temperature departure.

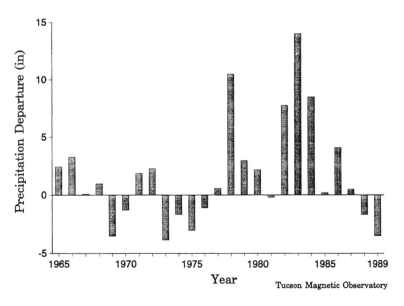

Figure 6 Twenty-five year record of annual mean precipitation departure.

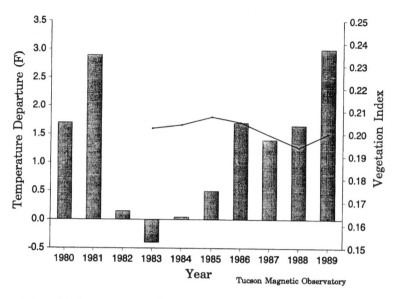

Figure 7   Ten year record of temperature departure and vegetation index.

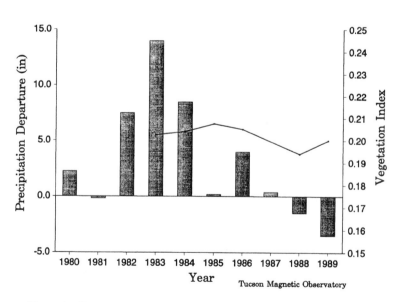

Figure 8   Ten year record of precipitation departure and vegetation index.

(Figures 7 and 8). The small decline in the vegetation indices roughly parallels the decline in precipitation at the sites. No obvious correlation is apparent between temperature and vegetation index values.

As a further index of aridity, Thornthwaite water balance indices (Thornthwaite and Mather, 1955) were calculated for 1981 to 1989. This index takes into account the actual and potential evapotranspiration as well as soil moisture storage to calculate water surplus or water deficit values in inches. Cumulative values for April, May, and June were calculated for each of the study years and plotted together with vegetation index means (Figure 9). No discernable pattern was evident between the two indexes.

Vegetation Indices and Water Levels

The municipal well water level files were edited so that they contained location information in a format that permitted their input to the GIS database. The 64 wells that fit the criterion of having at least 4 years of data collected corresponding to the years of satellite image coverage were utilized. These data points, which were provided in latitude/longitude, were converted to UTM coordinates to allow for analysis and comparison with the satellite data. Again using an empirical approach, several distance buffers away from the course of the channel were evaluated. A distance buffer of 1000 m provided the closest spatial match with the riparian vegetation distribution on the level sliced vegetation index maps. This resulted in a selection of 30 wells that had multitemporal well level data for 1983 to 1989 and were within 1000 m of the channel of the Tanque Verde and its major tributaries (and therefore included all city

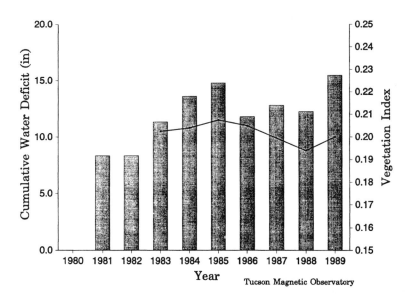

Figure 9   Thornthwaite water balance index and vegetation index.

wells within the riparian zone). The well level data for these 30 wells were then added to the attribute table associated with the wells and used to calculate the difference in well level from year to year over the study period.

To determine the spatial distribution of wells within the 1000-m buffer in the Tanque Verde subset, a plot was created that displays both city and private well locations. To assess the relationship between yearly change in water level for Tucson City water wells and the change in yearly vegetation index levels, a series of maps (Figures 10 to 14) were generated. While the 1983 to 1984 and 1984 to 1985 maps (Figures 10 and 11) show significant amounts of both increasing and decreasing vegetation index values, the trend shifts with the 1985 to 1986 map (Figure 12) to a pattern dominated by declining vegetation index values (Figures 13 and 14). Figure 15 shows the overall pattern of change for the study period (1983 to 1989) which may be a clearer representation of overall change than the year-to-year data. Also plotted on these maps is the yearly change in static water level for city wells falling within the 1000-m zone along major channels. There is an obvious predominance of wells exhibiting increasing water levels on the 1983 to 1984 map (a result of a period of increased precipitation). From 1984 to 1985 onward, the water level in the majority of city wells show a net decrease. Figure 16 displays the 1983 to 1989 vegetation index change along with the locations of all city and non-city wells within the 1000-m buffer for reference. As previously noted, non-city well water level data were too inconsistent for analysis purposes.

By overlaying the desert scrub-mesquite class boundaries derived from the SPOT land cover classification with the 1983 to 1989 vegetation index change map (Figure 17), it was possible to stratify vegetation index changes in natural plant communities from other vegetated areas. The 1983 to 1989 change map presents a very clear picture that vegetation in a narrow corridor bounding the stream channels has in almost every instance decreased.

Regression Analysis

The use of a 1000-m buffer provides a good starting point for delineation of a riparian zone but such diverse factors as geomorphology, zoning, and land ownership combine to modify the shape and extent of the riparian zone within this arbitrary spatial designation. As a result, many of the wells located within the southwestern portion of the Tanque Verde subset are actually in areas dominated by urban/residential land cover. To ensure that the analysis was constrained to vegetation and ground water interactions, 13 wells were visually selected as truly lying within the floodplain of Tanque Verde Creek. These wells were located by their UTM coordinates on the vegetation index images using the image processing software.

Single pixel and five pixel array values were extracted at each well location from the six available Landsat MSS images. Pixel DN values were then con-

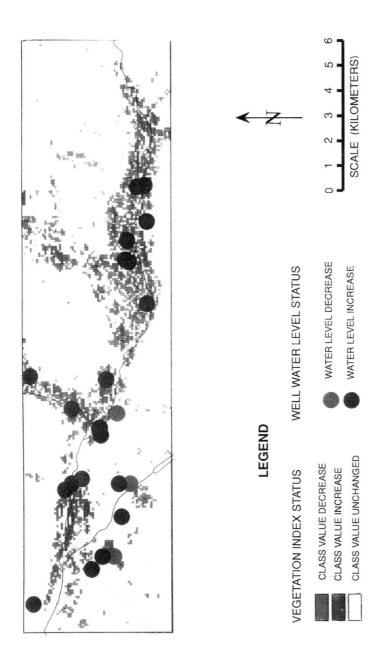

Figure 10 1983–1984 vegetation index and well level change map.

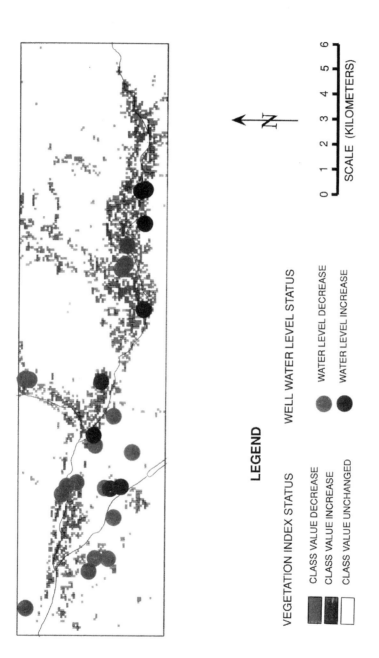

Figure 11   1984–1985 vegetation index and well level change map.

Figure 12 1985–1986 vegetation index and well level change map.

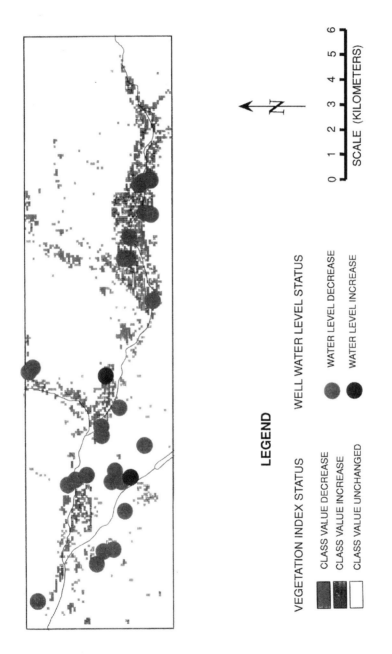

Figure 13   1986–1988 vegetation index and well level change map.

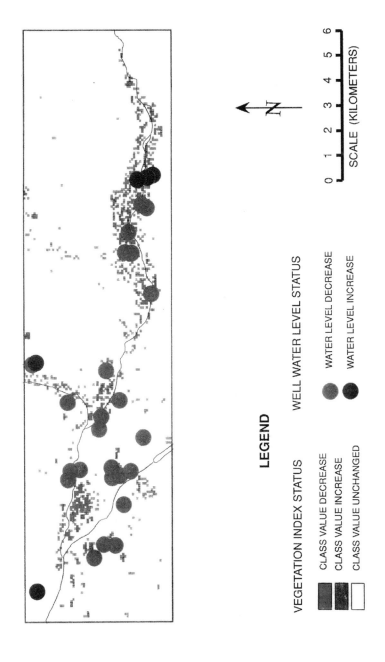

Figure 14   1988–1989 vegetation index and well level change map.

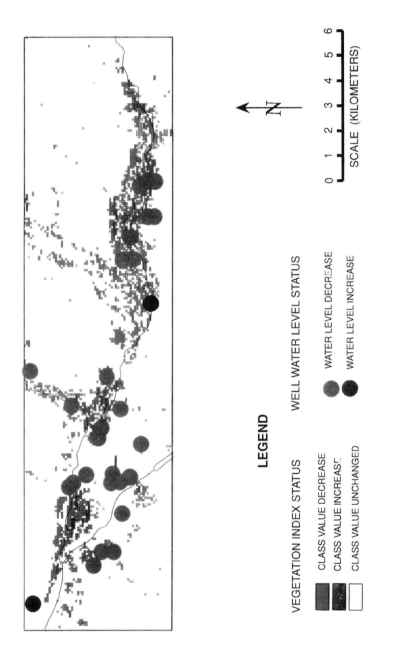

Figure 15   1983–1989 vegetation index and well level change map.

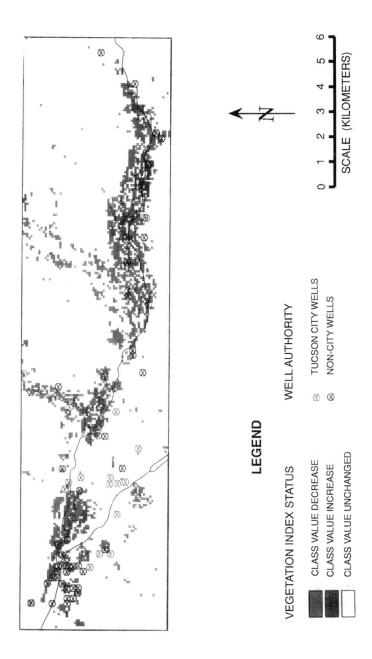

Figure 16 1983–1989 change map with wells within the 1000-m buffer.

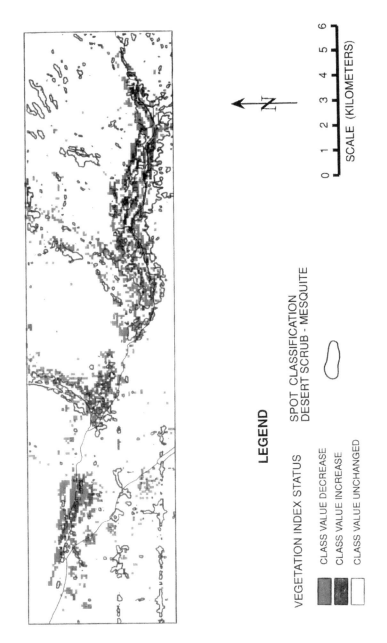

LEGEND

VEGETATION INDEX STATUS

CLASS VALUE DECREASE

CLASS VALUE INCREASE

CLASS VALUE UNCHANGED

SPOT CLASSIFICATION
DESERT SCRUB - MESQUITE

SCALE (KILOMETERS)

0 1 2 3 4 5 6

**VEGETATION INDEX CHANGE (1983-1989) AND DESERT SCRUB-MESQUITE CLASS,
TANQUE VERDE WASH**

Figure 17 Example of the level sliced vegetation index map—1984.

verted to vegetation indices and vegetation index change values were computed between yearly values and for the entire range of study dates (1983 to 1989). Change values for well level data, which had previously been computed in ARC/INFO, were entered into a statistical package along with the vegetation index change values. The data were then examined using regression analysis to explore relationships between yearly changes in groundwater levels and vegetation indices.

Comparisons were made directly between variables of the same year, as well as 1 and 2 year offsets of vegetation indices to account for potential lag effects between the changes in groundwater levels and the response of vegetation to changing conditions. In all cases no statistical correlation existed between changing vegetation index and changing water level variables. Similar data extraction techniques were utilized for the single date SPOT vegetation index to explore the effect of increased spatial resolution. Similar results were obtained although it should be noted that this was single year data only.

CONCLUSIONS

Although a direct statistical relationship between changes in vegetation indices, water level, and climatic variables could not be established, the multitemporal satellite data does provide an important historical view of vegetation condition within the Tanque Verde riparian area. Multitemporal maps of changing vegetation condition and well water levels reveal a definite decline in both after 1985. A variety of spatial and temporal scale considerations may have played a role in the lack of statistical relationships between the variables. Lack of correlation is at least partially a result of the coarse spatial resolution of the Landsat MSS data (80 m) coupled with local variations in terrain and subsurface hydrology. Certainly, a denser network of meteorological and hydrological gauging stations, more frequent measurements from these stations, and higher frequency multispectral satellite data may have provided greater insight into the role and correlation of these variables.

No chance exists to use remote sensing data to finely dissect the past to provide instant, quantitative, and definitive answers. This study was significant in that it does provide, for the first time, maps that depict the nature and spatial extent of changing vegetation patterns in one of Arizona's most important urban riparian habitats. The results of this study illustrate the potential of the available 20 year archival Landsat MSS data base to yield important information on the historical dynamics of urban riparian habitats.

With proper attention to preprocessing of the images to normalize variations in sensor response and atmospheric effects, these satellite data can be input and analyzed in a GIS environment. Use of GIS to analyze changes in vegetation index categories provides useful quantitative information concerning the percentage of vegetation class change throughout a given area and provides a means for producing hard copy outputs which illustrate the qualitative relationship between vegetation class changes and the fluctuations in the local water table. Use of classified satellite images to help stratify areas undergoing change due to stress vs. those being converted to other land uses further helps in extracting useful information from this valuable archival data set.

ACKNOWLEDGMENTS

The authors wish to thank the City of Tucson and Tucson Water for initiating this investigation and providing financial and logistical support. Dr. Charles F. Hutchinson, Dr. Jiang Li, Mr. Greg Saxe, Mr. Douglas Kliman, and Mr. James L. Walsh of the Arizona Remote Sensing Center provided invaluable support and input to this work. This work was performed while the senior author was a Senior Research Associate at the Arizona Remote Sensing Center.

REFERENCES

Chavez, P. S., Jr., An improved dark-object subtraction technique for atmospheric scattering correction of multispectral data, *Remote Sensing Environ.*, 24, 459–479, 1988.

Chilar, J., St.-Laurent, L., and Dyer, J. A., Relation between the normalized difference vegetation index and ecological variables, *Remote Sensing Environ.*, 35, 279–289, 1991.

Hewitt, M. J., III, Synoptic inventory of riparian ecosystems: the utility of Landsat thematic mapper data, *Forest Ecol. Manag.*, 33/34, 605–620, 1990.

Holm, R. G., Moran, M. S., Jackson, R. D., Slater, P. N., Yuan, B., and Bigger, S. F., Surface reflectance factor retrieval from Thematic Mapper Data, *Remote Sensing Environ.*, 27, 47–57, 1989.

Huete, A. R., A Soil-Adjusted Vegetation Index (SAVI), *Remote Sensing Environ.*, 25, 295–309, 1988.

Marsh, S. E., Walsh, J. L., Lee, C. T., Beck, L. R., and Hutchinson, C. F., Comparison of multi-temporal NOAA-AVHRR and SPOT-XS satellite data for mapping land-cover dynamics in the West African Sahel, *Int. J. Remote Sensing*, 13 (16), 2997–3016, 1992.

Meinzer, O. E., Plants as indicators of ground water, *U.S. Geological Survey Water Supply Paper 577*, U.S. Government Printing Office, Washington, D.C., 1927, 95 pp.

Price, J. C., Calibration of satellite radiometers and the comparison of vegetation indices, *Remote Sensing Environ.*, 21, 15–27, 1987.

Robinove, C. J., Computation with physical values from Landsat digital data, *Photogramm. Eng. Remote Sensing*, 48(5), 781–784, 1982.

Robinson, T. W., Phreatophytes, *U.S. Geological Survey Water Supply Paper 1423*, U.S. Government Printing Office, Washington, D.C., 1958, 85pp.

Stromberg, J. C., Tress, J. A., Wilkens, S. D., and Clark, S. D., Response of velvet mesquite to groundwater decline, *J. Arid Environ.*, 23, 45–58, 1992.

Thornthwaite, C. W. and Mather, J. R., The water balance, in *Publications in Climatology*, Laboratory of Climatology, Drexel Institute of Technology, Centerton, NJ, 8, 1–104, 1955.

Tucker, C. J., Red and photographic infrared linear combinations for monitoring vegetation, *Remote Sensing Environ.*, 8, 127–150, 1979.

Tucker, C. J., Vanpraet, C., Boerwinkel, E., and Gaston, A., Satellite remote sensing of total dry matter production in the Senagalese Sahel, *Remote Sensing Environ.*, 13, 461–474, 1983.

Tucker, C. J., Vanpraet, C., Sharman, M. J., and Van Ittersum, G., Satellite remote sensing of total herbaceous biomass production in the Senegalese Sahel: 1980–1984, *Remote Sensing Environ.*, 17, 233–249, 1985.

CHAPTER 8

Remote Sensing of Sediments and Wetlands in Lake Erie

John G. Lyon

ABSTRACT

Two research projects have demonstrated the value of remote sensor measurements of coastal and lacustrine resources. Benefits included a better understanding of water characteristics as measured by remote sensors, a methodology for measuring some characteristics of nonpoint sources of sediment, and a methodology for quantifying wetlands over time. Results provide insight on transport of sediment from bays or estuaries to the nearshore zone, and the loss of wetlands due to natural and human events.

INTRODUCTION

The two studies presented here focus on problems faced by Lake Erie ecosystems. These problems include nonpoint sources of pollution, and loss of wetlands. Two methodologies were used to provide information on the problems. Each approach illustrates how remote sensor data can be used to evaluate these problems, and compliment traditional measurement technologies.

To address the problem of nonpoint sources of sediment required information from a combination of sources including remote sensors and model simulations. Categorizations of satellite data allowed evaluation of total suspended sediment concentrations in Sandusky Bay/Estuary area. Four satellite images displayed trends in the measured on-site data and distribution of water colorants. Satellite data were compared to the results of a hydrodynamic and water quality model. Satellite derived images of relative or absolute suspended sediment concentrations could be interpreted as a tracer for some water quality and hydrodynamic conditions. Results from satellite data and models had similar concentrations of sediment and distribution of those concentrations. Comparisons also indicated new areas to be evaluated in continuing remote sensor and modeling research.

An earlier version of this chapter was published in the National Oceanic and Atmospheric Administration (NOAA), *Lake Erie Estuarine Systems: Issues, Resources, Status, and Management*, U.S. Department of Commerce, 1989, pp. 125–142.

The change in quantity of wetlands vs. Lake Erie water levels was determined from measurements on historical aerial photographs (1935 to 1980). Analysis
indicated there has been a loss of approximately 700 ha of wetlands before 1950.
The loss was a result of impoundments on the Huron River, and decreased nour-
ishment of the Pointe Mouillee delta and estuarine wetlands from riverine sources.
Rising Lake Erie water levels were found to negatively influence the remaining
amount of wetlands in undiked areas (1950 to present). The total amount of wetlands followed the water level conditions, and since 1950 the totals have remain
about the same at any particular water level. This methodology of measurements
from historical aerial photographs and determination of local hydrological conditions has proved useful in other Great Lakes for quantifying change in these
estuarine and lacustrine wetlands.

BACKGROUND

Improved measurement technologies are required to supply data on rivers,
estuaries, and lakes. Remote sensing data can provide a valuable contribution to
inventory resources, or to assist the modeling of water resource characteristics.
The combination of traditional and remote sensor measurement technologies and
modeling methods can potentially supply information in a rapid, relatively lower
cost, and less people-intensive manner than traditional approaches alone.

Methodologies employing satellite data have proven useful for Great Lakes
applications. Great Lake problems often involve large area measurements, and
necessitate repetitive coverage to address complex issues such as nonpoint
sources of pollution. Other measurement problems require inventory and monitoring of estuarine and coastal wetlands. Remote sensor methods can supply historical and current inventories, and allow evaluations of long-term comparisons.
They also supply measurements of water resource or terrestrial characteristics
adjacent to wetlands resources, that are often involved in maintaining a wetland.

These examples illustrate some of the problems of Lake Erie and its adjacent rivers and estuaries. Each addresses a problem and employs a remote sensor technology to supply data for analysis. The examples demonstrate remote
sensor contributions to assessment of quantities of wetlands in Lake Erie, and
to the use of satellite data to evaluate suspended sediment concentrations as a
adjunct to modeling sediment transport through coastal areas to the open lake.

Remote Sensing of Lake Erie

Previous studies of Lake Erie have demonstrated the capability to measure
water and wetland variables in combined remote sensing and on-site sampling
experiments. These experiments were conducted by a variety of users, and for
a variety of reasons. They can be grouped as: (1) evaluations of nonpoint sources
of pollution; (2) use of tracers to follow water currents; and (3) evaluations of
water quality.

Many efforts have evaluated remote sensor data as input to nonpoint studies. Monteith et al. (1981) used land cover statistics from computer categoriza-

tions of satellite data to compare water quality sampling and composition of watershed land cover classes. Similar approaches have worked to evaluate tillage practices from residue cover in fields. Categorized Landsat Thematic Mapper data demonstrated good separability of no-plow land, chisel-plowed land, and mollboard plowed fields (Schaal, 1986). This method of measuring tillage practices promises to be an accurate and cost-effective approach. It is hoped that the monitoring of fall and spring tillage conditions could be used to quantify implementation of erosion-reduction practices. Further work is being conducted by Terry Logan, John Lyon, and Andy Ward of OSU, and Gary Schaal of Ohio Department of Natural Resources. Their efforts are focused on improving our knowledge of measuring differences in tillage practices from spaceborne sensors, and are funded by NASA through the OSU Center for the Commercial Development of Space.

Erosion of shorelines and bluffs also contributes substantial amounts of sediment to the lake through coastal processes. This phenomenon has implications for water quality due to the contribution of glacial deposits which are also a source of phosphorus. Several studies on the Canadian side of Lake Erie focused on these phenomenon (Bukata et al., 1975, 1976; Coakley, 1976; Haras and Tsui, 1976). The studies have mapped the location of erosion prone shores to provide for management (Haras, Bukata, and Tsui, 1976). Black and white infrared photographs were used as a basemap, and engineering data were recorded as an overlay. The resulting book is a detailed source of coastal conditions in both photo and narrative forms.

Other studies have used remote sensing to characterize the transport of suspended sediments. They have also used sediment as a tracer of water circulation patterns. Lake Erie has several prominent, coastal features. To study their genesis and understand the opportunities for protection and management, satellite and aircraft data were found to be useful. Bukata et al. (1975, 1976) identified the common orientation of longshore drift and movement, and deposition areas of entrained materials. Of particular note was the location where opposite currents meet and deposition has formed coastal landforms. These include Pointe Pelee (where a vortex or gyre of sediment-laden water has been recorded from the air and space), and Turkey Pointe and Pointe Rondeau where offshore transport can be seen from satellite data.

Other authors have used NOAA Environmental satellite measurements of lake thermal and sediment conditions to supply detail on surface circulation patterns. Sea surface temperature (SST) studies of Lake Erie demonstrated some of the prevailing areas of offshore transport as identified in the studies mentioned. In the Long Point Bay and estuarine wetlands area in Ontario, transport and water quality conditions were sampled with Landsat and Coastal Zone Color Scanner data by LeDrew and Franklin (1985). Similar transport has been identified on Advanced Very High Resolution Radiometer data (AVHRR) by Lyon et al. (1988). NOAA Environmental satellite data has also been valuable in evaluating lake ice cover, and has been used to examine ice spectra and ice movement by the Great Lakes Environmental Research Laboratory in Ann Arbor, MI.

Water quality evaluations have also been conducted with remote sensors. NASA participated in some data acquisition and analysis in the 1970s. Studies included airborne scanner measurements of Cleveland harbor (Raquet et al., 1977), and remote sensing inputs to quantification of nonpoint sources of pollution in the Maumee Bay area (Shook et al., 1975). Some Daedalus airborne multispectral scanner data were flown of the Lake St. Clair, Detroit River, and the extreme Western Basin of Lake Erie. Lyon is currently working on this data set under Ohio Sea Grant funding (NA84AA-D-00079, R/EM-8). The expected benefits included a radiometric model of light interaction with water, adaptations of the model for use with satellite data, and production of maps of general water bathymetry (1 m increments) and general bottom type using remote sensor data, on-site sampling, and the radiometric model.

The studies above demonstrate the use of remote sensor data in studies of Lake Erie. However, these projects generally restrict their analyses to one date or time only. This approach will supply no temporal data, and will provide only limited assistance in analyses of frequently changing water characteristics. This makes the results of remote sensor experiments difficult to use in any operational application.

To further the application of remote sensor data in water resources requires multiple date evaluations. Remote sensor data are acquired by satellites each day, or by aircraft on a periodic basis. Data potentially provide high frequency sampling of water and wetland spectra. It remains to address each data type and format, and to attempt to input data from each in an experiment.

Suspended Sediment and Nonpoint Sources

To limit lake and estuary eutrophication it is necessary to reduce phosphorus inputs or loads. Contributions from urban sewage and phosphate detergents have been limited at great cost. To further reduce causes of eutrophication it is necessary to limit nonpoint sources of pollution such as sediment eroded from farm fields (Schaal, 1986; Logan, 1987).

Effective management of activities that contribute to sediment and phosphorus loadings requires more information than is currently available. Fundamental questions about sediment transport, resuspension, deposition, and forcing functions need to be answered (Verhoff, 1980; Bedford and Abdelrhman, 1987). Quantifying sediment transported from farm regions to the lake is difficult. Traditional measurement technologies are inaccurate, qualitative, time consuming, and people intensive. New approaches answer these questions by effectively measuring the impact of farm erosion, and sediment loadings on Lake Erie.

This project focused on measuring suspended sediment concentrations using on-site data collection and remote sensor techniques. We have developed a methodology for modeling surface suspended sediment concentrations carried by river runoff to Lake Erie estuarine and coastal areas. A combination of synoptic satellite data, on-site sampling, and sediment concentration results from models potentially fulfilled a need for more information to understand and calculate real loadings to the Great Lakes.

Quantities of Wetlands

Waves interact with long-term fluctuations of Great Lake water levels to influence the hydrology of wetlands on Lake Erie (Beeton and Rosenberg, 1968; Bruce, 1984). Long-term water fluctuations are related to climatic change and occur within 10 to 30 years (IGLLB, 1973). Extremely high lake levels were experienced during the early 1950s, the early 1970s, and the mid-1980s. Extremely low lake levels were experienced during the 1930s and early 1960s. The interval between periods of high and low lake levels varies, as does the length of high or low water periods (IGLLB, 1973). These changes in lake level will continue to occur, and their effects on coastal resources have received a minimum of study.

Fluctuations in Great Lake water levels have been found to influence the extent of coastal wetlands (Harris et al., 1981; Jaworski et al., 1979; Lyon, 1980). High water levels alter the hydrological conditions in wetlands which can kill vegetation, or can destroy barrier beaches which protect the wetlands from wave action (Harris et al., 1981). In a study of seven different varieties of Great Lakes wetlands, a decrease of 29% of wetland area was found between the lowest and highest lake levels that were studied (Jaworski et al., 1979). It is clear that long-term fluctuations in water level can increase or decrease the total area of wetlands. Information concerning the effects of lake levels on wetlands and beaches is necessary for resource management. Likewise, a methodology for quantifying local effects is required to supply data for resource management.

To determine the change in the quantity of wetlands we analyzed current and historical aerial photos. The objectives included: (1) measurement of the extent of wetlands from historical aerial photographs, and (2) use of the measurements to understand cause and effect relationships between wetlands and water levels.

The utility of the mapping approach has been demonstrated in a study of the Straits of Mackinac area of Michigan. A lower quantity of wetland and beach area was found along the Lake Michigan coast of the straits in years of high water levels (Lyon, 1981; Lyon and Drobney, 1984; Lyon et al., 1986). From the study of the Straits of Mackinac and other locations, there appears to be a direct relationship between lake levels and the presence of wetland plant communities and beaches. To further test the hypothesis that water levels influence the presence of wetland areas required investigation of longer-term influences over several years and several different lake levels, and different Great Lakes. A determination of the long-term effect of water level fluctuations on wetland communities in the Pointe Mouillee region was a further test of the concept.

METHODS

Suspended Sediments

June 1 to July 15, 1981 was studied due to the occurrence of four large storms. Research indicated that the major transport of sediment to the open lake occurs under storm flow conditions. An additional, important factor was the availability of water sampling (Richards and Baker, 1982) and satellite data during this period (Figure 1).

Satellite data were processed into products that could be interpreted for relative or absolute surface suspended sediment concentrations. Data products were developed in two ways (Lyon et al., 1988): (1) satellite data were computer categorized to produce a map of sediment concentration classes or types; and (2) the resulting class mean brightness values from satellite data were used in comparisons with on-site sampling of suspended sediment concentrations (mg/l). Analysis included regressions of brightness values with the on-site sampling data, and rank correlation analysis to demonstrate the capability of the categorizing algorithm to select spectrally distinct classes.

Suspended sediment concentration class types were developed from computer categorizations of satellite data. This procedure selected homogeneous areas of suspended sediment concentration classes with a clustering algorithm. The resulting sediment classes effectively divided Sandusky Bay into concentration classes and provided a map of their locations (Figures 2 and 3; see color section).

Two NOAA-AVHRR scenes were evaluated for their capability to supply spatial distribution of suspended sediment concentrations in the form of images. The scenes were acquired on June 26 to 27, 1981 (one and two days after on-site sampling) (Figure 4). All five data channels were used to make the computer categorized products mentioned above.

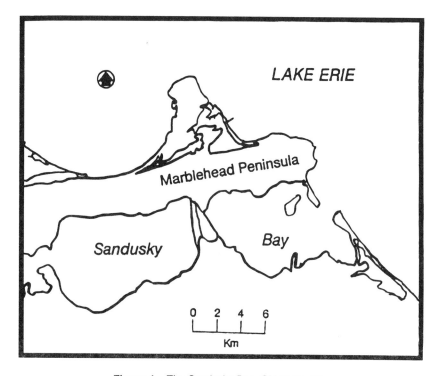

Figure 1   The Sandusky Bay, OH study site.

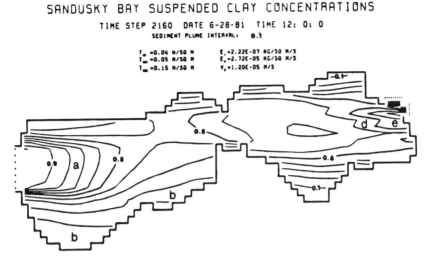

SANDUSKY BAY SUSPENDED CLAY CONCENTRATIONS

TIME STEP 2160   DATE 6-28-81   TIME 12: 0: 0

SEDIMENT PLUME INTERVAL: 0.1

T₌ -0.04 H/SQ H   E,·2.22E-07 KG/SQ H/S
T₌ -0.05 H/SQ H   E,·2.72E-05 KG/SQ H/S
T₌ -0.15 H/SQ H   V,·1.20E-05 H/S

Figure 4   Results of hydrodynamic and water quality model simulations for the same time shown in Figure 2. The contour lines show proportional concentration of total suspended solids relative to 595 mg/l (after Lee, 1986).

Individual concentration measurements of sediment and satellite brightness values were used to generate regression models of the relationship. Twelve individual sediment sampling sites were compared to the average brightness of the red and near-infrared reflectance of computer categorized class. Linear or multiple variable linear regression model were developed from brightness values and on-site sampling. Histograms and tests of normality indicated the distributions were normal, and that linear models were appropriate for the concentrations involved. For Landsat images, on-site sampling stations were located on the lineprinter maps of the categorized scene. The brightness values of twenty-five picture elements or pixels (0.16 km²) surrounding the sampling station were averaged and compared to sediment concentrations measured in the water. This has been shown to be a suitable area to average the spectral response of water. Landsat channels 5 and 6 were supplied the most detail on sediment concentration conditions.

For AVHRR images, a single pixel (1 km²) encompassing the sampling station was located, and the spectral characteristics of the categorized product were used for comparison with water samples. AVHRR channels 1 (red) and 2 (near infrared) provided the most useful data on sediment, and these data were used for later analysis.

A rank correlation, nonparametric Spearman Rho correlation test proved valuable for demonstrating that the categorization algorithms could recognize sediment class from satellite-measured brightness values. The test indicated that the categorizing algorithm distinguished sediment concentration classes based on their brightness, and there was a nonrandom association between on-site measures of sediment and computer-generated classes.

Quantities of Wetlands

The historical distributions of wetland and beach areas were measured from aerial photographs acquired during periods of low water (1935, 1964), and high water (1973, 1978, 1980). Use of these photographs allowed a 41 year period for analysis, and several examples of each water level from different decades. The exact dates, scales, and water levels are included in Table 1. This intensive sample of wetland areas and water levels was completed in the Pointe Mouillee area. For the inventory, only wetlands which were undiked and subjected to Lake Erie water levels were measured. Diked wetlands have been managed by flooding and accurate assessment of historical, hydrological conditions would be difficult (Figure 5). Boundaries of wetland and beach areas were interpreted and traced on to Mylar from enlarged aerial photographs. Area measurements were made with a planimeter, and corrected for actual photo scale to produce determinations of area.

Historic water level data were obtained for the area from the class 1 gage stations at Fermi Power Plant, Gibralter, Monroe, Michigan, and Toledo, Ohio (NOAA, 1978). When possible, daily levels were used instead of monthly averages, and the closest gage in operation was used for the water level value. It is important to note that tides have a very small effect on the Great Lakes water levels. Wind or pressure generated events (e.g., seiches or storm surges), and long-term changes in lake water levels from rainfall create lake level fluctuations which influence hydrology of wetlands (IGLLB, 1973).

Table 1   Results of Regression Analyses of Landsat Digital Brightness Values for Bands 5(red) and 6 (near infrared) with On-Site Total Suspended Solids (n = 12), and Results of Spearman Rho Tests Comparing the Nonrandom Assortment of On-Site Total Suspended Solids with Categorized Water Colorant Class from Landsat and AVHRR Multiple Channel Data.

Date	Data source	Linear regression $r^2=$, $p>$		Spearman rho $r=$, $p>$
June 10	Landsat categorized raw data, bands	5 & 6 5	0.80, 0.0015 0.80, 0.0002	0.86, 0.0124
June 12	On-site samples collected	—		—
June 25	On-site samples collected	—		—
June 26	AVHRR categorized raw data, bands	1 & 2 1	0.87, 0.0003 0.80, 0.0002	0.86, 0.0007
June 27	AVHRR categorized raw data, bands	1 & 2 1	0.90, 0.0003 0.81, 0.0004	0.78, 0.0072
June 28	Landsat categorized raw data, bands	5 & 6 5	0.90, 0.0001 0.72, 0.0009	0.93, 0.0001

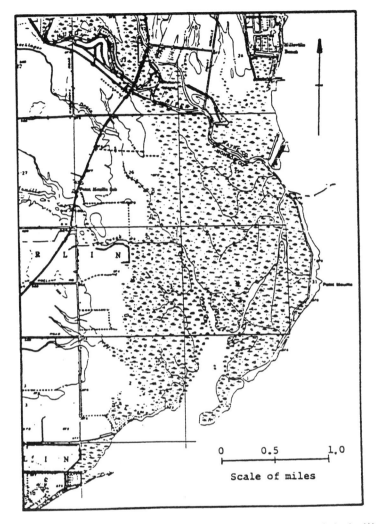

Figure 5   USGS map of the Pointe Mouillee area from 1942. The area is in the Western
Basin of Lake Erie. Original map scale was 1:62,500.

RESULTS

Suspended Sediments

Suspended sediment concentration measurements and satellite digital data
products of Sandusky Bay, the nearshore region and open part of Lake Erie were
compared from June to July 1981. Calculations were completed using traditional,
on-site sampling data, and raw brightness values and suspended sediment concentration categorization products from satellite sensors.

Work yielded good agreement between satellite data and on-site sampling
(Lyon et al., 1988). A 4-day sequence of satellite and suspended sediment sam-

pling demonstrated the basic concept and the proposed methodology. AVHRR and Landsat scenes from June 10, 26, 27 and 28, 1981 present the distribution of suspended sediment classes in Sandusky Bay (Figures 2, 3, and 4). Linear regression models of Landsat digital brightness values and suspended sediment concentrations showed a strong relationship (Table 1). Nonparametric Spearman Rho tests also demonstrated a strong relationship between suspended sediment classes derived from AVHRR and Landsat satellite data, and on-site measurements.

These calculations and final results were compared with results from hydrodynamic and water quality models. General concentration class boundaries and general concentration levels simulated by the model were also demonstrated independently from categorized or enhanced satellite data.

Comparisons between satellite data and the results of hydrodynamic and water quality model runs (Lee, 1986; Yen, 1987) indicate several areas of agreement. Figure 2 presents the July 28 Landsat categorized scene and the relative sediment concentrations contours from the model. Similar contour boundaries are indicated in Figure 4, and it is apparent that both data sources could produce similar results. All data sources indicated the plume (A) in the upper bay. It is important to note that both AVHRR and Landsat categorized scenes show this plume (Figures 3 and 4). Regressions of Landsat brightness value and on-site sampling from June 25 indicated this plume had clay concentrations ranging from 600 to 490 mg/l. The satellite categorized scene and model results also identified lower concentration areas along the south shore of the upper Bay (B). They also identified lower concentration areas along the north shore of the upper and lower bay (C, 466 mg/l).

In the lower bay there were several areas of similarity. Large scale eddys can be identified (D) where bay water and open lake water mix. Storm surges, seiches, and lake modes of resonance all can cause backflow and influx of open lake water into the lower bay (Lee, 1986). The satellite categorizations display this as a change in water colorant classes over a short distance. The hydrodynamic and water quality model predicted some of this change.

Quantities of Wetlands

Results demonstrated all of the following: (1) the recent, historical extent of wetlands in the study area (1935, 1011 ha, 2496 acre); (2) the great fluctuation between low water and high water years, for example 1964 (173.64 m water level, 828 ha, 2045 acre) and 1973 (174.67 m water, 332 ha, 819 ac and 1980, 174.54 m, 203 ha, 502 acre); (3) area measurements indicated a decrease of approximately 700 ha of wetlands and beaches from 1935 to 1950. The dates of photo coverage, scales, and water levels included: (1) 1935, 1:30,162, 173.29–173.43m; (2) 1964, 1:20,556, 174.37m; (3) 1973, 1:41,605, 174.67m; (4) 1978, 1:24,130, 174.37m and (5) 1980, 1:24,771, 174.54m. The 1935 photos had no date and it has been impossible to establish it. Hence, a range in water levels from May through September 1935 were employed.

Analysis of the five sets of photos indicates there is a similar linear relationship between water levels and total undiked wetlands (Lyon et al., 1985). In general, higher lake levels result in lower quantities of wetlands. This result has been also demonstrated for Lake Erie wetlands at Long Point, Ontario (Whillians, 1985). Other studies have also found this relationship to be true (Lyon, 1981; Lyon et al., 1986).

An interesting result of this work was the loss in wetlands between 1935 and 1950. Figures 6 to 9 show the change in wetlands during this period (Green, 1987). While Lake Erie water levels have been generally high since the 1970s, lake levels alone do not explain this loss. Lake levels have varied from high to low since they have been recorded, and earlier accounts document the natural fluctuations of lake levels. Because neither weather conditions nor lake levels have changed drastically since the 1920s it is conceivable that the loss of sediment input to the Huron River delta or Pointe Mouillee has resulted in the decrease of total wetlands from 1935 to 1973.

This loss presumably resulted from the damming of the Huron River and entrapment of silt, sand, and larger size particles (Greene, 1987). The Pointe Mouillee area is the delta and estuarine wetlands of the Huron. The delta has been present since the retreat of the Wisconsin glaciation over 4000 years ago. The delta was largely intact until the 1920s or 1930s. Maps and accounts before that time record a delta of large size (Lyon et al., 1985). However, the early part of this century saw construction of seven or more dams on the Huron River, with documented decreases in transport of sediments to the delta area.

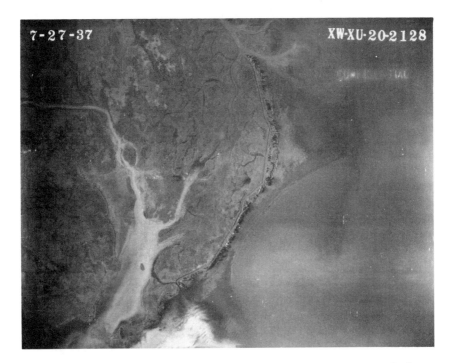

Figure 6   Aerial photo of wetland areas from 1937 (water level was 174.00 m).

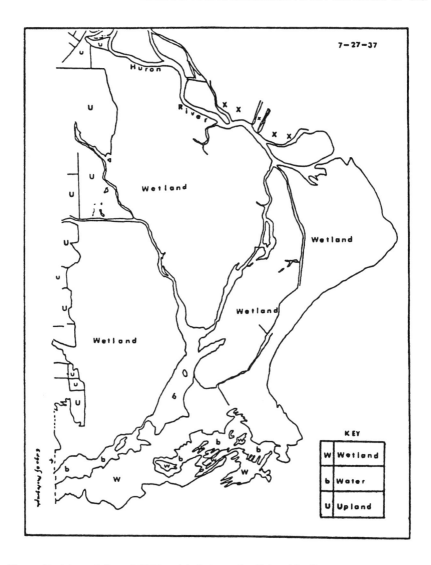

Figure 7   Interpretation of 1937 aerial photographs, Pointe Mouillee. Areas outside the study area are shown by letter x.

DISCUSSION

Suspended Sediments

Quantifying sediment transported from farm regions to the lake is difficult. Traditional measurement technologies are inaccurate, qualitative, time consuming, and people intensive. New approaches can potentially answer these questions by effectively measuring the impact of farm erosion and pollutant loadings on Lake Erie.

Figure 8   Aerial photo of wetlands area from 1950 (water level was 173.92 m).

Best Management Practices (BMP) such as conservation tillage have been introduced to farms near Lake Erie tributaries to reduce farm-related erosion. To measure the implementation and future effects of BMPs requires identification of sediment transport characteristics and the fate of the sediment. It is necessary to understand the various nonpoint source loading mechanisms, and to parameterize these loadings for use as input in management models.

The absence of an organized use of satellite and modeling methodologies results from the difficulty of analyzing several different data types and models. Often this combination of data is unavailable due to the cost of on-site sampling, the infrequency of aircraft or satellite coverage, and the difficulty of operating hydrodynamic and water quality models. The methodology proposed here can potentially overcome some of these difficulties to provide a basis for a future, operational methodology.

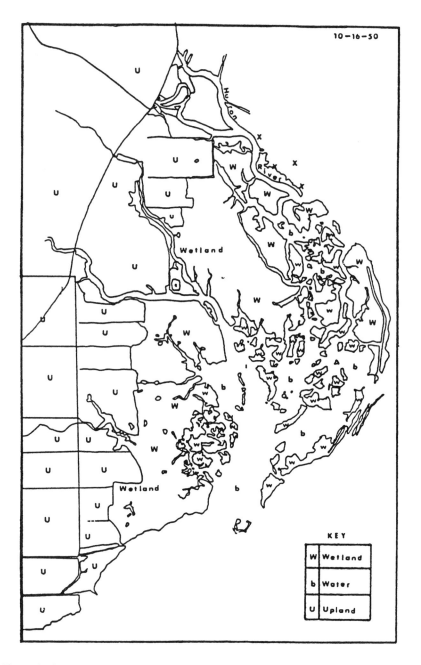

Figure 9   Interpretation of 1950 aerial photographs, Pointe Mouillee. Areas outside the study area are shown by letter x.

Quantities of Wetlands

Record high Great Lakes water levels prompt questions concerning their influence on coastal resources and structures. One of the concerns is whether lake water levels result in different quantities of wetlands and beaches. The results of this and other studies indicate a decrease in available wetland and beach areas with increasing water level. However, the variable nature of precipitation input to the lakes results in fluctuating water levels. This too results in variable quantities of wetlands.

The relatively constant presence of wetland and shore areas through time indicated that early seral communities have been maintained by fluctuations of water levels and flooding. Fluctuating water levels are a perturbation similar to fire in prairie, boreal forest, or chaparral ecosystems. Low lake levels are historically followed by high lake levels which flood previously dry areas and kill nonwetland shrub and trees. Wetland plants recolonize these areas and wetland areas are maintained by the periodic disturbance of high water levels.

CONCLUSIONS

Estuary and nearshore areas experience a variety of changes, and have an influence on open lake systems. Additional inputs of datas are needed to better manage resources and model their behavior. Remote sensor information can be very useful as an adjunct to traditional sources of data. The case studies presented here demonstrate the utility of remote sensing for both inventory and modeling. Both case studies and previous work demonstrated the value of remote sensor inputs to studies of coastal and lacustrine systems.

There is great interest in the use of remote sensing technologies, and a need for operational applications of aerial and satellite data along with traditional measures. Potentially, a remote sensing approach can increase the accuracy of model determinations and reduce the costs associated with on-site sampling. Availability of daily remote sensor data can assist in providing more accurate estimates of water resources.

ACKNOWLEDGMENTS

This research project was sponsored by the Ohio Sea Grant Program with funds from the National Oceanic and Atmospheric Administration, Office of Sea Grant, Grant no. NA81AA-D-00095 and NA84AA-D-00079 (R/EM-2, R/EM-7, R/EM-8), and from appropriations made by the Ohio State Legislature. Additional support was provided the Center for Lake Erie Research, and the Department of Civil Engineering at The Ohio State University. This work was the natural outgrowth of Michigan Sea Grant projects awarded during 1977 to 1981 (04-M01-134, R/CW-7 and -3), and three Conservation Fellowships awarded by the National Wildlife Federation.

REFERENCES

Bedford, K. and Abdelrhman, M., Analytical and experimental studies of the Benthic Boundary layer and their applicability to near-bottom transport in Lake Erie, *J. Great Lake Res.*, 13, 628–648, 1987.

Beeton, A. and Rosenberg, H., Studies and research needs in regulation of the Great Lakes. Proceedings of the Toronto Water Regulation Conference, 1968, 311–340.

Bruce, J., Great Lakes levels and flows: past and future, *J. Great Lakes Res.*, 10, 126–134, 1984.

Bukata, R., Haras, W., and Bruton, J., The application of ERTS-1 digital data to water transport phenomena in the Point Pelee-Rondeau Area. *Verh. Internat. Verein. Limnol.*, 19, 168–178, 1975.

Bukata, R., Haras, W., Bruton, J., and Coakley, J., Satellite, airborne and ground-based observations of storm-induced suspended sediment transport off Point Pelee in Lake Erie. Proceedings of International Conference on Human Environment Conservation, Warsaw, Poland, 1976.

Coakley, J., The formation and evolution of Point Pelee, Western Lake Erie, *J. Earth Sci.*, 13, 136–144, 1976.

Greene, R., Lake Erie water level effects on wetlands as measured from aerial photographs, Master's Thesis, Dept. of Civil Engineering, The Ohio State Univ., Columbus, OH, 1987, 75 pp.

Haras, W., Bukata, R., and Tsui, K., Methods for recording Great Lakes shoreline change, *Geosci. Canada*, 3, 174–184, 1976.

Haras, W. and Tsui, K. (Eds.), Canada/Ontario Great Lakes shore damage survey, coastal zone atlas, Min. of Natural Resources-Ontario and Environment Canada, Ottawa, Ontario, 1976, 637 pp.

Harris, H., Bosley, T., and Roznik, F., Green Bay's coastal wetlands—a picture of dynamic change, Proceedings of the Wabesa Wetlands Conference, Wabesa, WI, 1981, 340–351.

International Great Lakes Levels Board (IGLLB), Regulation of Great Lakes water levels. Report to the International Joint Commission, Windsor, Ontario, 1973, 294 pp.

Jaworski, E., Raphael, C., Mansfield, P., and Williamson, B., Impact of Great Lakes water level fluctuations on coastal wetlands, Institute of Water Resources, Michigan State University, MI, 1979, 351 pp.

Lathrop, R. and Lillesand, T., Use of thematic mapper data to assess water quality in Green Bay and Central Lake Michigan, *Photogramm. Eng. Remote Sensing*, 52, 671–680, 1986.

LeDrew, E. and Franklin, S., Surface current analysis from Landsat and CZCS Imagery at Long Point Bay, Lake Erie, Ontario, Canada. Proceedings of the Eighteenth International Symposium on Remote Sensing of Environment, Paris, France, 1984.

LeDrew, E. and Franklin, S., The use of thermal infrared imagery in surface current analysis of a small lake. *Photogramm. Eng. Remote Sensing*, 51, 565–574, 1985.

Lee, D., The development of a multiclass size sediment transport model and its application to Sandusky Bay, Ohio, Master's Thesis, Dept. of Civil Engineering, The Ohio State University, OH, 1986, 175 pp.

Logan, T., Diffuse (non-point) source loading of chemicals to Lake Erie. *J. Great Lakes Res.*, 13, 649–658, 1987.

Lyon, J., Remote sensing analyses of coastal wetland characteristics: the St. Clair Flats, MI, Proceedings of the Thirteenth International Symposium on Remote Sensing of Environment, Ann Arbor, MI, 1979, 1117–1129.

Lyon, J., Data sources for analysis of Great Lakes wetlands, Proceedings of the Annual Meeting of the American Society of Photogrammetry, St. Louis, MO, 1980, 512–525.

Lyon, J., The influence of Lake Michigan water levels on wetland soils and distribution of plants in the Straits of Mackinac, Michigan, Ph.D. Dissertation, School of Natural Resources, University of Michigan, MI, 1981, 132 pp.

Lyon, J. and Drobney, R., Lake Level effects as measured from aerial photos, *ASCE J. Surveying Eng.*, 110, 103–111, 1984.

Lyon, J., Gauthier, R., Greene, R., and Jaworski, E., Assessment of historical wetland regeneration from aerial photos, Landsat MSS and TM Data, Proceedings of U.S. Army Corps of Engineers Fifth Remote Sensing Symposium, Ann Arbor, MI, 1985.

Lyon, J., Drobney, R., and Olson, C., Effect of Lake Michigan water levels on wetland soil chemistry and distribution of plants in the Straits of Mackinac, *J. Great Lakes Res.*, 12, 175–183, 1986.

Lyon, J., Bedford, K., Yen, J., Lee, D., and Mark, D., Determinations of suspended sediment concentrations from multiple day Landsat and AVHRR Data, *Remote Sensing Environ.*, 24, 9, 1988.

Monteith, T. and Jarecki, E., Land cover analysis for the United State Great Lakes Watersheds, Great Lakes Basin Commission, Ann Arbor, MI, 1978.

National Oceanic and Atmospheric Administration (NOAA), Great Lakes Water Levels, 1860–1975. U.S. Department of Commerce, Riverdale, MD, 1978, 187 pp.

Raquet, C., Salzman, J., Coney, T., Svehla, R., Shook, D., and Gedney, R., Coordinated aircraft/ship surveys for determining the impact of river inputs on Great Lakes Water-Remote Sensing Results, PLUARG Technical Report, U.S. Task D, International Joint Commission, Windsor, Ontario, 1977, 200 pp.

Richards, R. and Baker, D., Assimilation and flux of sediments and pollutants in the Sandusky River Estuary, Sandusky Bay and the adjacent nearshore zone of Lake Erie, Supplemental Report, Water Quality Lab., Heidelberg College, Tiffin, OH, 1982, 125 pp.

Saylor, J. and Miller, F., Studies of Large-Scale Currents in Lake Erie, 1979–80, *J. Great Lakes Res.*, 13, 487–514, 1987.

Schaal, G., Residue studies, Ohio Dept. of Natural Resources, Div. of Soil and Water Conservation, Columbus, OH, 1986, 200 pp.

Shook, D., Raquet, C., Svehla, R., Wachter, D., Salzman, J., Coney, T., and Gedney, D., A preliminary report of multispectral scanner data from the Cleveland Harbor Study, NASA Technical Memorandum, TM X-71837, 1975, 42 pp.

Strong, A., Remote sensing of algal blooms by aircraft and satellite in Lake Erie and Utah Lake, *Remote Sensing Environ.*, 3, 99–107, 1974.

Verhoff, F., River Nutrient and Chemical Transport Estimation, *ASCE J. Environ. Eng.*, 106, 591–608, 1980.

Whillians, T., Related long-term trends in fish and vegetation ecology of Long Point Bay and Marshes, Lake Erie, Ph.D. Dissertation, Univ. of Toronto, Toronto, 1985, 252 pp.

Whitlock, C., Kuo, C., and Leroy, S., Criteria for the use of regression analysis for remote sensing of the sediment and pollutants, *Remote Sensing Environ.*, 12, 151–168, 1982.

Yen, J. C., Remote sensing and a three-dimensional numerical model analysis of suspended sediment distributions in Sandusky Bay/Lake Erie nearshore region, Master's Thesis, Dept. of Civil Engineering, The Ohio State Univ., OH, 1987, 204 pp.

CHAPTER 9

Use of a Geographic Information System Database to Measure and Evaluate Wetland Changes in the St. Marys River, Michigan

Donald C. Williams and John G. Lyon

ABSTRACT

A digital database was constructed by photo interpretation, mapping, and digitizing seven dates of aerial photography on the St. Marys River, MI. The database was used in conjunction with geographic information system software to examine historical changes in wetland area. Total wetland area between 1939 and 1985 ranged from 7200 to 7317 ha over a 46-year period of high and low water. There was greatest variation in areas of emergent wetland and scrub-shrub wetland, which appeared to be responding primarily to changes in water level.

INTRODUCTION

The St. Marys River (Figure 1) is the outlet for the oligotrophic water of Lake Superior to the lower Great Lakes. Because of the amount of wetland along the river, and the relatively low phytoplankton productivity of this water, wetlands have been shown to play a dominant role in the primary productivity of the St. Marys ecosystem (Duffy et al., 1987; Liston et al., 1986). The Detroit District, U.S. Army Corps of Engineers (USACE) proposed to extend winter season navigation through the locks at Sault Ste. Marie, MI (USACE, 1987). The USACE undertook further wetland study in conjunction with environmental impact analysis because of the importance of the wetlands to the ecosystem in this connecting channel.

Local distribution of wetlands, their species composition, productivity, and growth rates had been studied in selected parts of the St. Marys River (Liston et al., 1980; Jude et al., 1986; Liston et al., 1986). The area of St. Marys River in the USACE study (Figure 2), primarily Lake Nicolet, contains a diversity of

From *Hydrobiologia* 219: 83–95, 1991, M. Munawar & T. Edsall (eds), Environmental Assessment and Habitat Evaluation of the Upper Great Lakes Connecting Channels. © 1991 Kluwer Academic Publishers. Printed in Belgium. Reprinted by permission of Kluwer Academic Publishers.

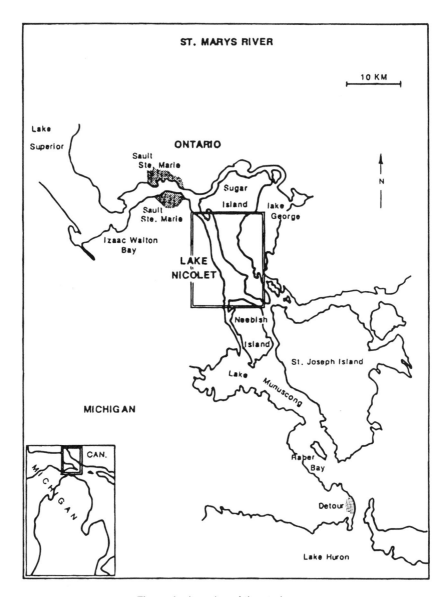

Figure 1   Location of the study area.

wetlands. To determine and evaluate wetland distribution changes it was necessary to classify them, map their current distribution, and measure their areas. A
historical inventory of these wetlands was necessary to determine their changes
over time. For ease of analysis, the information from the inventory was placed
in a geographical information system (GIS). From measurements derived from
this system it was possible to address the relative abundance of wetlands, and
consider past, present, and potential sources of change in the wetlands.

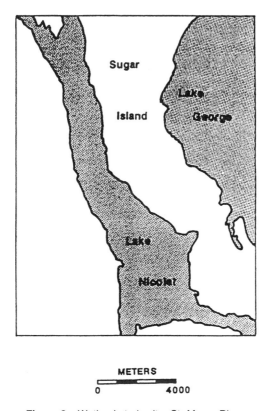

METERS

0   4000

Figure 2 Wetland study site, St. Marys River.

METHODS

The USACE obtained photographs for the summer and fall seasons of 1939, 1953, 1964, 1978, 1982, 1984, and 1985. The film types included black and white, black and white infrared, color, and color infrared. The scales of the photos ranged from 1:12000 to 1:58000 (Table 1). Wetland types and areas were interpreted from these photos by the National Wetland Inventory (NWI) of the Fish and Wildlife Service (FWS), which provided wetland maps, tabular data, and digital files for use in the study.

The aerial photos were acquired at seven different water levels. Like the Great Lakes, the connecting channels are subject to a wide range of water levels resulting from large variations in water balance in the Great Lakes basin. Water levels can range plus or minus one meter from the long-term average. The aerial photos obtained for this study documented wetland changes in association with water levels that varied within a range of 1.04 m, from 176.72 to 177.20 m. This allowed an analysis of the influence of St. Marys River water level elevations on the wetlands. The specific dates of the photos and the yearly average water levels are given in Table 1. The water levels were recorded at the U.S. Slip gauge on the St. Marys River.

Table 1   Images for Photo Interpretation and Database Creation and Corresponding Yearly Average Water Levels

Year	Date	Scale	Emulsion	Water level (m)
1939	7/2	1:20000	black and white	176.72
1953	7/19	1:16000	black and white infrared	177.19
1964	6/28	1:16000	black and white	176.16
1978	6/29	1:12000	black and white	176.83
1982	10/25	1:58000	color infrared	176.66
1984	9/18	1:12000	color infrared	177.00
1985	10/19	1:24000	natural color	177.20

Wetland areas were quantified for each date of aerial photo by a NWI contractor experienced with photo interpretation and identification of wetlands. The project area was also visited by the contractor. Interpretations were based on NWI conventions (USFWS, 1987) which provide specific instructions for applying the FWS classification system (Cowardin et al., 1979). Initial interpretations were completed with an analog stereo plotter using the 1984 images. Wetland boundaries were interpreted in and along the river up to one-half mile inland from the shore. The 1984 study area limits were maintained for the other years. The plotter established geometric control and was used to correct any inaccuracies found in the photos. The interpreter viewed stereo pairs of photos, and outlined the wetland boundaries. The other dates of photography were compared to the 1984 maps and wetland boundary adjustments were made with a zoom transfer scope. The boundaries were plotted on U.S. Geological Survey 7.5-min maps of the project area.

The wetland types were characterized as emergent, aquatic bed, scrub-shrub, forested, unconsolidated shore, and unconsolidated bottom wetland types based on the FWS classification system (Cowardin et al., 1979), and can be grouped as riverine and palustrine wetland ecological systems. However, the unconsolidated bottom type probably includes many areas with submergent vegetation, since much of the bottom of the St. Marys River is vegetated outside the navigation channel (Liston et al., 1986).

Large scale (1:24000) maps were produced by the NWI for the seven 7.5-min U.S. Geological Survey quadrangles covering the study area. A separate map was produced for each year of interpreted aerial photography. Wetlands were mapped on the following quadrangles: Sault Ste. Marie South, Baie De Wasai, Oak Ridge, and small parts of Munuscong and Munuscong NE. The wetland boundaries for each wetland class were converted into area measurements, and reported in tenths of acres (0.04 ha) by the USFWS wetland analytical mapping system (WAMS) software. Area data were summarized in wetland maps and tables of wetland areas for each year and quadrangle.

The maps and tabular summaries provided by the NWI were appropriate for completion of most analyses. However, since the tabular summaries were completed on a quadrangle by quadrangle basis, it was difficult to locate and quantify local changes. For these more detailed analyses, digital files were evaluated.

The digital files of each year for each quadrangle were created by NWI using a GIS. The techniques are those being used to construct a national georeferenced wetland data base. WAMS was used to digitize the mapped information and place it in a "common ground" geographic reference system. Digital files from WAMS were imported to the map overlay and statistical system (MOSS). Using MOSS, the files were converted from latitude/longitude to the universal transverse mercator (UTM) projection, rasterized, and converted to ELAS format, which is readable on the Detroit District's GIS system (USFWS, 1988).

Analyses of the digital files concentrated on one quadrangle, Oak Ridge, because it contained approximately 46% of the total wetlands in the study area and a representative cross section of wetland classes. Digital files of four dates of interpreted photos, 1939, 1964, 1978, and 1982, were used for comparisons.

Comparisons were made between scenes using the ERDAS GIS software program MATRIX. Wetlands at two dates were compared by creation of an $n \times n$ matrix, where n was equal to the number of classes in the scene. In this case eight classes were used for each scene, the seven FWS classes (unconsolidated bottom, aquatic bed, emergent, unconsolidated shore, scrub-shrub, forested, and upland) plus other areas outside the study area.

These digital analyses allowed exact location of wetland increases and decreases. The MATRIX computation produced an 8×8 matrix. The diagonal elements indicated the cells in which no class change occurred between dates, and the off-diagonal elements represented the cells in which there were changes to other wetland and nonwetland classes. The matrix results were then mapped into a digital file which showed where changes occurred.

RESULTS

Tabular Data

Historically, the quantity of total wetlands in the project area has remained nearly the same (Figure 3). The tabular data from the interpretation of historical aerial photos from the 46-year period indicated that the total area of wetlands (including unconsolidated bottom) ranged from 7200 to 7317 ha. The maximum difference was 1.6%. During that period, water levels varied by 1.04 m. The total wetland in 1939 was 7317 h at water elevation 176.72 m, and in 1985, 7247 ha at water elevation 177.20 m. The wetland areas were similar with the lake levels 0.48 m higher in the recent photos. Comparisons between any two of the four dates of photographs indicated that changes in overall wetland areas were small, ranging between 0.5 and 1.6% over the 45-year period.

Related local studies of wetland plant populations have demonstrated little change in quantities of plants from year to year (Liston et al., 1986). These year-to-year studies do not explain evidence of past losses at the outer edges of emergent wetland stands. A significant part of these small variations in totals appears to be explained by changes in water levels. This was probably the reason for the small increase in total wetlands in the St. Marys study area between 1953 and

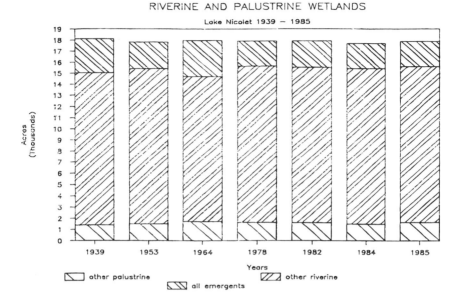

Figure 3   Emergent and other palustrine and riverine wetland areas in the St. Marys River study area, 1939–1985.

1984. In the latter year, water levels were substantially lower: 1953, 7200 ha at 177.19 m; 1984, 7247 ha at 177.00 m.

The most clear relationship to water levels was found in the emergent wetland class. Figure 4 shows strong evidence of a relationship between increasing water levels and the decreasing emergent wetland areas between 1939 and 1985. A regression equation was derived relating emergent vegetation areas to average annual water levels (Dixon, 1985). This equation was significant at the 0.05 level. The multiple R-square value for the regression was 0.64. These parameters indicate that a strong relationship exists between the emergent wetlands variability and water level elevation in the St. Marys River. There was approximately a 32% change in emergent wetland area between high and low water (1347 vs. 917 ha). The overall results suggest that some emergent wetland losses can be expected during periods of high water.

There was a parallel relationship between water level and areas of the scrub-shrub wetland class. Areas of scrub-shrub wetland decreased during periods of high levels and increased during periods of low levels. This relationship is shown in Figure 5. The regression equation derived for the relationship between average annual water level and area of scrub-shrub wetland was also significant at the 0.05 level, and the multiple R-square value was 0.69. While the relationship was statistically significant, there was only a 16% change in area of scrub-shrub wetland from high to low water (390 vs. 328 ha), so that area of scrub-shrub vegetation did not vary as widely under changing water regimes as did emergent wetland.

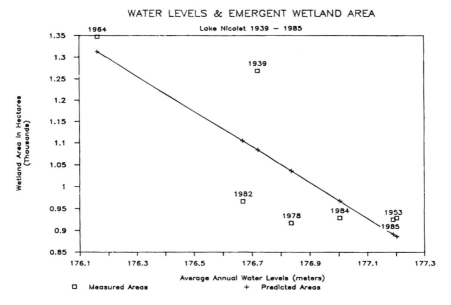

Figure 4   Relationship of emergent wetland area to water levels, 1939–1985.

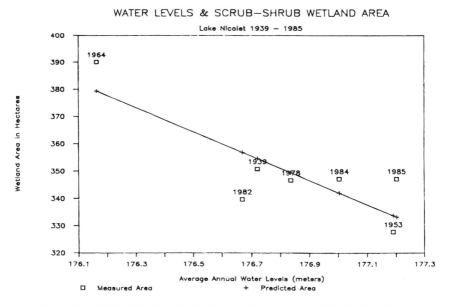

Figure 5   Relationship of scrub-shrub wetland area to water levels, 1939–1985.

Digital Data

The digital data analyses of the Oak Ridge wetlands showed summary results similar to the tabular data. In these analyses, unconsolidated bottom was not included as wetland because this class may or may not be vegetated, and because it includes a large part of the study area. Dates with the largest difference in water levels were compared to investigate changes primarily due to water level. Analysis of the 1964 and 1982 files indicated that 6.2% (731 ha) of the quadrangle study area remained wetland between the low water year, 1964, with levels of 176.16 m, and the high water year, 1982, with levels of 176.66 m. Wetlands were lost in 2.39% of the area (281 ha), and gained in 0.65% of the area (77 ha). The net change was therefore a loss of wetland in 1.7% of the study area or a loss of wetland in 204 ha with an increase in water level of 0.50 m.

Additional comparisons were made to determine short- and long-term changes at similar water levels. 1939 and 1982 were used to indicate changes over the long term. Between 1939 and 1982, 6.0% or 710 ha of the Oak Ridge study area remained in wetlands. Wetlands were lost in 1.8% of the mapped area (209 ha), and gained in 0.8% of the area (97 ha). The net change in total wetlands was therefore 1.0% of the quadrangle study area, or 113 ha. This took place over a 43-year period with a decrease of 0.06 m in average annual water level.

Very small changes took place over the short term. Between 1978 and 1982, 5.8% or 689 ha of the Oak Ridge study area remained as wetland. There was a loss of 0.6% (71 ha) and a gain of 0.7% (81 ha). Hence, there was no significant net change in wetland area between those 2 years in the Oak Ridge study area. There was, however, a decrease of 0.17 m in water levels between the 2 years.

Further analysis of the digital files indicated the types of changes that were occurring in the Oak Ridge wetlands. Inputs and outputs from each wetland type were tabulated for the water level, long-term, and short-term comparisons described above. These changes included movement between wetland classes, and losses or gains of wetland to upland. The results are given in Figures 6, 7, and 8. The major changes due to water levels (1964 to 1982) appear to have been (1) losses of emergent vegetation, and (2) gains in unconsolidated bottom. The major long-term changes appear to have been (1) losses of emergents, (2) losses of scrub-shrub, and (3) large gains of forested wetlands and unconsolidated bottom. The only changes in the short term (1978 to 1982) appear to have been small gains in emergent wetlands. These apparent gains may also be due to the early date of the 1978 photography, which was well before the vegetation in the St. Marys had fully emerged (Liston et al., 1986).

The digital analyses allowed mapping of the specific sites of changes in wetland type. The sites of change indicated in the digital analyses are shown on the maps displayed in Figures 9, 10, and 11. These show the wetland areas in the Oak Ridge quadrangle. Rather than the specific community changes given in graphs above, the maps show areas of stable wetland, and areas where losses and gains have occurred. Maps are given again showing changes associated with water level variation, long-term changes, and short-term changes.

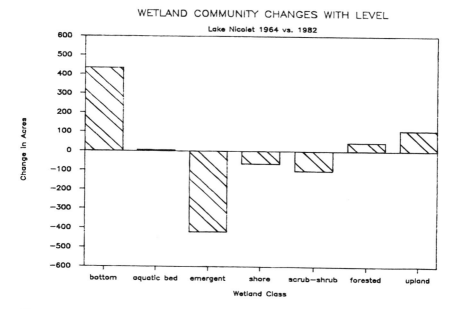

Figure 6   Changes in wetland class related to changes in water level from 1964 to 1982.

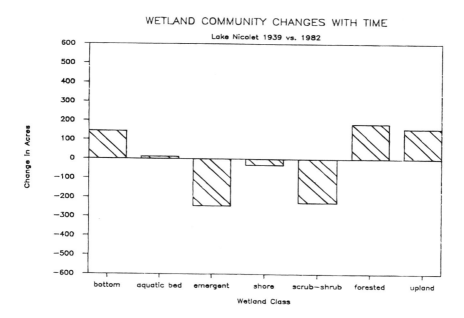

Figure 7   Changes in wetland class from 1939 to 1982.

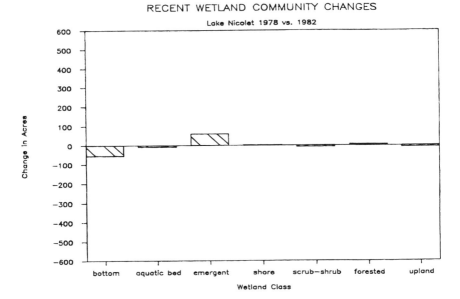

Figure 8   Changes in wetland class from 1978 to 1982.

DISCUSSION

This analysis indicates that GIS provides a convenient method for documentation and analysis of wetland data. The National Wetland Inventory data are currently processed in a GIS system that provides for quantitative analysis of wetland trends and changes. The availability of digital files on a map-by-map basis allows the use of digital computers so that wetland changes may be tracked and analyzed.

The information developed during this study has provided an opportunity to determine whether the wetlands in the St. Marys River exhibit historical changes like other Great Lakes wetlands. The above analyses have indicated that a significant inverse relationship exists between water level and the extent of emergent wetland. This relationship has been shown to consistently occur in Great Lakes coastal wetlands. Lower water levels result in greater quantities of coastal emergent wetlands (Lyon and Drobney, 1984; Payne, et al., 1985; Lyon et al., 1986; Jaworski et al., 1979; Greene, 1987). The relationship may be due to the geometry of the nearshore profile (Bukata, et al., 1988). According to this analysis, the inverse relationship also holds in the case of the wetlands of the Great Lakes connecting channels.

On the other hand, judging from the significance of the regression derived, and the R-square value of the regression, this relationship is less clear than in the case of coastal wetlands of the Great Lakes proper (Table 2). The additional interpretation of wetland trends in the St. Marys made possible by the digital files and the matrix analysis may provide some insight as to differences in response to water levels.

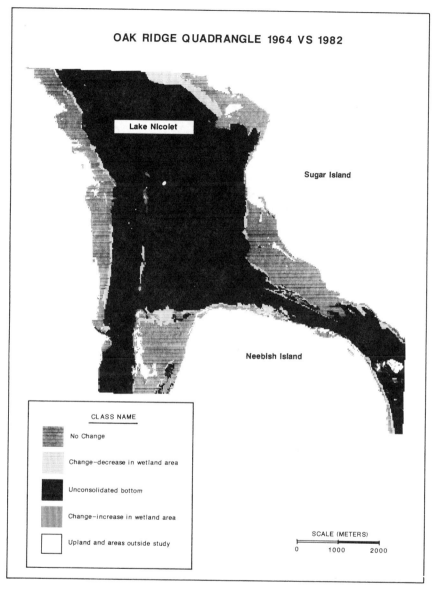

Figure 9 Change in wetland area related to water level changes, 1964 vs. 1982, Oak
Ridge quadrangle, Lake Nicolet, St. Marys River. Unconsolidated bottom excluded from wetland area.

It is interesting to note that the long-term trend of gains in unconsolidated
bottom, losses in emergent wetland, and stability or gains in forested wetland
found in the St. Marys River, is consistent with long-term trends found in New
England wetlands (Larson and Golet, 1982). There is general acceptance of the
concept of "pulse stabilization" in Great Lakes coastal wetlands. That is, these

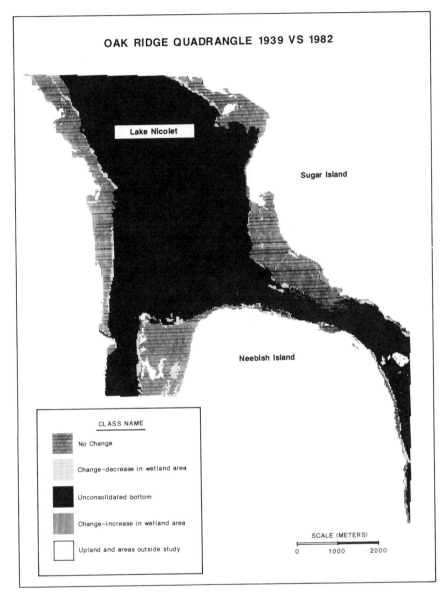

Figure 10   Change in wetland area between 1939 and 1982, Oak Ridge quadrangle, Lake Nicolet, St. Marys River. Unconsolidated bottom excluded from wetland area.

wetlands do not undergo long-term succession because of periodic disturbance from changing water levels (Lyon, 1981; Lyon et al., 1986; Harris, 1977). The GIS-derived data suggest that in the case of the St. Marys River wetlands, this may not be entirely the case. Possibly because of their more limited exposure than the coastal wetlands, the wetlands in the St. Marys River may be undergoing slight successional changes.

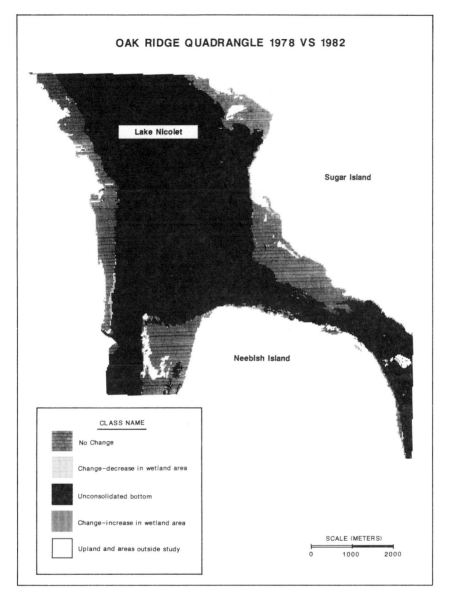

Figure 11   Change in wetland area between 1978 and 1982. Oak Ridge quadrangle, Lake Nicolet, St. Marys River. Unconsolidated bottom excluded from wetland area.

The major conclusions of the wetlands analysis are that: (1) there have been no significant changes in the total amounts of wetland found in the project study area; (2) the changes that have occurred, especially in the emergent wetland class, appear to be primarily related to changes in water levels, although ice, recreational and commercial vessels, currents, and other factors probably have some effects; and (3) quantitative evaluations using digital files provide some

Table 2   Coefficients of Determination (R^2) and Significance
Levels (p) of Regression Equations; Water Levels on
Great Lakes Wetland Areas

Wetland location	Author(s)	R^2	p
Lake Michigan	Lyon (1981)	0.93	<0.001
Lake Huron	Payne et al. (1985)	0.75	<0.025
Lake Erie	Greene (1987)	0.87	<0.005
St. Marys River	—	0.64	<0.05

indication that there are long-term successional trends which are not directly evident in Great Lakes coastal wetlands.

ACKNOWLEDGMENTS

Mr. Ross Lunetta, now of the U.S. Environmental Protection Agency, Las
Vegas, Nevada Laboratory, and Ms. Robin Gebhard, Chief Cartographer, National
Wetland Inventory, St. Petersburg, Florida, made important contributions to the
successful use of NWI digital files in this study. Mr. Les Weigum, Chief of the
Environmental Analysis Branch, Detroit District Corps of Engineers, provided
generous support toward completion of this project.

REFERENCES

Bukata, R. P., Bruton, J. E., Jerome, J. H., and Haras, W. S., A mathematical description
    of the effects of prolonged water level fluctuations on the areal extent of marsh-
    lands, Inl. Wat. Direct., Ntnl. Wat. Res. Inst., Can. Inl. Wat., Burlington, Ontario,
    Sci. Ser., 1988, 166.
Cowardin, L. W., Carter, V., Golet, F. C., and LaRoe, E. T., Classification of wetlands
    and deepwater habitats of the United States, U.S. Dept. Inter., Fish Wild. Serv.,
    Report No. FWS/OBS-79/31, Washington, D.C., 1979, 103 pp.
Dixon, W. J. (Ed.), BMDP Statistical Software, Univ. California Press, Berkeley, CA,
    1985, 734 pp.
Duffy, W. G., Batterson, T. R., and McNabb, C. D., The St. Marys River, Michigan: an
    ecological profile. U.S. Fish Wild. Serv. Biol. Rep., 85(7.10), 1987, 138 pp.
Greene, R. G., Effects of Lake Erie water levels on wetlands as measured from aerial
    photographs; Pointe Mouillee, Michigan, M.S. Thesis, Dept. Civil Engin., Ohio
    State Univ., OH, 1987, 70 pp.
Harris, H. J., Bosley, T. R., and Roznik, F. D., Green Bay's coastal wetlands: a picture
    of dynamic change, in Wetlands Ecology, Values and Impacts, Dewitt, C. B., and
    Solway, E., Eds., Proc. Waubesa Conf. Wetlands, Madison, WI, June 2–5, 1977,
    337–358.
Jaworski, E., Raphael, C. N., Mansfield, P. J., and Williamson, D. B., Impact of Great
    Lakes water level fluctuations on coastal wetlands. Off. Res. Devel., Inst. Res.,
    Michigan State Univ., 1979, 351 pp.

Jude, D. J., Winnell, M., Evans, M. S., Tesar, F. J., and Futyma, R., Drift of zooplankton, benthos, and larval fish, and distribution of macrophytes and larval fish during winter and summer, 1985. Subm. to Detroit District, Corps of Engineers, MI, 1986, 174 pp. + appendices.

Larson, J. S. and Golet, F. C., Models of freshwater wetland change in southeastern New England, in *Wetlands, Ecology and Management,* Gopal, B., Ed., Ntnl. Inst. Ecol. and Internat. Sci. Publ., Jaipur, India, 1982, 512 pp.

Liston, C. R. and McNabb, C. D. (principal investigators) with Brazo, D., Bohr, J., Craig, J., Duffy, W., Fleischer, G., Knoecklein, G., Koehler, F., Ligman, R., O'Neal, R., Siami, M. and Roettger, P., Limnological and fisheries studies of the St. Marys River, Michigan, in relation to proposed extension of the navigation season, 1982 and 1983. Sub. to Detroit District, Corps of Engineers, Detroit, MI and U.S. Fish Wildl. Serv., Twin Cities, MN, Issued as U.S. Fish Wildl. Serv., Off. Biol. Serv. Rep. OBS/85(2), 1986, 764 pp. + appendices.

Lyon, J. G., The influence of Lake Michigan water levels on wetland soils and distribution of plants in the Straits of Mackinac, Michigan. Ph.D. Dissertation, Univ. Michigan, MI 1981.

Lyon, J. G. and Drobney, R. D., Lake level effect as measured from aerial photos, *ASCE J. Surv. Engin.,* 110, 103–111, 1984.

Lyon, J. G., Drobney, R. D., and Olson, C. E., Jr., Effects of Lake Michigan water levels on wetland soil chemistry and distribution of plants in the Straits of Mackinac, *J. Great Lakes Res.* 12(3), 175–183, 1986.

Payne, F. C., Schuette, J. L., Schaeffer, J. E., Lisiecki, J. B., Regalbuto, D. P., and Rogers, P. S., Evaluation of marsh losses: Maisou Island complex. Prep. for: Wildl. Div., Michigan Dept. Ntrl. Resources, 1985, 67 pp. + appendix.

USACE (U.S. Army Corps of Engineers), Draft Environmental Impact Statement: Supplement II to the Final Environmental Impact Statement for Operations, Maintenance, and Minor Improvements of the Federal Facilities at Sault Ste. Marie, Michigan (July 1977), Operation of the Lock Facilities to 31 January ± 2 Weeks. Detroit District, Corps of Engineers, 1987, 176 pp. + appendices.

USFWS (U.S. Fish & Wildlife Service), Photointerpretation conventions for the National Wetlands Inventory, May 1, 1987, 51 pp.

USFWS (U.S. Fish & Wildlife Service), St. Marys River, Michigan, geographic information system applications. Final report, January 19, 1988, 1988, 5 pp.

Airborne Multispectral Scanner Data for Evaluating Bottom Sediment Types and Water Depths of the St. Marys River, Michigan

John G. Lyon, Ross S. Lunetta, and Donald C. Williams

ABSTRACT

Airborne multispectral data and models of the light interaction with water were employed for general assessment of river bottom soil types and water depths in the study area. This required analyses of scanner data using multivariate pattern recognition techniques. Subsequent radiometric modeling and analyses of field data resulted in determination of class identities and creation of bottom type and water depth thematic maps from scanner data. Accuracy assessments indicated that the effort produced very good identification of five general bottom sediment types (85%) and general water depths (95%).

INTRODUCTION

Large engineering projects often require environmental analyses and production of Environmental Impact Statements (EIS). One portion of this effort is the assessment of impacts of a project or action on environmental resources such as wetlands, wildlife, and fisheries. Evaluations of these resources are expensive using traditional measurement techniques. The resources vary in space and in time, and it is difficult to sample their variability. As a result, increased attention has been focused on advanced technologies to evaluate resource character-

By permission, The American Society for Photogrammetry and Remote Sensing: Airborne multispectral scanner data for evaluating bottom sediment types and water depths of the St. Mary's River, Michigan. J.G. Lyon, R.S. Lunetta and D.C. Williams, *Photogrammetric Engineering & Remote Sensing*, July 1992, 58(7): 951–956.

istics on a per-area basis. Technologies such as remote sensing can provide area-wide information and a potentially unique source of data for analyses of environmental resources.

The Detroit District Corps of Engineers of the U.S. Army (USACE) is charged with operation and maintenance of the locks at Sault Ste. Marie, Michigan. They are concerned with the environmental resources, their natural variability, and any potential affects of lock operation and navigation on the St. Marys River and Lake Superior system (USACE, 1979, 1988). The locks and St. Marys River receive a large volume of iron ore, coal, and wheat, and transport of these bulk and other cargoes is critical to the regional and national economy.

The importance of environmental resources in the St. Marys River area has also been recognized, and the government and a number of groups have many questions that must be answered. Hence, there is a continuing need to evaluate environmental resources in the St. Marys River as they are influenced by federal actions or projects. This need for information is mandated by the National Environmental Policy Act (NEPA), the Fish and Wildlife Coordination Act, and other laws and regulations.

The large size of the river system, extreme weather, and rural surroundings make data acquisition difficult and expensive. Remote sensing and aerial photography were initially proposed to supply a variety of data to augment other point measurements. The value of remote sensing was its capability to provide data over large areas, and to measure spectral and spatial variables that are ordinarily unavailable from traditional sampling methods.

Studies have demonstrated a unique capability to measure aquatic and water resource variables using combined remote sensing and field sampling experiments (Klemas et al., 1974; Lathrop and Lillesand, 1986; Lyon et al., 1988). To further the application of remote sensor data for measurement and modeling of water resources requires efforts to employ deterministic models in addition to the statistical approaches which have been so successful. Radiation Transfer (RT) models have the potential to simulate various water resources, and also to fulfill the need for a deterministic model based on physical and/or chemical processes (Scherz and Van Domelen, 1975; Weidmark et al., 1981; Butkata et al., 1983; Suits, 1984; Hollinger et al., 1985; Bukata et al., 1988). It remained to develop a simple RT model for the St. Marys River, and to determine the accuracy of this combined remote sensor data and deterministic modeling approach to develop bottom type and water depth map products for use in environmental analyses.

METHODS

A variety of methods were combined to develop the final products and test their accuracy. These efforts included the acquisition of remote sensor data and the processing of the data into thematic maps of bottom types and water depths, as well as an independent accuracy assessment of the quality of the maps.

Airborne Scanner Data

The need for a detailed spectral and spatial data set was met by use of multispectral remote sensor data. The U.S. Environmental Protection Agency (EPA) made an aerial overflight of the St. Marys River study area on October 19, 1985 during very good weather conditions. The exact configuration of the sensor flown by the EPA Environmental Monitoring Systems Laboratory in Las Vegas, NV was a Daedalus 1260 scanner that acquired data in 12 spectral windows or bands from 0.38 to 14.00 μm, with a 10-m resolution element or pixel. 1:24,000-scale color aerial photographs were also taken with a Wild RC-8 camera. Concurrently, water variables were sampled in the field by boat at 36 stations. Data included total suspended solids, Secchi Disk depth, temperature, chlorophyll a, and depth to bottom.

The EPA Laboratory processed the raw aircraft scanner data, including unpacking the data from high density format into a computer compatible density of 1600 BPI. They also made radiometric calibrations, corrections of scan angle distortions, and preliminary geometric corrections of any pixel size distortion due to aircraft yaw, pitch, and roll.

The processed scanner data were then used to generate both image and computer categorization products. Data analysis required two general steps. The first step included geometric corrections, selection of a sub-set of spectral bands for analysis (feature selection), and statistical pattern recognition (Moik, 1980) of water colorants classes. Resulting products were a database of bottom sediment types and water depth classes ready for subsequent analyses in step two of the methods presented here (Hollinger et al., 1985).

The second step focused on development of bottom type and water depth thematic maps. The approach employed Secchi Disk measurements, water class statistics from categorization, and the RT model for calculating the spectral radiance or brightness from the water column and bottom (Scherz and van Domelen, 1975). The simple RT model estimated incoming light condition and fore- and backscattering of light in the water column, and was used to relate the contributions of varying water depths to the resulting brightness values measured by a remote sensor. Elements of this approach are consistent with more complex approaches for deterministic modeling of water depth and water quality (Bukata et al., 1978; 1983, Jain et al., 1981; Philpot, 1981; Suits, 1984; Hollinger et al., 1985).

Step 1: Identification of Bottom Types

The first objective was to identify the general bottom sediment types in the study area. This was necessary for the second objective as well, because different bottom types commonly exhibit great differences in reflectance. For example, sand substrates tend to have higher reflectances than silt, clay, or organic soil materials. These differences can result in incorrect determinations of water depth, and it was necessary to separate or stratify our data into bottom type classes. As a result, determination of water depths for each bottom type class or stratum could be made in a direct manner (Weidmark et al., 1981).

In the Lake Nicolet portion of the St. Marys River there were several distinct bottom sediment types. These included the following river bottom sediment categories or types: sand, silt/clay and silt/sand combinations, and sand/silt and sand-rock/silt combinations. Each bottom type exhibited distinct spectral reflectance in several spectral data bands, and they could be discriminated spectrally. Hence, unsupervised computer categorizations or pattern recognition experiments were employed to separate bottom types, and thus achieved stratification of classes by bottom type to allow water depth analyses to be completed as a separate effort (Jain et al., 1981; Weidmark et al., 1981).

Several procedures were followed to develop the desired thematic map products. Initially, a feature selection experiment identified an optimal subset of the 12 bands for processing (Moik, 1980). Daedalus 1260 bands 1 (0.38 to 0.42 μm) and 2 (0.42 to 0.45 μm) were eliminated due to atmospheric and/or water opaqueness at short, visible wavelengths. Bands 9 (0.80 to 0.89 μm) and 10 (0.92 to 1.10 μm) were also eliminated due to sensor anomalies. The bands mentioned here were also eliminated because they supplied little or no detail on water resources useful for this analysis.

The remaining six bands were used in the experiment. Bands 3, 4, and 5 (0.45 to 0.50, 0.50 to 0.55, and 0.55 to 0.60 μm, respectively) provided data for bottom type categorizations and measurement of water depths. Band 6 (0.60 to 0.65 μm) was valuable for identification of shallow water areas and their depth conditions. Near- and thermal-infrared channels 7 and 11 (0.65 to 0.69 and 8.00 to 14.00 μm) were used to identify upland classes in the scanner scene, and to separate them from the water resource classes of interest.

As a first step, upland or terrestrial areas in the scene were "masked out" or removed from the analysis. The goal was to reduce the size of the data set by eliminating data unimportant to the project. It also reduced the contribution of spectrally distinct terrestrial data to this spectrally based pattern recognition experiment aimed at identifying aquatic resources. To do this, a two-band data set was copied from the original; it consisted of near- and thermal-infrared bands 7 and 11. This two-band scene was categorized by clustering; six land and water categories were selected a priori; and a thematic map of terrestrial and aquatic resources was made. All terrestrial spectral classes were subsequently recorded to class zero, and all aquatic spectral categories were recorded to class one. The original four-band image was multiplied by the resulting terrestrial/aquatic image of "1s" and "0s." The result was a scene composed of only areas of the water resource classes of interest.

The data set used in categorizations was this "masked" four-band scene described above. Preliminary categorizations on a smaller area indicated that a clustering algorithm with 50-classes yielded the best product. In addition, a minimum cluster distance of two was determined to optimize the recognition of bottom types. Categorizations were completed using these conditions and the clustering algorithm of the ERDAS system.

The final categorized products required geometric corrections. To correct the geometry, Ground Control Points (GCPs) were selected from road intersections, aids to navigation, and other distinct points on both maps and image products. GCPs were digitized from 1:24,000-scale USGS maps and referenced in UTM coordinates. The row and column indices of GCPs were taken from coordinates of the scanner data, and were compared to UTM coordinates in the transformation. A root-mean-square (RMS) error of 5 was achieved with a cubic-convolution algorithm (Moik, 1980).

Final products were made from the geometric corrected, categorized scanner data images. The resulting products included model determinations and maps of general bottom type and water depth. The individual spectral classes selected by clustering ($n = 50$) were identified as to bottom type. Classes were assigned or stratified into a bottom type class by a combination of approaches, including (1) evaluation of field data, (2) interpretation of the color photographs from the overflight, (3) aerial photos from other dates, and (4) two-axes graphical plots of cluster means for each class. Resultant images displayed the location of sand, combinations of silt/clay and silt/sand, and combinations of sand/silt and sand-rock/silt types.

Water quality data were acquired during the overflight, and samples exhibited the oligotrophic quality characteristics of the Lake Superior headwaters. In the study area, localized concentrations of suspended solids were less than 1.0 mg/l and no higher than 3.0 mg/l, and chlorophyll a was approximately 0.3 mg/l. Due to these low concentrations, it was assumed that their contributions to remote measurements were small and probably would not alter the results. Hence, it was deemed unnecessary to model these water quality characteristics for this particular overflight.

Step 2: Water Depth Determination

The modeling effort employed on-site Secchi Disk measurements to estimate the extinction or attenuation of light by the water column. The extinction coefficient of water was estimated using Beer's Law and approximations made by Scherz et al. (1974). The basic assumption was that the energy returned by the Secchi Disk to the observer in a boat was approximately 10% of the initial radiance on the water surface (Scherz et al., 1974). From this assumption, it followed that we could calculate

$$L(\text{radiance on Secchi Disk, } Ysd)$$
$$= L(\text{radiance initially under the water surface})/e^{aY},$$

where a is the extinction coefficient and Y is the depth of water. As the energy reaching the Secchi Disk (sd) at depth "Ysd" is approximately 10% of L (at the water surface), the extinction coefficient (a) can be calculated. It follows that

$$e^{aYsd} = L(\text{surface})/LYsd = 1/0.10 = 10.0$$

Rearranging the equation results in a solution:

$$aYsd = 2.3,$$
$$a = 2.3/Ysd$$

The *Ysd* measured during the overflight was 10 ft (3 m). The resulting extinction coefficient was calculated as $a = 2.3/10.0$ or 0.23. This extinction coefficient value was within the range observed by other investigators in the same or similar waters (Bukata et al., 1978, 1983; Liston et al., 1986).

Once the extinction coefficient was established, the RT model was used to calculate the radiance of the water column. The model incorporated the fore- and backscattering and extinction of light in uniform layers of depth, the bottom reflectance, and suspended and dissolved substance influences on radiance. These components were addressed by modeling the attenuation of light with depth as a series of five layers (Scherz and Van Domelen, 1975; Scherz et al., 1977; Suits, 1984). Each layer of equal depth or thickness was characterized by a separate calculation of the contribution of light from fore- and backscattering and extinction of light by water. The results were used to develop a "lookup table" of brightness values for the five water depth classes for each bottom type.

The entries in each lookup table consisted of five brightness values that may be expected for each given depth, 1/5 *Ysd* or 2 ft (0.8 m). The model calculated the percentage of each layer's contribution and was used to determine the light that would be measured from a given depth. Individual lookup table entries for a certain depth range were calculated from

$$L(Y\ at\ "i") = [L\ (\text{shallow water}) - L\ (\text{deep water})]$$

(proportion of light returned from layer *"i"*).

Contributions of light from these depth layers were calculated by Scherz and Van Domelen (1975) and Scherz et al. (1977) using the relationship

$$L_{Yi} = L\ (\text{incident})\ t^{(2i-2)}\ F^{(2i-3)}$$

where L_{Yi} is the radiance from a given layer *"Y,"* *"t"* is the transmission of light through a layer, *F* is the combination of unit backscatter that the light will experience, and *"i"* is the given layer of thickness *"Y."*

Using this approach, Scherz et al. (1977) determined that clear water will return radiance approximately in the proportion of 0.590 or 59.0% of the incident radiance from the first layer, 24.0% from the second, 9.5% from the third, 7.1% from the fourth, and 0.4% from the fifth layer of water. The depth or thickness of each layer is again one-fifth of the Secchi Disk depth. For this analysis we assumed $BV \cong L(\text{sensor})$. This is a suitable assumption due to (1) the small study area and low altitude overflight, (2) the early morning and clear atmospheric conditions during the overflight, and (3) the uniform conditions of path radiance would presumably contribute uniformly to the L(sensor).

The BV data from a given bottom type and its classes were searched to identify "shallow water" and "deep water" examples. The mean BV data in the four bands were used to identify the high and low BV and supply a range. The proportions from the Secchi Disk calculations were used to identify the portion of the BV range that represented the water depths we sought to identify. The BV in each band, for each water depth class (2.0 to 4.0, 4.0 to 6.0, 6.0 to 8.0, >8.0, and >10.0), was calculated and "stored" in a lookup table.

The above calculations allowed us to determine the brightness values (BV) associated with different depths of water over the same bottom type. By selecting a very shallow or high BV example, and a very deep or low BV example of each bottom type, we were able to develop a lookup table consisting of mean brightness values for each bottom type from the four spectral bands. The resulting lookup tables were used to assign general depth classes in 2-ft increments (0.8 m) to 10 ft (3 m) for each class in the categorized scanner scene.

The final product was developed from (1) lookup tables and use of (2) field sampling data from the overflight, sampling data from 1987, and sampling data from other studies (Liston and McNab, 1983).

The lookup table also incorporated a correction for the actual water depth calculated. The final product needed to be in low water datum (LWD), as were NOAA charts, to allow comparison. This was completed by subtracting the difference of the actual water level during overflight and the LWD.

To verify the quality of the bottom type and water depth thematic maps, an accuracy assessment was completed. In August 1987, 137 bottom samples and water depths were collected during a 3-day period in the St. Marys River area. Ninety-seven of the bottom samples were submitted to the USACE soils laboratory in Cincinnati, OH and soil types of the bottom samples were measured by feel. Seventy-four of these soil and 76 of the depth samples were taken in the Lake Nicolet study area.

RESULTS AND DISCUSSION

A combination of techniques—including unsupervised categorization of scanner data, use of a radiometric model, and field data analysis—yielded thematic maps of general bottom soil type and water depth (Figures 1 and 2 in color section). These products compared well to field sampling data (Figures 3 and 4). Verification of the accuracy of bottom types was performed with sampling data from the 1987 field season (USACE, 1989). Class accuracies are presented in Table 1. The total accuracy was 63 of 74 samples or 85%.

Verification of water depths was completed by comparing, point by point, selected areas of (1) the USACE water depth maps and (2) samples from the 1987 field data collection for verification. The individual class accuracies are recorded in Table 2. The total accuracy was 72 of 76 samples or 95%.

These results were partially due to the very low concentrations of water colorants in the Lake Superior headwaters of the St. Marys River (Bukata et al., 1978, 1988). Absence of water colorants made this a valid approach, and others have experienced similar results (Lathrop and Lillesand, 1986; Hutchinson,

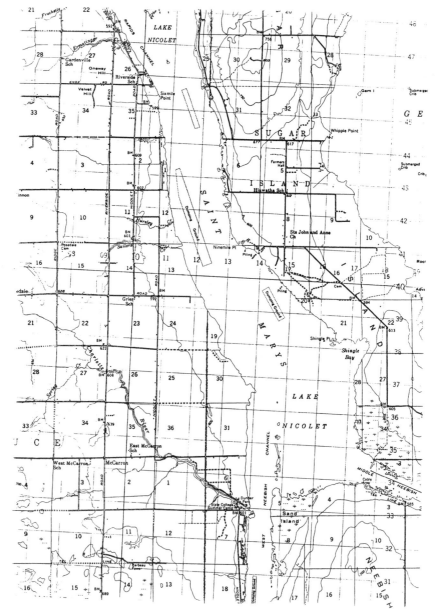

Figure 3 Location map for the St. Marys River and Lake Nicolet, Michigan study area.

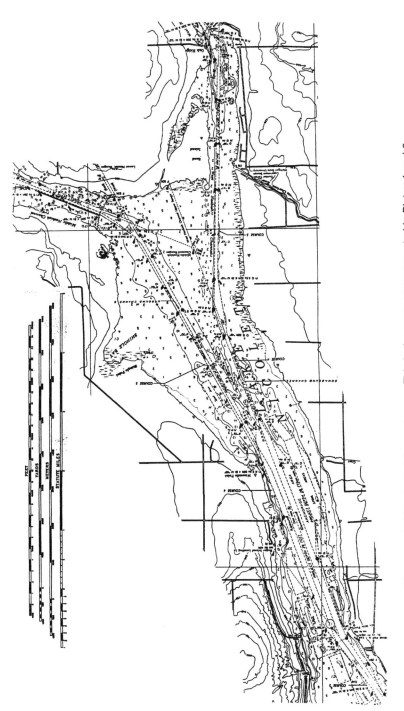

Figure 4 NOAA chart of the study area. This is the same area presented in Plates 1 and 2.

Table 1 Class Accuracies of the Remote Sensor Derived Bottom Categories and the Same Categories Identified in the Field

Derived categories	Remote sensor derived thematic map categories				
	Sand	Silt/Clay	Silt/Sand	Sand/Silt	Sand-Rock/Silt
Sand	18/21, 85%				
Silt/Clay		9/14, 64%			
Silt/Sand		5/14, 36%	12/13, 92%	1/11, 9%	
Sand/Silt	3/21, 15%		1/13, 8%	10/11, 91%	3/15, 25%
Sand-Rock/Silt					12/15, 75%

Table 2 Class Accuracies of the Remote Sensor Derived Water Depth Categories and the Same Categories Identified in the Field

	2.0–4.0 ft	4.0–6.0 ft	6.0–8.0 ft	>8.0 ft	>10.0 ft
2.0–4.0 ft	5/6, 83%				
4.0–6.0 ft	1/6, 17%	12/13, 92%			
6.0–8.0 ft		1/13, 8%	17/17, 100%	1/17, 6%	
>8.0 ft				16/17, 94%	1/23, 4%
>10.0 ft					22/23, 96%

1989). The same results would not be possible in waters containing concentrated phytoplankton or suspended sediments.

The remote sensor approach may be valuable for several types of environmental analyses. The resulting thematic maps of water depths and bottom types over large areas can supply data that are difficult and much more expensive to obtain as compared to products resulting from other technologies. The combination of limited field sampling and remote sensor data potentially could supply similar information in other clear water areas.

CONCLUSIONS

In the St. Marys River, the use of a combination of categorized airborne scanner, radiometric, and field data yielded thematic maps of water depths and bottom types. The approach used here, of identifying the individual contribution of bottom sediment types and stratifying the data set by bottom type and then determining the brightness value of water depths, allowed for accurate measurements. This combination of remote sensing measures and modeling was very useful in meeting requirements for information to evaluate the characteristics of environmental resources.

ACKNOWLEDGMENTS

This research was sponsored by the Detroit District of the U.S. Army Corps of Engineers and by the NOAA Office of Sea Grant through grants NA81AAD-0095, R/EM-2, -3, -7, and -8. Data were processed at the Computer Mapping Center, Detroit District, and Airborne scanner data were supplied by the EPA Environmental Monitoring Systems Laboratory. We appreciated the assistance of Tom Mace, Mason Hewitt, Gene Meier, and Mark Olsen. The assistance of Carl Argiroff, William Willis, Les Weigum, Roger Gauthier, and Tom Freitag of the Detroit District is also gratefully acknowledged.

REFERENCES

Bukata, R., Bruton, J., and Jerome, J., Use of chromaticity in remote measurements of water quality, *Remote Sensing Environ.* 13, 161–177, 1983.

Bukata, R., Jerome, J., Bruton, J., and Benett, E., Relationship among optical transmission, volume reflectance, suspended mineral concentration, and chlorophyll a concentration in Lake Superior, *J. Great Lakes Res.*, 4, 456–461, 1978.

Bukata, R., Jerome, J., and Bruton, J., Relationships among Secchi Disk depth, beam attenuation coefficient, and irradiance attenuation coefficient for Great Lakes water, *J. Great Lakes Res.*, 14, 347–355, 1988.

Hollinger, A., O'Neil, N., Dunlop, J., Cooper, M., Edel, H., and Gower, J., Water-depth measurement and bottom type analysis using a two-dimensional array imager, Proceedings of the Nineteenth International Symposium on Remote Sensing of Environment, 1985, 553–563.

Hutchinson, W. H., *Application of a Radiometric Model and Remote Sensor Data for Evaluation of Water Depths*, M.S. Thesis, Dept. of Civil Engineering, The Ohio State University, OH, 1989, 85 pp.

Jain, S., Zwick, H., Weidmark, W., and Neville, R., Passive bathymetric measurements of inland waters with an airborne multispectral scanner, Proceedings of the Fifteenth International Symposium of Remote Sensing of Environment, 1981, 947–951.

Klemas, V., Otley, M., Philpot, W., and Rogers, R., Correlation of coastal water turbidity and circulation with ERTS-1 and Skylab imagery, Proceedings of the Ninth International Symposium on Remote Sensing of Environment, 1974, 1289–1317.

Lathrop, R. and Lillesand, T., Use of thematic mapper data to assess water quality in Green Bay and Central Lake Michigan. *Photogramm. Eng. Remote Sensing* 52, 671–680, 1986.

Liston, C. and McNab, C., *Limnological and Fisheries Studies of the St. Marys River, MI, in Relation to the Proposed Extension of the Navigation Season, 1982 and 1983*, U.S. Fish and Wildlife Service, Rept. No. FWS/OBS 86(3), St. Paul, MN 1983, 764 pp.

Lyon, J., Bedford, K., Chien-Ching, J., Lee, D., and Mark, D., Suspended sediment concentrations as measured from Landsat and AVHRR data, *Remote Sensing Environ.*, 25, 107–115, 1988.

Lyon, J. and Drobney, R., Lake level effects as measured from aerial photos, *ASCE J. Surveying Eng.* 110, 103–111, 1984.

Lyon, J., Drobney, R., and Olson, C., Effects of Lake Michigan water levels on wetland soil chemistry and distribution of plants in the Straits of Mackinac, *J. Great Lakes Res.*, 12, 175–183, 1986.

Moik, J., *Digital Processing of Remotely Sensed Data.* NASA, Beltsville, MD, 1980, 340 pp.

Philpot, W., *Remote Sensing of Optically Shallow, Vertically Inhomogeneous Waters: A Mathematical Model.* Sea Grant Pub. DEL-GS-12-81, University of Delaware, Newark, DE, 1981, 12 pp.

Scherz, J., *Assessment of Aquatic Environment by Remote Sensing*, Rept. No. 84, Institute for Environmental Studies, University of Wisconsin, Madison, WI, 1977, 235 pp.

Scherz, J. and Van Domelen, J., Water quality indicators obtainable from aircraft and Landsat images, and their use in classifying lakes, Proceedings of the Tenth International Symposium on Remote Sensing of Environment, 1975, 447–460.

Suits, G., A versatile directional reflectance model for natural water bodies, submerged objects, and moist beach sands, *Remote Sensing Environ.*, 16, 143–156, 1984.

U.S. Army Corps of Engineers, *Final Supplement to the Operation and Maintenance Environmental Impact Statement for the Federal Facilities at Sault Ste. Marie, MI, Addressing Limited Season Extension of Operation*, Detroit District, MI, 1979, Sections I-VI and Appendices A–L.

U.S. Army Corps of Engineers, *Supplement II to the Final Environmental Impact Statement, Operations Maintenance and Minor Improvements of the Federal Facilities at Sault Ste. Marie, MI (July 1977), Operation of the Lock Facilities to 31 January +/– 2 Weeks*, Detroit District, 1988, Sections A–H.

Weidmark, W., Jain, S., Zwick, H., and Miller, J., Passive bathymetric measurements in the Bruce Peninsula region of Ontario, Proceedings of the Fifteenth International Symposium on Remote Sensing of Environment, 1981, 811–822.

Section III

Environmental Engineering Applications of GIS and Remote Sensing

CHAPTER 11

Introduction to Environmental Engineering Applications

John G. Lyon and Jack McCarthy

GIS and remote sensor data can greatly facilitate modeling in scientific or engineering endeavors. The level of success depends on a number of factors. As mentioned, GIS information can be input directly to the model as primary data. GIS results can be used to estimate the values of model coefficients, thereby making the coefficients more realistic as compared to nature. GIS results can be compared with model results as their independent check or verification.

Whether GIS and remote sensor technologies will help a given project depends on the objectives and on the given application. The skills and experience of the users are also important. The following chapters present important, real-world applications that combine a variety of technologies and methods of analysis. GIS plays a role as an element in the experimental design, yet it makes a valuable contribution in each application. The quality of the results directly relates to the skill in using GIS and remote sensor technologies, and also in combining those results with results from other more traditional measurement and modeling technologies.

This section has several papers that address the capabilities of GIS and related technologies. These chapters examine the theme of surface water resources and their management. GIS and remote sensor technologies are particularly important to these sort of evaluations because: (1) they supply data and information on land cover in study areas, (2) they implement Best Management Practices (BMP), and they identify impact areas in support of water quality studies, and (3) GIS and remote sensor technologies are particularly valuable tools in evaluations of nonpoint sources of pollution. This is because nonpoint sources of pollution are usually distributed spatially over the landscape. Consequently, they have no discrete point source location and hence are nonpoint in origin.

Nonpoint pollution is of vital importance to water resource and land management activities. GIS and remote sensor technologies offer the promise of measurements, and database or modeling capabilities in support of nonpoint pollution research and management studies. The landscape scale of the problem demands both the spatial and spectral input data from remote sensing, and the database management and modeling capabilities of GIS.

Remote sensing techniques supply a variety of information that can be analyzed or interpreted for conditions related to nonpoint pollution. For example, remote sensor technologies: (1) can identify the presence of crop residue and inventory the tillage activities over large areas; (2) can identify areas of active erosion; (3) can identify riparian areas that have been impacted; and (4) can help evaluate a variety of other related variables.

GIS techniques offer special capabilities that can greatly enhanced the results of nonpoint pollution studies or modeling. GIS can do the following: (1) extrapolate or scale point-sampling results to area-based estimates; (2) evaluate the spatial distribution of crops or other land cover changes; (3) evaluate the mix of land covers in a given watershed, and identify impervious classes based on land cover and soils; and (4) supply other data and information.

The chapters in this section examine the nonpoint surface water pollution using spatial analyses. They also examine in detail some of the techniques applied and the strengths of these approaches. Chapter 12 evaluates how spatial data, GIS, and other techniques can help to supply information on nonpoint sources over large areas.

The DEM and nonpoint chapter examines the issue of nonpoint pollution from other perspectives. A detailed examination of sensitivity of model analyses is conducted using the digital elevation model (DEM) or point elevation data. The characterization of the behavior of a given variable and its influence on the results of model simulations is called a sensitivity analysis. Sensitivity analyses of some form are advisable in the calibration and simulation steps of a modeling experiment. This chapter presents a good analysis using an application of great import for the reader.

GIS and remote sensor technologies are particularly valuable to planning and sitting of projects or of utility corridors. Sewer and related storm flow pipelines need to be routed. The plan can be graphed and delivered as a plan or map product. The product must be evaluated, modified, and presented in final form. The products will probably be used in the design and/or construction phases of a project and hence a whole variety of needs must be served.

This sort of application also illustrates how a number of competing factors or variables must be integrated to arrive at an optimal solution. In routing a utility corridor, the following factors must be addressed: (1) the presence of cultural and wetland resources, (2) ownership of property, topography, (3) location of floodplains, and (4) other variables. This level of complexity is beyond easy human interpretation or integration of the many details. The use of GIS technologies and a decision model that are weighted to represent the optimal condition or characteristics of a site can expedite this process.

The sewer system modeling chapter displays how GIS can provide valuable information and data for planning and sitting. The details of selection criteria have been addressed and presented in graphical or map form for important variables. Their integration provides for a routing that best represents the realities of the decision-making process and the groups involved.

Oil pollutant management is one example of a landscape scale event that can be modeled or, the event planning can be facilitated, by the use of GIS and related technologies. Response to spills is largely an emergency action. GIS and related technologies can help prepare for contingencies and estimate where impacts will likely occur.

The chapters presented in this section examine the use of GIS and related technologies for evaluations of oil spill response activities, and for evaluation of real time data inputs to GIS for spill management and control. They demonstrate and discuss the advantages of GIS for addressing the contingency and the spatial distribution of the spill and response to the spill.

CHAPTER 12

GIS Targets Agricultural Nonpoint Source Pollution

James Hamlett, Tawna Mertz, and Gary Petersen

INTRODUCTION

Although it is difficult to pinpoint the exact origin of nonpoint source pollution, the relative contribution from watersheds and subwatersheds can be assessed using GIS. Pollution susceptibility models for surface and subsurface waters combine hydrogeologic parameters, such as soil type, topography or slope, depth of water table, net recharge, aquifer characteristics, and conductivity of the media, with pollutant characteristics and demographic information. These screening or indexing models make relative comparisons within and among regions. Such assessments have become necessary with 1987 amendments to the federal Clean Water Act (CWA). The CWA, administered by the EPA, requires states to develop assessments of nonpoint source water quality problems and management plans to remediate nonpoint pollution problems (Hession et al., 1992).

Sixty percent of all sediment delivered to the nation's waters is from agricultural land. Best estimates indicate that sediment and sediment-associated contaminants cause damages of more that $4 billion (1980 dollars) annually (USDA, 1989). Wadleigh (1968) and the Council on Environmental Quality (1980) estimated total annual sediment contribution to U.S. streams to be 3.6 billion tons. These sediments often contain pesticide residues and nitrogen and phosphorus from fertilizers. In Pennsylvania, agriculture's contribution to nonpoint source pollution is second only to mine drainage. Pennsylvania farms located within the Susquehanna River basin contribute nearly one-third of the nitrogen and one-quarter of the phosphorus pouring into the Chesapeake Bay (EPA, 1983).

Degradation of the water quality within the Chesapeake Bay led to a cooperative approach to pollution problems among the governments of Virginia, Maryland, Pennsylvania, the District of Columbia, the Chesapeake Bay Commission, and the U.S. EPA, via the Chesapeake Bay Agreement of 1983. The Chesapeake Bay Agreement of 1987 set specific water quality goals, including a 40% reduction of nitrogen and phosphorous entering the Bay from 1985 to 2000 (Chesapeake Bay Commission, 1987). Because resources for combating non-

Based on an article published by GIS World, April 1993. With permission.

point source pollution are limited, states are required by the EPA to develop a list of high priority, or critical, water bodies on a watershed basis (Hession et al., 1992). High priority regions receive preference for funds to help alleviate nonpoint sources of pollution.

Groundwater pollution from domestic, agricultural, commercial, and industrial sources has only recently entered the spotlight. Across rural America, groundwater is considered the prime source (more than 95%) of drinking water supply (USGS, 1990); on a national basis, nearly half of the population depends on groundwater. More than one-third of Pennsylvanians depend on groundwater for their water needs (USGS, 1986). Contamination of groundwater results from the generation of potential pollutants and the transport of these pollutants from the source to the receiving aquifer. The federal Safe Drinking Water Act requires all community water systems and nontransient, noncommunity water systems to monitor their drinking water (40 CFR Part 141, 142, and 143) for pesticides and synthetic organic chemicals.

GIS, combined with modeling capabilities, offers an efficient means of identifying and ranking the nonpoint pollution potential of areas for both surface water and groundwater. Government agencies like the Soil Conservation Service (SCS) and Pennsylvania's Department of Environmental Resources (PADER) are starting to rely on models to help them direct programs and limited monies available for reducing nonpoint source pollution.

GIS APPLICATIONS TO NONPOINT SOURCE POLLUTION

Existing databases can be used when applying GIS to assess the potential nonpoint source pollution problems. Data may range from specific field cropping practices from county level land-use data to statewide data compiled through the U.S. Geological Survey's (USGS) Land Use Data Analysis program. National pollution data can be obtained from sources such as the Toxic Release Inventory, EPA's Storet database, USGS's WATSTORE, as well as state sources. Soils information is available through the SCS's STATSGO database. Demographic data, transportation network information, hydrogeology, topography, and many other types of data are also available from various sources and at different scales. These data provide the foundation for GIS-based resource analyses.

Availability of data at various scales becomes an important issue as one considers the formulation of nonpoint source pollution assessment/screening models. Ideally, a physically based model at a detailed scale is preferable when attempting to assess the extent and magnitude of agricultural nonpoint pollution. The use of such models, however, requires a substantial quantity of high-resolution data (both spatially and temporally) that is often not readily available and that requires a tremendous investment in time and resources to obtain. Few watersheds currently have this extensive data available. Screening models, on the other hand, use less physically based algorithms and rely more on loading functions related to general watershed characteristics (e.g., land use/land cover, topography, soils, etc.), nutrient and pesticide usage, and proximity of source areas to water bodies. These screening models generally are used at regional scales and, hence, require less detailed data. However, reliable data sets at regional/statewide scales are necessary if reasonable comparisons of potential nonpoint pollution are to be realized. The detail of the model screening or assessment is highly dependent on the extent of data needed by a particular model, the extent and scale of data that are readily available, and the compatibility of data scales. In short, the resolution of the modeling assessment is only as good as the limiting data layer.

When identifying or targeting critical watershed areas on a regional basis, data layers must be fairly small scale (1:1,000,000 or 1:2,500,000). Generally, a reasonable database at this scale can be compiled (obtained or developed) from readily available programs (e.g., USGS land use data, USGS topographic data, hydrogeology, SCS STATSGO soils information, Census data, etc.). Such regional databases have been compiled and used in various locations, including Pennsylvania, Montana, Rhode Island, and the Chesapeake Bay basin. National databases (at scales such as 1:5,000,000 and smaller) have also been used for initial resource screening studies. These screenings provide a first-level sorting of regions, relatively large river basins, or counties that have potential resource problems. Watershed site-specific assessments, which require detailed data (i.e., 1:24,000 scale), remain somewhat limited because the magnitude of data (i.e., various types of data as well as quantity) are quite limited and expensive to acquire or collect. Assessment efforts may be approached by considering availability of data and developing a multitiered screening/assessment methodology, whereby small-scale (national scale) data are used to identify general regions or basins where pollution potential seems likely. Statewide or regional data (1:100,000 or 1:250,0000) can then be used to further target the critical watersheds warranting initial detailed (large scale at 1:24,000) model-based assessments. Resources are efficiently and effectively targeted through this approach.

At Penn State, a GIS-based model has been developed that ranks the agricultural nonpoint pollution potential of 104 watersheds in Pennsylvania. Enhanced spatial resolution is currently being used to explore subwatersheds within several problem regions in Pennsylvania. The groundwater pollution potential for 478 groundwater subbasins in the state is also being assessed by combining GIS with modeling capabilities. Because the Penn State models use widely available databases, the federally funded National Center for Resource Innovations (NCRI), in conjunction with SCS and other cooperating groups, adopted the Penn State approach as a base for evaluating agricultural nonpoint pollution potential in the Chesapeake Bay drainage region. This project is ongoing.

STRUCTURE OF A REGIONAL NONPOINT POLLUTION MODEL

The relative agricultural nonpoint pollution potential of surface and groundwater basins in Pennsylvania is being assessed through ongoing GIS-based projects. Increased understanding of the potential contaminant generation (by source and locale), the relative risk of transport pathways that conduct pollutants to aquifers and streams, and targeting of possible areas for control programs/practices requires more detailed studies. As potential problem areas are identified, educational efforts and controls can be developed and implemented, thereby improving protection of water resources within the Commonwealth. The following screening assessment could be used as a guide for other states/regions interested in conducting similar watershed/subbasin pollution potential screenings.

The GIS approach utilized at Penn State consists of acquiring or developing statewide databases (scale of 1:250,000 where possible), building statewide nonpoint pollutant screening models (loading function based) that use these data, and processing the GIS data with these models so as to identify "critical" pollutant source areas (watersheds) within Pennsylvania. Primary efforts are directed at compiling the database and developing both surface water and groundwater pollution susceptibility screening models that are reasonable, given the type and extent of readily available data. Efforts toward conducting a surface water screening model were initiated in 1990 with an initial ranking completed in 1992. The susceptibility of subsurface aquifers to both nitrate leaching and pesticide pollution are currently being investigated. All three studies were undertaken to help PADER identify potential "critical" areas for agricultural nonpoint pollution. In each case, PADER was interested in completing the screening studies in a relatively short time frame (1 to 2 years). Figure 1 schematically illustrates the overall framework of the agricultural pollution potential screening being conducted in Pennsylvania.

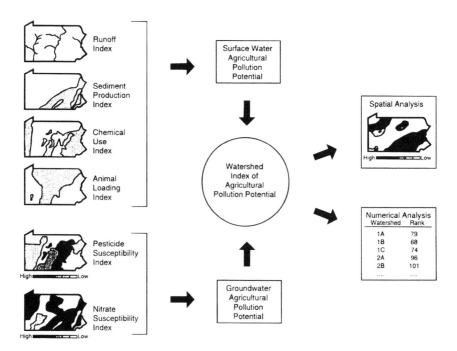

Figure 1   Schematic of agricultural pollution potential screening in Pennsylvania.

Facilities

The Office for Remote Sensing of Earth Resources (ORSER), Environmental Resources Research Institute, and the Land Analysis Laboratory at Penn State are cooperatively developing the database using state-of-the-art technology. ORSER computer facilities include a Local Area VAX cluster supporting graphic workstations and PC workstations with high resolution color monitors. Two SUN workstations are also available for data processing. The Land Analysis Lab has similar computational facilities. Programmers at ORSER and the Land Analysis Lab have extensive experience handling GIS data and programming and provide the necessary support for research faculty and graduate students working on related GIS projects. Several third-party and public domain software packages that handle spatial information are used in these ongoing resource studies. These include: ERDAS (Earth Resources Data Analysis System, 1994), ARC/INFO (Environmental Systems Research Institute, 1990), the Land Analysis System (LAS) from NASA (National Aeronautics and Space Administration and U.S. Geological Survey, 1990), and GRASS (U.S. Army Corps of Engineers, 1993).

Through these projects and others, a statewide database is being developed for Pennsylvania. For each data layer, we collate existing map data units from various sources to produce parameter layers in the GIS. When digital spatial information are unavailable, maps are acquired, digitized, mosaicked, and registered to produce the desired data layer. All data layers are processed to 100-m grid cells in an Albers equal area projection. The state is covered by a grid of 2860 rows and 4950 columns, resulting in nearly 14.2 million grid cells. This database provides a sound base for a multitude of resource-based, statewide studies.

COMPONENTS OF THE MODELS

Surface Water Assessment

The surface water ranking model focused on 104 major watersheds in Pennsylvania and was most concerned with the potential production and transport of sediment, nutrients, and pesticides from agricultural sources. The transport of these constituents with runoff and sediment-attached forms was considered by estimating the runoff and sediment delivery from the distributed cells (100 m by 100 m) used in the model. The 1-year time frame of the project forced the researchers to use existing data. Seven data layers (scale of 1:250,000) were used in this initial study. If spatial information was not available, maps were digitized and registered to produce the data layer. The data layers were adapted from a number of existing data sets; similar data sets are available in most states.

1. Watershed Boundaries were adapted from Pennsylvania's Bureau of Water Resources Management State Water Plan.

2. Land Use data gleaned from the USGS Land Use Data Analysis program were available at a map scale of 1:250,000. This information did not differentiate between agricultural pastureland and cropland, so the researchers stratified the layer, designating areas with 1 to 15% slope as cropland and areas with 0 to 1% and >15% slope as pasture.

3. Animal densities were obtained from 1987 U.S. Census Bureau data. Type and number of animals were available by zip code. Nitrogen and phosphorus production were calculated by developing nutrient loadings for the different animals.

4. Topography was extracted from the USGS 1:250,000 Digital Elevation Data. Latitude/longitude data were converted to 100-m grid cells and the slope was calculated for each cell.

5. Soils data were acquired through the linkage of two databases. The Soil Conservation Service, National Cartographic Center's State Soil Geographic Data Base (STATSGO) provided details on types and percentages of individual soils at a scale of 1:250,000. This information was joined with the Natural Resources Inventory Soils-5 database, which added information on the proportionate extent of the component soils and physical properties of soils. Soil properties, such as erosion potential and hydrological soil group, were then calculated for the model.

6. Precipitation layers were developed in collaboration with research scientists at ZedX, Inc., of Boalsburg, PA, a commercial firm that markets specialized databases for agricultural decision making. Two databases from the climatological station summaries published by the National Climatic Data Center in Asheville, NC, were integrated in a three-step process. The result was the maximum daily precipitation that could be expected each month, for a one-in-five year event, spatially interpolated to create a map with a 1 sq km resolution.

7. Rainfall-runoff factor was also developed in conjunction with ZedX, Inc. Twenty-two-year average annual and regional cumulative erosion index values were combined with topographic factors to create the spatial interpolation needed for this data layer.

An Agricultural Pollution Potential Index (APPI) was developed from the data layers to rank the 104 watersheds according to their relative nonpoint pollution potential. Four indices were used: (1) the sediment production index assessed potential erosion and sediment delivery to streams; (2) the runoff index evaluated the watershed's surface runoff potential; (3) the animal loading index ranked potential manure production within the watersheds; and (4) the chemical use index predicted pollution potential from commercial chemical applications.

Groundwater Assessment

The GIS database for Pennsylvania is being augmented to allow identification of groundwater subbasins most susceptible to groundwater pollution potential. Models assess two types of nonpoint source pollution: pesticides and nitrate contamination. Data were (or are being) acquired from the U.S. Department of Commerce, PADER, The Pennsylvania Department of Agriculture, and The Pennsylvania State University. Data layers used in these screenings, registered to the 100-m resolution digital database, include: precipitation, hydrogeology, soils, topography, geology, animal type and distribution, land cover, groundwater subbasins, pesticide use by crop, minor civil subdivisions, depth to water, and average net recharge areas. Attribute data for soils, geology, groundwater quality, chemical-nutrient use, and cropping are also incorporated.

Nitrate Pollution Potential

The indexing model to rank relative nitrate pollution potential of groundwater subbasins considers the potential for nitrate pollutant generation and the relative susceptibility of nitrate movement from the surface, through the soil and vadose zone, to the subsurface aquifers. Initial indexing was conducted using the available statewide data (at a scale of 1:250,000), the modified DRASTIC model (Aller et al., 1987), and ARC/INFO and ERDAS software.

DRASTIC uses a numerical ranking system to assess the groundwater pollution potential for hydrogeologic settings. The DRASTIC model evaluates the pollution potential of a hydrogeologic setting with seven factors: (1) depth to water; (2) net recharge; (3) aquifer media; (4) soil media; (5) topography; (6) impact of the vadose zone; and (7) hydraulic conductivity of the aquifer. Each factor possesses a range of different elements which are assigned a rating based on each one's pollution potential with respect to the other elements in the range. The ratings range from 1 to 10. Each factor is weighted, based on its potential contribution in comparison to the other factors. Factor weights range from 1 to 5. The depth to water was approximated using nearly 30,000 points of well data from across Pennsylvania; data was collected by the USGS. Net recharge was estimated using the Percolation Index (PI) developed by Williams and Kissell (Follet et al., 1991). Statewide soils data and the average annual precipitation data comprised the data layers used to develop net recharge. The aquifer media was based on the Pennsylvania geology map and related literature; a DRASTIC rating of aquifer media was determined from these. The statewide soils map (STATSGO) and attribute data provided the soil media layer. Because the top layer is generally disturbed by tillage and surface activities, texture of the second soil layer was used as the soil media texture. Petersen et al. (1991) previously determined topographic information. Impact of the vadose zone and hydraulic conductivity of the aquifer were generated based on the ratings of the aquifer media and the water yielding capabilities of rocks of Pennsylvania (Geyer and Wilshusen, 1982). Deichert (1993) provides more information on these procedures.

Nitrate-N levels of approximately 3000 wells located across Pennsylvania were available from the USGS database. A statewide map of nitrate-N levels was compiled from these data and compared with the DRASTIC rating of potential nitrate pollution. Given the resolution of the available nitrate data and the GIS-based calculations, a reasonable comparison of nitrate pollution potential was achieved. As nitrogen-use data (loadings at the land surface) and a finalized geologic map and attribute data become available, efforts will be made to link these data with the model and update the DRASTIC susceptibility rating.

Pesticide Pollution Potential

The Pennsylvania statewide ranking of groundwater susceptibility to pesticides is a first-tier screening procedure to determine the relative susceptibility of groundwater to pesticide contamination from agricultural sources. This model is intended for regional or statewide screenings to aid environmental planning and/or regulation. The screening is accomplished by evaluating the hydrogeologic vulnerability and the leachability characteristics of the pesticides used. The groundwater susceptibility methodology can be applied for any region with the availability and quality of data dictating the effectiveness of the modeling assessment. Major steps in the process are: (1) to obtain or develop pesticide-use data; (2) to develop a pesticide leachability coverage using the pesticide-use data combined with a leachability model and a land cover layer; (3) to develop a hydrogeologic vulnerability coverage; (4) and to combine the leachability and hydrogeologic vulnerability coverages to create a groundwater susceptibility to pesticides screening model. This process allows a relative ranking of subbasins to groundwater pesticide pollution potential. Figure 2 graphically displays the screening methodology being used. The model considers data from the statewide database, including depth of soil, soil type, geology, depth to water, precipitation, percolation, topography, crop type, and pesticide type and use. The assessment involves two parts: (1) a soil and hydrogeologic pollution attenuation potential; and (2) a rating of leachability of pesticides used with the subbasins.

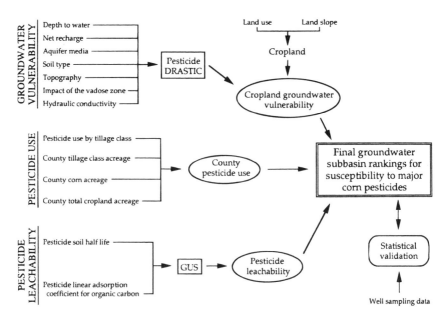

Figure 2   Procedure to determine relative susceptibility of groundwater pesticide contamination.

Pesticide DRASTIC (Aller et al., 1987) provides the basis for determining the relative hydrogeologic vulnerability of the groundwater subbasins to pesticide contamination. A pesticide screening model described by Gustafson (1989), called the groundwater ubiquity score (GUS), is used to create a relative ranking of pesticides according to their potential leachability. GUS assesses pesticide leachability based on persistence and mobility of the pesticides, with highly persistent and highly mobile pesticides being the most leachable. GUS relies on physical properties of the pesticides that are independent from specific sites and conditions. Statewide pesticide use data are estimated from land use/cover data, crop acreage available from agricultural statistics, and pesticide-use data from the National Agricultural Statistics Service (USDA, 1993). This information is linked with GUS and the modified Pesticide DRASTIC to provide an overall assessment of groundwater subbasin pesticide pollution susceptibility.

Pesticide Vulnerability Assessment Procedure

As a specific application of the GIS model for groundwater susceptibility to pesticides, development of a pesticide vulnerability assessment procedure (VAP) is planned for Pennsylvania groundwater subbasins that serve as a water source for public water supplies. Federal regulations allow a public water system to be exempt from monitoring requirements for pesticides and synthetic organic chemicals if they satisfy certain vulnerability criteria. Two types of waivers that exempt public water systems from monitoring are: (1) an area-wide use waiver, issued if regulated pesticides have not been used, manufactured, transported, stored, or disposed of within the surface/groundwater subbasin where the water system is located; and (2) a susceptibility waiver issued to public water systems that satisfy specific vulnerability criteria.

The VAP will be developed in consultation with pesticide specialists from the Pennsylvania Department of Agriculture (PDA), PADER, the EPA, and Penn State. The VAP uses a GIS to manage and manipulate the spatial data required to delineate groundwater subbasins within Pennsylvania that satisfy waiver criteria. Delineation of 478 groundwater subbasins will be coupled with public water system latitude/longitude data to determine the location of the water systems within each subbasin. Groundwater subbasins are first evaluated according to area-wide use waiver criteria. Area-wide use waivers can be recommended for those pesticides that are not and have not been used in a given subbasin. Subbasins that do not qualify for the area-wide use waiver are then evaluated considering the susceptibility waiver criteria. Susceptibility waiver evaluation is intended for regional decision support. A spatial decision support system (SDSS) will be developed to supplement the VAP.

By combining additional data layers from the GIS database with the public water system locations, other geographic areas can be identified that satisfy the area-wide use waiver criteria for each of the regulated pesticides included in the screening. Digital land-use data will be combined with pesticide-use data to pinpoint potential pesticides present in the groundwater subbasins. Estimated pesticide use by type and rates will be determined from various sources, including county, agricultural census, and/or Resources For the Future (RFF) data. Pesticide manufacturing, storage, and spill data may also be available from sources like the PDA, PADER, EPA, and the Toxic Release Inventory. This information will help determine whether individual community water supply systems meet the designated criteria for waiver.

The susceptibility waiver methodology will be developed similar to the ground-water pesticide pollution screening procedure outlined above. The modified pesticide DRASTIC index and GUS will be combined to couple the hydrogeologic pollution potential and pesticide leachability screenings. The susceptibility indexing provided by a modified Pesticide DRASTIC/GUS procedure will not necessarily indicate subbasins that can be issued susceptibility waivers; rather, it will indicate those subbasins that may potentially be issued a waiver or that are definitely not qualified for a waiver. Additional knowledge of the subbasin area and pesticide usage/levels within the subbasin or individual public water systems will be necessary to ensure proper issuance of susceptibility waivers.

RESULTS AND DISCUSSION

Surface Water

The 104 watersheds were ranked according to their potential for nonpoint pollution of surface waters. This initial study focused on the pollution potential of agricultural lands, such as cropland, pasture, orchards, animal feeding areas, and farmsteads. Results (illustrated in Figure 3), show that watersheds with the greatest agricultural pollution potential are distributed somewhat throughout Pennsylvania. This scattered distribution results from the dispersed locations of farms having high animal densities and chemical use, combined with the effects of steeper topography that increase the possibility that pollutants will be transported from land to stream systems. Although the high hazard watersheds are distributed, there are three general areas of the state that could logically be targeted for agricultural pollution control programs. These include the southeastern region, which has intensive cropland, animal production systems, and widespread chemical use; the northeast portion, where cropland acreage is generally on steeper terrain and where farms incorporate livestock production systems; and the south central and southwestern section, where terrain and cropping combine to increase the pollution potential.

Groundwater

The statewide database used for assessing groundwater nitrate pollution potential is not completed, but an initial relative ranking of nitrate pollution potential of groundwater in Pennsylvania was conducted and the results are depicted in Figure 4. Results show that regions under the most intense agriculture,

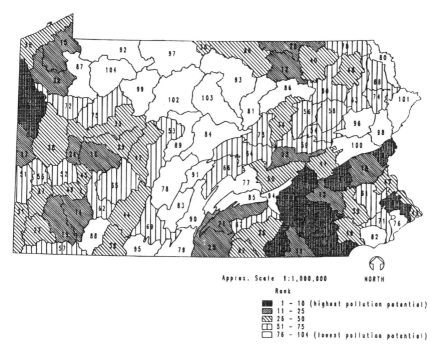

Figure 3 Ranking of Pennsylvania surface watersheds by agricultural index. (From Hamlett et al., *J. Soil Water Conserv.*, 47(5), 399–404, 1992. With permission.)

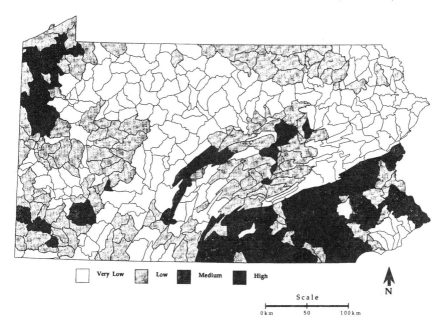

Figure 4 DRASTIC-based nitrate pollution potential ranking of Pennsylvania subbasins (considering only agricultural lands as nitrate sources).

with the shallowest groundwater aquifers, and those with karst topography are most susceptible to nitrate pollution. Initial model results, on a relative basis, compare favorably with measured nitrate concentrations in well water. Additional nitrate assessments and revisions are ongoing.

The pesticide groundwater assessment and the vulnerability assessment for community water supply systems are currently underway and results are not yet available. However, the necessary data are being compiled and initial modeling approaches have been determined. As the data are finalized and incorporated into the models, the susceptibility of groundwater to pesticide pollution will be assessed.

USE OF MODEL BY PADER

Cost share monies are available to help farmers implement Best Management Practices (BMPs) to reduce farm runoff. Because these funds are limited, Pennsylvania needs to target watersheds with the greatest potential runoff problems. Results from the original model, ranking the potential nonpoint source pollution of Pennsylvania's watersheds, combined with other water quality-related data (including acid mine drainage, in-stream water quality, designated uses of streams, etc.), are used by PADER as a first-cut screen to focus control programs and implementation resources.

Greater data resolution and more specific pollutant models, however, are needed to precisely pinpoint sources of nonpoint pollution within critical watersheds. Penn State, working with PADER, recently took the extra step and began examining in greater detail critical watersheds identified by the first model. New land cover data, with enhanced resolution, is allowing researchers and policy makers to explore the nonpoint source pollution potential of subwatersheds within high-risk watersheds.

ORSER's GIS-based model may also affect management of nonpoint pollution in the Chesapeake Bay drainage basin. A consortium of agencies and investigators under the sponsorship of the SCS and leadership of the federally funded National Center for Resource Innovations, are using the agricultural pollution potential model developed by Penn State researchers to evaluate agricultural nonpoint pollution potential in the Chesapeake Bay drainage region. Additional components are being added to the model and the individual indexes are being strengthened with improved data and loading algorithms. Potential data sets for the model are being compiled for the entire bay basin. States in the drainage basin are assisting in the compilation of the databases at a level that can be applied to the nonpoint pollution potential model. This Chesapeake Bay decision support system (CBDSS) is intended to provide resource managers with a tool to assess current and changing resource conditions, focus program-based technology, research, personnel and funding, and to reassess program effectiveness at regional and subregional levels. When completed, the CBDSS will provide a system for spatial inputs and for analyses through a series of indexes/models for water quality/pollution potential assessments. As improved data and understanding become available, the CBDSS will likely be updated to allow continued capabilities in assessing resource conditions.

THE FUTURE

GIS-based computer models provide scientists with an important communication tool—visual graphics. Policymakers now consider GIS an essential tool for setting priorities and the pictures produced by models help those policymakers visualize the targeting process. The flexibility of GIS-based models, with the capability of adding or deleting layers, allows the models to respond to the changing needs of different agencies and groups. As the technology improves, resolution (of both data and models) can be enhanced to examine problems on a more detailed scale. At the other extreme, small-scale data like that used for national and regional assessments, will continue to be valuable as environmentalists and scientists take a more holistic view and assess problems from a global perspective. Improved data for an enhanced scale of study and efforts in screening approaches at the enhanced scale will continue to be needed. Additionally, as modeling and computation capabilities improve, considerable attention will be directed to developing expert systems that rely on GIS technology and a menu of assessment models that can be combined to assess the extent of resource conditions.

Additional layers that may someday be incorporated into GIS-based models evaluating nonpoint source pollution include urban nonpoint pollution potential and mining effects. More detailed farm information, such as farm management strategies, structural practices and economics, would improve the targeting process conducted by government and resource agencies. Future uses of GIS-based models by private and government agencies will continue to expand as decision makers begin to understand—and demand—GIS.

REFERENCES

Aller, L., Bennett, T., Lehr, J. H., Petty, R. J., and Hackett, G., DRASTIC: a Standardized System for Evaluating Ground Water Pollution Potential Using Hydrogeologic Settings, EPA 600/2-87/035, 1987.

Chesapeake Bay Commission, *The Chesapeake Bay Agreement of 1987*, 1987.

Council on Environmental Quality, *Environmental Quality—1980: the Eleventh Annual Report of the Council on Environmental Quality*, U.S. Government Printing Office, Washington, D.C., 1980.

Deichert, L. A., *Evaluation of Nitrate Ground Water Pollution Potential in Pennsylvania Using DRASTIC and NLEAP models*, M.S. Thesis, The Pennsylvania State University, PA, 1993, 96 pp.

Earth Resources Data Analysis Systems, Inc., *ERDAS Field Guide*, Atlanta, GA, 1994.

Environmental Systems Research Institute, Inc., *ARC/INFO User Documentation*, Redlands, CA, 1990.

Follett, R. F., Kenney, D. R., and Cruse, R. M. (Eds.), *Managing Nitrogen for Ground Water Quality and Farm Profitability*, Soil Science Society of America, Inc., Madison, WI, 1991.

Geyer, A. R. and Wilshusen, J. P., *Engineering Characteristics of The Rocks of PA*, Envir. Geolo. Rpt. 1. PA Geological Survey, Harrisburg, PA, 1982.

Goss, D. W. and Wauchope, D., The SCS/ARC/CES pesticide properties database: II. Using it with soil data in a screening procedure, in *Pesticides in the Next Decade: The Challenges Ahead*, Weigman, D. L., ed., Virginia Water Resources Research Center, Blacksburg, VA, 1990, 471–493.

Gustafson, D. I., Groundwater ubiquity score: a simple method for assessing pesticide leachability, *Environ. Toxicol. Chem.*, 8, 339–357, 1989.

Hamlett, J. M., Miller, D. A., Day, R. L., Petersen, G. W., Baumer, G. M., and Russo, J., Statewide GIS-based ranking of watersheds for agricultural pollution prevention, *J. Soil Water Conserv.*, 47(5), 399–404, 1992.

Hession, W. C., Flagg, J. M., Wilson, S. D., Biddix, R. W., and Shanholtz, V. O., *Targeting Virginia's Nonpoint Source Programs*, ASAE Paper No. 92-2092, American Society of Agricultural Engineers, St. Joseph, MI, 1992.

National Aeronautics and Space Administration and U.S. Geological Survey, *Land Analysis System User documentation*, Goddard Spaceflight Center, Greenbelt, MD, and EROS Data Center, Sioux Falls, SD, 1990.

Petersen, G. W., Hamlett, J. M., Baumer, G. M., Miller, D. A., Day, R. L., and Russo, J. M., *Evaluation of agricultural nonpoint pollution potential in Pennsylvania using a geographical information system*, Final Report ME89279. Environmental Resources Research Institute, The Pennsylvania State University, PA, 1991, 60 pp.

U.S. Department of Agriculture, *The Second RCA Appraisal—Soil, Water and Related Resources on Nonfederal Land in the United States: Analysis of Conditions and Trends*, USDA, 1993.

U.S. Army Corps of Engineers, *GRASS 4.1 User's Reference Manual*, CERL, Champaign, IL, 1993.

U.S. Environmental Protection Agency, *Chesapeake Bay Program: Findings and Recommendations*. USEPA, Region III, PA, 1983.

U.S. Geological Survey, *National Water Summary, Ground-Water Quality*, Pennsylvania, PA, 1986.

U.S. Geological Survey, *Information Exchange on Models and Data Needs Relating to the Impact of Agricultural Practices on Water Quality: Workshop Proceedings*, Cosponsored by USGS, Agricultural Research Service, and Soil Conservation Service, Office of Water Data Coordination, USGS, VA, 1990.

Wadleigh, C. H., *Wastes in Relation to Agriculture and Forestry*, U.S. Department of Agriculture, Misc. publication 1065, 1968.

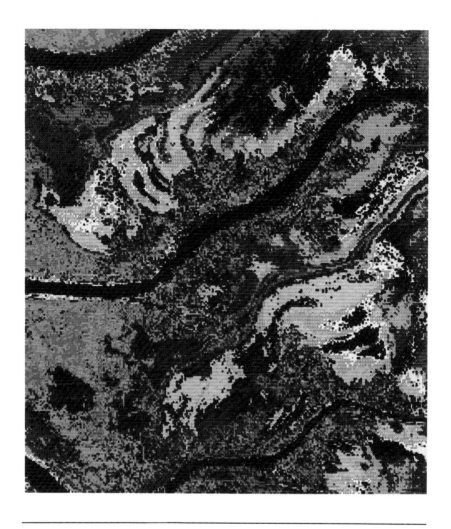

CHAPTER 5, FIGURE 3 Multispectral scanner image of the St.Clair Flats.

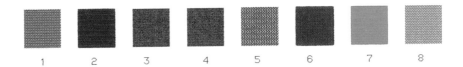

The variable name is : SW Harsens Wetlands

Value	Class Name	Number of points	%
1	Submerged Aquatics (Eel Grass)	1890	27.78
2	Shallow Marsh (sedges)	2660	39.09
3	Cropland	362	5.32
4	Cropland	0	0.00
5	Deep Marsh (Bur reed)	653	9.60
6	Shallow Marsh (Cattails)	930	13.67
7	Buttonbush (>6 in. water)	256	3.76
8	Deep Marsh (Cattails)	53	0.78

CHAPTER 5, FIGURE 5 Bur-reed in deep marsh.

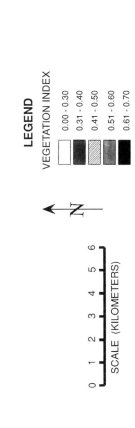

LEGEND

VEGETATION INDEX

0.00 - 0.30
0.31 - 0.40
0.41 - 0.50
0.51 - 0.60
0.61 - 0.70

N

SCALE (KILOMETERS)

0   1   2   3   4   5   6

LEVEL SLICED VEGETATION INDEX, TANQUE VERDE WASH, 1984

CHAPTER 7, FIGURE 2   Example of the level sliced vegetation index map (1984).

SPOT Derived Land Cover Classification, Tanque Verde Subset, 1988.

Bare Soil
Urban Cover
Residential Cover
Desert Scrub (mixed cover)
Desert Scrub (mesquite)
Dense Cover Vegetation

N

SCALE (KILOMETERS)

0 1 2 3 4 5 6

CHAPTER 7, FIGURE 4   SPOT unsupervised classification map.

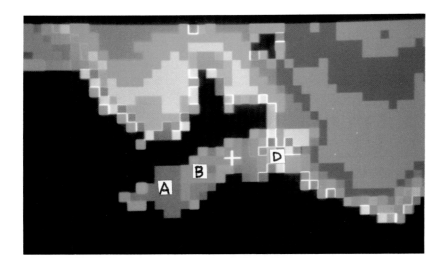

CHAPTER 8, FIGURE 2 Categorized water colorant classes from the June 27 AVHRR scene. Description of letter symbols is given in the text.

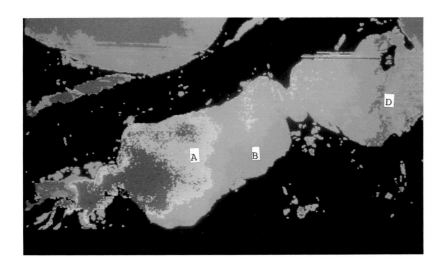

CHAPTER 8, FIGURE 3 Categorized water colorant classes from the June 28, 1981 Landsat scene. Description of letter symbols is given in the text.

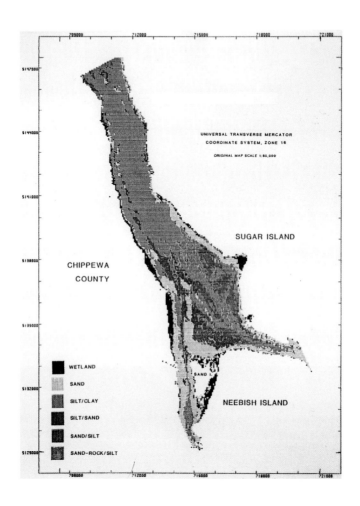

Legend (within figure):

WETLAND
SAND
SILT/CLAY
SILT/SAND
SAND/SILT
SAND-ROCK/SILT

UNIVERSAL TRANSVERSE MERCATOR
COORDINATE SYSTEM, ZONE 16

ORIGINAL MAP SCALE 1:80,000

CHIPPEWA
COUNTY

SUGAR ISLAND

NEEBISH ISLAND

SAND I.

CHAPTER 10, FIGURE 1   Thematic map of river bottom types.

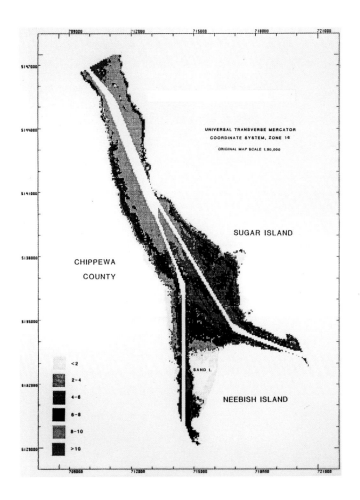

CHAPTER 10, FIGURE 2 Thematic map of water depth classes.

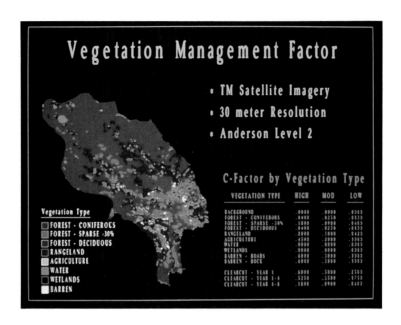

CHAPTER 21, FIGURE 3 Vegetation management factor.

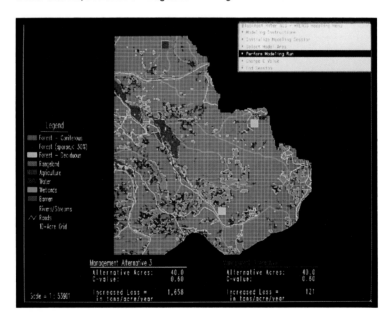

CHAPTER 21, FIGURE 10 Management alternatives.

CHAPTER 13

Water Resource Engineering Application of Geographic Information Systems

Uzair M. Shamsi

ABSTRACT

This chapter assesses the hydraulic capacity of the Chartiers Creek Interceptor of the Allegheny County Sanitary Authority (ALCOSAN) to convey the dry and wet weather flows under the existing and anticipated ultimate future development conditions. The assessment was performed by developing a hydraulic model of the interceptor system using U.S. EPA's computer program *storm water management model (SWMM)*. The model consisted of 56 tributary sewer subareas, 75 sewer segments, 96 manholes, 34 orifices, and 34 weirs/outfalls. The input data for the model were extracted from a *geographic information system (GIS)* of the watershed. The watershed GIS consisted of data layers of digitized subareas, soil associations, and census tracts; present and future land use; and USGS digital elevation models (DEMs).

INTRODUCTION

Established in 1946, ALCOSAN is by far the largest of all the sewerage systems in Allegheny County, Pennsylvania and serves more than half of the local municipalities, through a series of interceptor lines and a sewage treatment plant. The Chartiers Creek watershed, shown in Figure 1, is one of the largest watersheds in Allegheny County enveloping all or portions of 23 municipalities, including the city of Pittsburgh, 12 boroughs and 10 townships. More than 90% of the watershed is served by Chartiers Creek Interceptor. Of the 56 tributary areas shown in Figure 1, 22 have combined sewers. Most of the separate sewer tributary areas contribute wet weather inflow to the interceptor. The service area includes urban, suburban, and rural areas. The area is approximately 148 sq miles of rolling hills, whose development ranges from

Portions of this paper are adapted from Shamsi, U.M. and Schneider, A.A., GIS forecasts sewer flows, *GIS World*, March 1993, and Shamsi, U. M. and Schneider, A. A., GIS based hydraulic model pictures the interceptor future, in *New Techniques for Modelling the Management of Stormwater Quality Impacts*, William James, Ed., Lewis Publishers, 1993. With permission.

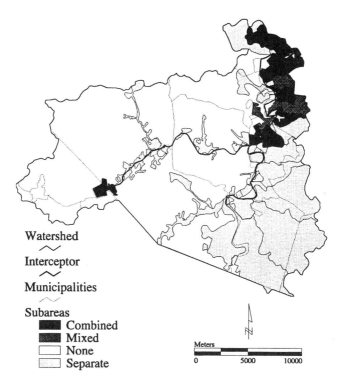

Figure 1   GIS map of subareas.

very dense commercial, residential, and industrial areas; sparsely developed rural and agricultural areas; lightly to extensively strip mine disturbed areas; to varying size tracts of undeveloped forested areas. The elevation within the six 7.5-min USGS quadrangles covering the area ranges from 690 to 1500 ft above mean sea level.

The interceptor ranges in size from 24 to 45 in. The ALCOSAN sewage treatment plant is located on the east bank of the Ohio River. The interceptor flow is conveyed to the treatment plant through a 54-in. underground tunnel that passes under the Ohio River. Most of the service areas tributary to the interceptor have diversion chambers. If operating correctly, a diversion chamber should permit all dry weather flows and a limited amount of wet weather flows to enter the interceptor.

The watershed is expected to undergo significant development in the future. Recently, a new interceptor serving several developing communities has been connected to the ALCOSAN interceptor, raising concerns about the interceptor's hydraulic capacity to convey the future flows. This chapter addresses a case study (Shamsi, 1991) that assessed the hydraulic capacity of the Chartiers Creek Interceptor of ALCOSAN to convey the dry and wet weather flows under the existing and anticipated ultimate future development conditions.

WATERSHED GIS

This study developed and analyzed a GIS for the Chartiers Creek watershed in order to provide input to the watershed SWMM. The geographic data set that was created consisted of digital components that would allow preprocessing, analysis, and display using techniques commonly referred to as remote sensing and GIS analysis. An analysis of the terrain, hydrography, soil associations, land use, census properties, and locations of major trunk and interceptor sewers was conducted to develop the watershed GIS.

Technical Approach

The software used in this project was primarily ARC/INFO, a vector-based GIS, and ERDAS (an image processing and raster-based GIS). Additional programs were written whenever needed for data format changes or the creation of a product for which the methodology was not available in either of the commercial packages. This included the determination of percent slope averages, the creation of the percent impervious and developable/nondevelopable images, and the construction of tabular output.

The vector data format is a topologically constructed set of points, nodes, lines, and polygons which define locations, boundaries, and areas. A raster format is a regular grid of uniform size cells, with a data value associated with each cell. The reason for using both kinds of data processing and handling formats is to take advantage of the best features of each. Data entry of vector information is more easily performed using a file digitized into the vector format. All GIS layers except the elevation data were initially digitized in vector format from their respective base maps utilizing ARC/INFO. The resulting polygon topology was then converted to raster format for GIS analysis.

Satellite, elevation imagery, and overlays for GIS cross-tabulation are appropriately dealt with in raster format wherein every cell of a given layer/image is registered to the corresponding cell of every other layer. The GIS analysis of the watershed and subareas was done in raster format utilizing ERDAS software. The SPOT image and the DEMs were initially in raster format. Each data set or information layer was coregistered to common coordinates, so that every raster grid cell matched its corresponding cell in other layers.

GIS Analysis and Results

Land use classes were derived from a manual interpretation of the SPOT image. Each land use class was assigned a percent impervious value. These values were based on the soil conservation service (SCS) estimates for similar land use types. Certain open space land use types were assigned a zero percent impervious value. Table 1 summarizes the watershed land use classes and their percent impervious values.

Table 1 Watershed Land Use and Percent Imperviousness

Land use	% impervious	% of study area
High density residential	52	2.5
Medium density residential	28	21.5
Low density residential	16	4.2
Commercial	85	0.6
Industrial	72	2.9
Open wooded	0	33.2
Urban disturbed	0	0.9
Rural disturbed	0	6.8
Strip mine, quarry, oil well	0	5.2
Open space	0	3.0
Open, nonwooded	10	18.1
Major roadway	100	1.1

Summary of Themes by Subarea

The final GIS analyses were performed on raster information layers with the following themes: subareas, percent slope, slope greater than 20%, slope less than or equal to 20%, land use, impervious area, soils, and census tracts. The area of each subarea was the basic theme against which the various other themes were summarized. The ERDAS "Summary" function, as well as an in-house written function called "Sumave," were utilized. The following tables were produced:

- A subarea table showing area in acres, percent slope, and mean percent imperviousness of each subarea. These values were used in SWMM to compute subarea wet weather flows.
- A soil association table showing percent of various soil associations in each subarea prepared by digitizing SCS soil association maps. This table was used to determine subarea infiltration parameters for stormwater runoff modeling.
- A census tracts table showing percent of various census tracts in each subarea prepared by digitizing U.S. Census Bureau census tract maps. Census tract data for population, number of dwellings, number of persons per dwelling, market value of housing units, and family income were obtained from the 1980 census data. Subarea percentage of census tracts and census tract percentages in subareas were applied to total tract values to estimate weighted mean subarea values. Mean subarea values computed in this manner were used in the watershed SWMM to estimate subarea dry weather flows.

The following color, raster, ARC/INFO maps were produced:

1. Subareas (Figure 1)
2. Land use (Figure 2)
3. Census (Figure 3)
4. Soils (Figure 4)
5. Future development (Figure 5)
6. DEM (Figure 6)

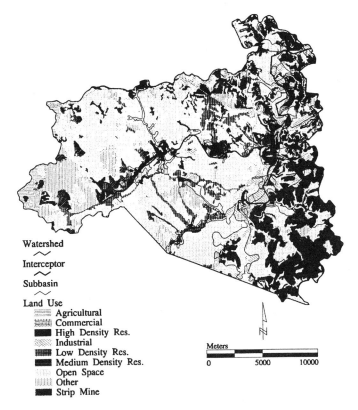

Watershed
〜
Interceptor
〜
Subbasin
〜
Land Use
    Agricultural
    Commercial
    High Density Res.
    Industrial
    Low Density Res.
    Medium Density Res.
    Open Space
    Other
    Strip Mine

Meters
0   5000   10000

Figure 2 GIS map of land use.

Determination of Future Developable Land

The present land use map (Figure 2) shows that substantial watershed area is still undeveloped, especially to the west of Chartiers Creek. In order to model ultimate future development in the watershed, six additional subareas were outlined as shown in Figure 5. These subareas were based on topography roughly following the natural surface water drainage basins on undeveloped land. By overlaying the vector land use layer and the vector "greater than/less than or equal to 20% slope" layer, a map coverage resulted showing existing developed areas, and future developable and nondevelopable areas. Based on the criterion that future development can occur only on 20% or less slopes of nondeveloped and nonstrip mined areas, the nonqualifying classes were eliminated resulting in a vector layer of future developable land shown in Figure 5.

The six future subareas were superimposed on the Watershed GIS future land use map to compute for each future subarea the area with slopes steeper than 20% and strip mines. This area and the area already developed were subtracted from the total future area to compute future developable area in each future sub-

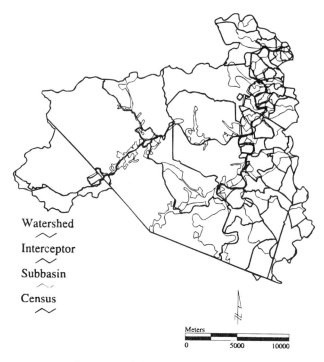

Figure 3 GIS map of census tracts.

area. Future subareas outlined in this manner, therefore, are based on the assumption that each developable parcel of land will be ultimately developed. Table 2 shows how the developable land area of future subareas was calculated.

HYDRAULIC MODEL OF THE INTERCEPTOR

The U.S. EPA's SWMM (Huber and Dickinson, 1991) was developed in the early 1970s and has been continually maintained and updated. It is perhaps the best known and most widely used urban runoff quantity/quality model. SWMM simulates dry weather sewage production and wet weather flows on the basis of

Table 2   Future Subareas

Subarea ID	Area (acres)	Area w/ steep slopes & strip mines (acres)	Area already developed (acres)	Developable area (acres)
600	1415.6	613.4	444.8	357.4
610	19,775.4	3632.5	2254.8	13,888.1
620	1240.4	87.0	73.4	1080.0
630	2399.3	539.6	46.3	1813.4
640	7245.0	1698.1	1037.5	4509.5
650	3920.6	511.7	1663.7	1745.2

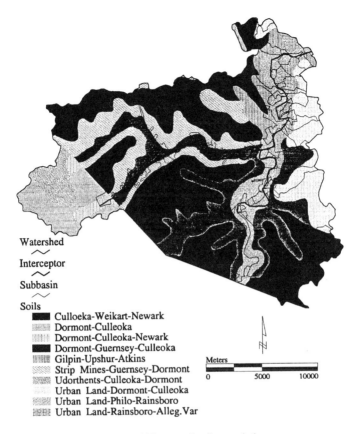

Figure 4 GIS map of soil associations.

land use, demographic conditions, the hydrologic conditions in the watershed, meteorological inputs, and conveyance/treatment characterizations. Based on this information, SWMM can predict combined sewage flow quantity and quality values. SWMM also provides the ability to perform detailed analyses of conveyance system performance under a wide range of flow conditions. As such, it is practically well suited to this study and is the model of choice for most combined sewer overflow feasibility studies.

The use of SWMM to model the Chartiers Creek watershed to assess the available capacity of the Chartiers Creek interceptor will be particularly advantageous for the following reasons:

1. SWMM represents the best means of producing estimates of current dry and wet weather flow rates from a service area as large and diverse as the Chartiers Creek watershed. Flow estimates can be prepared based on current land use conditions, topography, interceptor sewer characteristics, and selected meteorological conditions. The model can be calibrated against measured flow rates.
2. SWMM represents the best means of modeling the performance of the interceptor conveyance system under a range of dynamic flow conditions.

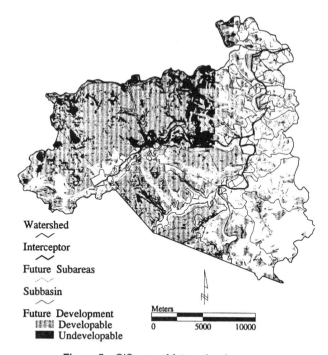

Watershed
〜
Interceptor
〜
Future Subareas
〜〜
Subbasin
〜

Future Development
Developable
Undevelopable

Meters
0   5000   10000

Figure 5   GIS map of future development.

Figure 6   Digital elevation model (DEM).

3. The use of SWMM provides a ready means of accounting for anticipated future development characteristics in an assessment of available capacity.
4. Using SWMM it will be possible to assess capacity in response to wet weather input. This characteristic can be very useful for subsequent analyses related to abatement of combined sewer overflows (CSOs).
5. The watershed SWMM developed under this study can be expanded to model water quality and to assess the effectiveness of several CSO abatement or treatment options.
6. Future CSO and system monitoring data expected to be collected in compliance with the Pennsylvania Department of Environmental Resources' CSO strategy can be used to increase the degree of calibration of the previously calibrated model.

Modeling Strategy

SWMM is flexible enough to allow different modeling approaches to the same area. An approach which adequately describes the service area and accurately computes and routes the flows at reasonable computing time and effort should be adopted. After reviewing the data, maps, and literature available to complete this study the following modeling strategy was adopted:

1. Run SWMM's Transport Block to generate sanitary flows (dry weather flows) in all the subareas.
2. Use Transport Block to model flow division (an approximate way to separate dry weather flows from combined flows) from the areas where detailed modeling of diversion chambers in the Extran Block is not feasible.
3. Enter the entire amount of sanitary sewage flow to the diversion chambers.
4. Run SWMM's Runoff Block to generate stormwater runoff (wet weather flow) in all the subareas.
5. For combined sewer subareas, enter the entire amount of runoff to the diversion chambers. For separate sewer subareas, enter only a fraction of total runoff to the diversion chambers to account for uncontrolled infiltration and inflows (I/I flows). This fraction will be determined by comparing the modeled and measured flows from separate subareas.
6. Import wet weather flows from Runoff Block into the Transport Block and combine them with the dry weather flows. Enter the combined flows to diversion chambers.
7. Use SWMM's Extran Block to regulate the dry weather flows entering the interceptor and route them through the interceptor.
8. Study interceptor response (i.e., capacity, surcharging, manhole flooding, etc.) from the Extran output.

Graphical User Interface

Most engineers have now become accustomed to the modern computer graphics features such as pull down menus, spreadsheet data input and editing, color plots, on-line help, etc., which are not currently available in SWMM. SWMM does not display the drainage network thus making detection and correction of connectivity-related errors very difficult. SWMM's ASCII format output is long, boring, and difficult to interpret. A graphical user interface called model turbo view (MTV) was, therefore, employed to overcome these deficiencies (10 Brooks Software, 1992). The graphical interface animates the underground phenomena by dynamically displaying three-dimensional plan views and profiles of the drainage system hydraulic gradient line (HGL). Such WYSIWYG features make the modeling process more interesting and interpretation of model results more convenient. Just as a picture is worth a thousand words in the business of fashion modeling, a graph is worth a thousand numbers in the business of hydraulic modeling. Some benefits of employing a graphical interface are:

1. Preparation of network schematics is not essential. Digitized plots of sewers and subarea boundaries can be used to create a drainage network diagram on a computer screen.
2. Flow and HGL data from the SWMM output can be displayed in either plan or profile view, providing an animated display of the HGL during the simulation time steps.
3. Zoom and pan features make even the large networks conveniently displayed on the screen.
4. Flow and HGL time-series plots can be displayed for any conduit or node in the hydraulic network.
5. Field collected flow and depth data can be displayed along with the model output for model calibration and verification.
6. Connectivity data errors are easily detected and can be edited while still in the program. Instabilities in the model output (often the most difficult errors to find), are also easily located.
7. Network graphics and model output in the plan and profile views can be generated using AutoCAD™ DXF format to help prepare the report.

PRECIPITATION AND FLOW MONITORING

Site selection for monitoring rainfall and flow was made in conjunction with the general arrangement of the watershed, delineated subareas, and the interceptor sewer system in order to maximize the extent and value of the data collected for use in calibration of watershed SWMM. Flow monitoring is necessary for several reasons. One reason is that hydraulic characteristics of the collection system may need to be confirmed or corrected. Another reason is that the flow monitoring provides necessary calibration and verification data to develop a mathematical model of the interceptor. Rainfall duration and intensity vary so much that parts of the collection system may be dry while others are deluged. Knowing that rain occurred only in a part of the collection system can have major implications on the hydraulic loading of the interceptor. Rainfall data coupled with flow monitoring data close the loop and establish cause and effect.

The determination of feasible locations for installing raingages was based on a combination of factors such as watershed topography, shape of the watershed, and availability of space. The objective was to distribute the raingauges so that most of the watershed storms could be captured. This objective requires that each rain gauge cover approximately an equal watershed area without leaving any watershed area uncovered. Since the eastern half of the watershed accounts for most of the service area, three rain gauges were uniformly distributed in this part. The western half of the watershed was covered by one gauge because it covers a relatively small service area.

The determination of feasible locations for installing flow monitors was based on a combination of factors such as type of flow (separate or combined), location with respect to the interceptor, and access. It was decided to install three flow monitors to sample a typical combined sewer service area, a typical separate sewer service area, and the interceptor. Flow monitoring was also conducted for 4 months.

MODEL CALIBRATION

Model calibration consists of adjusting model parameters (e.g., imperviousness, or roughness) until the predicted output agrees with measured observations. For example, the modeled hydrograph may be adjusted to agree with the measured hydrograph. Parameter estimates should fall within reasonable ranges of known values in order to enhance confidence in model results. This is easier for physically measurable parameters such as area, elevation and slope but harder for more abstract parameters such as imperviousness, depression storage, and roughness. As a result, the former are often fixed during the calibration and the latter are varied. It is often necessary to account for an unmodeled effect by varying a parameter beyond its normal range. In general, rainfall/runoff parameters and dry weather flow base infiltration rate estimating coefficients are adjusted to calibrate model output against measured conditions.

SWMM calibration was performed by comparing model outputs from Runoff, Transport, and Extran Blocks to measured flow rates. All three SWMM Blocks—Runoff, Transport, and Extran—were calibrated separately. Runoff and Transport Block input parameters were calibrated against measured wet and dry weather subarea flows, respectively. Extran Block input parameters were calibrated against measured interceptor flows both under dry and wet weather flow conditions. Only the Extran Block calibration is described here.

Extran block is based on physical properties of the interceptor conveyance system including sewers, manholes, and diversion chambers. Accurate information about the length, slope, and size of the sewers was obtained from the construction drawings. However, Manning's roughness coefficient for the sewers cannot be determined with full certainty and may be adjusted during the calibration process. Diversion chambers' geometry was determined from the construction or as-built drawings. However, SWMM's capability to model flow division in a diversion chamber is based on simplifying assumptions. For example, diversion chambers are treated as storage elements where water level rises or drops depending on the size of orifice opening for dry weather flow to the interceptor or the width and height of the overflow weir for combined sewer overflows. Conveyance within a chamber due to sloped bottom troughs is not accounted for. A diversion chamber is assumed to have a flat bottom with uniform cross section throughout its depth. Last but not least, partial openings due to orifice plates cannot be modeled. Thus, discrepancies between modeled and observed interceptor flows are likely. These discrepancies can be reduced by adjusting the parameters controlling the dynamics of flow in diversion chambers such as orifice coefficients, weir coefficients, and cross sectional area of the chambers.

Dry Weather Flow Calibration

Figure 7 shows the interceptor calibration for dry weather flows. The modeled interceptor dry weather flows were compared against the measured mean flows of May 28 and June 25 flows. The flow averaging was performed to account for two typical types of interceptor dry weather flows: (1) dry weather flows (June 25) found during prolonged dry weather periods, which peak at about 24 cfs. and (2) dry weather flows (May 28) found after substantial rainfall events, which peak at about 28 cfs. The latter type of flows include effect of wet weather infiltration on dry weather flows.

It can be seen from the plots in Figure 7 that modeled dry weather flows match the observed flows quite satisfactorily. The difference between the modeled and observed average daily flow and volume is less than 2%, which indicates a satisfactory calibration of the SWMM for dry weather interceptor flows. This degree of calibration was attained by the following SWMM parameter values:

1. Manning's "n" = 0.013
2. Diversion chamber orifice coefficient = 0.65
3. Diversion chamber weir coefficient = 3.00
4. Manhole/diversion chamber area = 50 ft^2

Wet Weather Flow Calibration

The interceptor tributary subarea flows are regulated in the diversion chambers. The diversion chambers allow a limited quantity of flow to enter the interceptor and divert any flows in excess of this amount to the receiving waters. If properly designed and maintained, this limited quantity is approximately 3.5 times the average dry weather flow rate. Improper design and lack of maintenance may cause the diversion chambers to allow more or less wet weather flows to enter the interceptor than set by this standard. Furthermore, some separate sewer tributary subareas have no diversion chambers and are referred to as having direct connections with the interceptor. Because these subareas are not always 100% water-tight, some quantities of wet weather flow are able to infiltrate into the interceptor. Thus, interceptor flows increase during the storm events and corresponding interceptor flows are called wet weather flows. In conclusion,

TIME	OBSERVED FLOW			MODELED
(HOURS)	June 25, 91	May 28, 91	MEAN	FLOW
	(CFS)	(CFS)	(CFS)	(CFS)
0	21.1	25.6	23.3	22.3
1	20.6	25.6	23.1	22.0
2	19.1	24.5	21.8	21.0
3	18.3	22.9	20.6	19.4
4	15.9	21.5	18.7	17.5
5	14.4	21.1	17.7	15.9
6	13.2	18.9	16.0	15.8
7	13.1	18.3	15.7	16.8
8	13.7	17.8	15.8	18.6
9	16.8	18.9	17.9	21.8
10	18.9	22.9	20.9	27.1
11	21.3	25.6	23.5	29.1
12	23.1	27.8	25.4	27.4
13	23.8	28.0	25.9	25.8
14	23.3	28.0	25.6	25.1
15	23.1	27.5	25.3	24.1
16	21.5	26.8	24.1	23.0
17	20.8	25.6	23.2	22.6
18	20.6	25.2	22.9	22.6
19	20.6	24.5	22.5	22.6
20	20.6	24.5	22.5	22.2
21	20.4	24.9	22.7	22.5
22	20.6	24.9	22.8	22.9
23	20.6	25.6	23.1	23.0
TOTAL	465.1	576.7	520.9	531.0
VOLUME	1,674,360	2,075,940	1,875,150	1,911,420
MEAN	19.4	24.0	21.7	22.1
% DIFFERENCE IN VOLUME				1.93
% DIFFERENCE IN MEAN FLOW				1.93

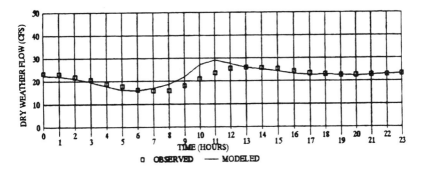

Figure 7 Dry weather flow calibration.

the interceptor wet weather flows include only that portion of the total subarea wet weather flow that is allowed by the diversion chamber to enter the interceptor, and any wet weather contribution of direct connection separate sewer subareas.

Figure 8 shows the interceptor calibration for wet weather flows. The modeled interceptor wet weather flows were compared against the measured flows of June 11, 1991. It can be seen from the plots in Figure 8 that modeled wet weather flows match the observed flows satisfactorily. The difference between the modeled and observed peak flow and volume is 5.5 and 7.7%, respectively, which indicates a satisfactory calibration. This degree of calibration did not require any further adjustment of the model parameters established under dry weather flow calibration.

INTERCEPTOR CAPACITY ANALYSIS

The following five type of flows were considered under present and anticipated future development conditions:

1. Dry weather flows
2. 2-year, 1-h wet weather flows
3. 25-year, 1-h wet weather flows
4. 2-year, 24-h wet weather flows
5. 25-year, 24-h wet weather flows

The hydraulic performance was evaluated in terms of capacity deficits indicated by surcharged sewers and manhole overflows. The model output was reviewed in MTVE to access the hydraulic performance of the interceptor under various operating conditions. Figure 9 shows an MTVE profile view of the Chartiers Creek Interceptor. SWMM ID, capacity (mgd), and sewer size (ft) are shown below each modeled interceptor segment. The profile views also show the HGL which represents a graph of the water surface elevation along the length of the interceptor. Under surcharged conditions the water surface, and hence the HGL, is above the sewers.

Present Conditions

The present flows were simulated by running the calibrated SWMM with the existing tributary subarea characteristics. Figure 10 shows a plot of maximum dry weather flows vs. interceptor capacity for all the interceptor segments modeled in SWMM. The maximum sewer flows reported here represent the peak flows simulated by the watershed SWMM. Figure 10 shows that the peak interceptor dry weather flow from the entire watershed is 35.7 cfs. Since the peak flows are less than the full flow capacity in all the interceptor segments, it is demonstrated that the interceptor has adequate capacity to transport the present day dry weather flows from the watershed.

TIME	RAINFALL	OBSERVED		MODELED	
(HOURS)	(INCHES)	FLOW	VOLUME	FLOW	VOLUME
		(CFS)	(CF)	(CFS)	(CF)
19	0.00	22.2	39,906	24.3	43,776
20	0.75	22.2	79,812	24.1	87,084
21	0.10	32.7	98,676	32.8	102,294
22	0.25	45.4	140,562	44.2	138,456
23	0.00	46.8	165,996	49.4	168,318
0	0.00	46.4	167,742	49.2	177,354
1	0.15	45.4	165,330	48.9	176,436
2	0.05	41.0	155,592	48.2	174,636
3	0.00	43.5	152,064	48.6	174,258
4	0.20	44.9	158,994	48.6	175,032
5		44.9	161,460	48.5	174,762
6		46.0	163,458	48.1	173,790
7		44.2	162,288	40.2	158,850
8		43.5	157,824	38.2	141,012
9		43.5	156,528	36.8	134,892
10		42.7	155,142	32.7	125,100
11		42.7	153,756	30.9	114,534
12		41.9	152,262	27.3	104,742
13		41.0	149,184	24.7	93,528
14		34.9	136,638	23.5	86,724
15		31.5	119,538	22.9	83,466
16		29.2	109,188	22.6	81,846
17		28.0	102,852	22.4	80,982
18		28.0	100,728	22.4	80,640
SUM	1.50		3,305,520		3,052,512
% DIFFERENCE IN VOLUME					−7.7
% DIFFERENCE IN PEAK FLOW					5.5

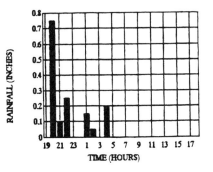
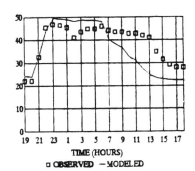

Figure 8 Wet weather flow calibration.

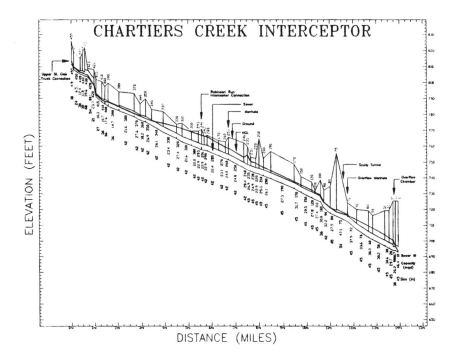

Figure 9   Interceptor profile and HGL.

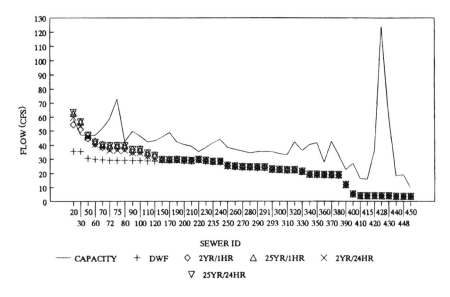

Figure 10   Present interceptor flows.

Figure 10 also shows plots of maximum wet weather flows under the 2-year/1-h, 25-year/1-h, 2-year/24-h, and 25-year/24-h storm conditions. The peak interceptor wet weather flows for the four design storms are 54.5, 62.7, 59.8, and 63.3 cfs, respectively. Peak wet weather flows are greater, under certain flow conditions, than the full flow capacity in the first three interceptor segments located at the watershed outlet. Table 3 summarizes the interceptor capacity deficits with respect to types of wet weather flows.

The above table shows that flows resulting from the short duration (1-h) storms are as critical as the long duration (24-h) storms since both types of design flows cause approximately the same amounts of capacity deficits and the same maximum wet weather flows. This phenomenon can be explained by the fact that high rainfall intensity associated with the short duration storms is known to cause more severe flooding conditions. It can be seen that, under all four types of wet weather flows, sewer 30 will have the maximum capacity deficit of 9.2 cfs. Sewers 20 and 50 will have capacity deficits of 4.6 and 0.1 cfs, respectively. Although interceptor capacity deficits due to surcharge interceptor segments did occur as described above, no manhole overflows (street flooding) were noted in the SWMM output. Thus, it can be concluded that the four types of wet weather flows surcharged the interceptor near the watershed outlet, but these flows were not large enough to cause manhole overflows. In other words, while the full pipe capacity of the interceptor for the above sewers is exceeded, surcharging is not severe enough to produce overflows.

Future Conditions

The future flows were produced by loading the calibrated SWMM with sub-area data descriptive of anticipated future land development conditions in the six future subareas (600, 610, 620, 630, 640, and 650). Figure 11 shows a plot of future maximum dry weather flows vs. interceptor capacity. It can be seen that the peak interceptor dry weather flow is 45.5 cfs, which is 27.5% greater than the present day dry weather flows. The peak flows are less than the full flow capacity in all the sewers except sewer 220. The capacity deficit based on full pipe flow for this sewer is 3.7 cfs. However, the model indicates that while the full pipe capacity in this segment is exceeded, no manhole overflows would be produced. Thus, despite a capacity deficit, the surcharge head of the inter-

Table 3   Interceptor Capacity Deficits Under Present Wet Weather Flows

Design flow	Deficient interceptor segments	Capacity deficit (cfs)
2-y/1-h	30	3.9
25-y/1-h	20	4.0
	30	9.2
2-y/24-h	20	1.1
	30	6.0
25-y/24-h	20	4.6
	30	9.1
	50	0.1

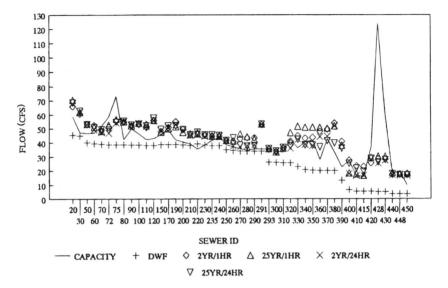

Figure 11 Future interceptor flows.

ceptor will enable sewer 220 to convey future dry weather flows. It is therefore concluded that the interceptor is adequate to handle future dry weather flows from the potential future service areas with the possibility of dry weather surcharging, provided that no major industrial development will take place and commercial and institutional growth will continue at the existing rate.

Figure 10 also shows plots of maximum future wet weather flows. The peak wet weather flows for the four design storms are 66.0, 70.6, 68.0, and 68.6 cfs, respectively. Peak wet weather flows are projected to be greater than the full flow capacity of the interceptor in all but a few interceptor segments where capacity was high already as indicated by the spikes on the capacity curve. A total of 12 modeled interceptor manholes overflowed during the course of future wet weather flows associated with 25-year/1-h rainfall events. Since the other three types of wet weather flows also demonstrate a similar pattern, it can be concluded that the interceptor does not have sufficient capacity to convey the future wet weather flows.

CONCLUSIONS

The study concluded that under the existing development conditions the interceptor was adequate to convey the dry weather flows, but it would result in a slight surcharge during wet weather conditions. Under the ultimate future development conditions, the interceptor was expected to convey the dry weather flows with slight surcharging, and wet weather flows with moderate to severe interceptor surcharging, manhole overflows, and street flooding.

REFERENCES

10 Brooks Software, Model Turbo View—EXTRAN and RUNOFF, User's Manual, Ann Arbor, MI, 1992.

Huber, W. C. and Dickinson, R. E., Storm Water Management Model. User's manual, Version 4, Environmental Research Laboratory, U.S. Environmental Protection Agency, Athens, GA, 1991.

Shamsi, U. M., Chartiers Creek Interceptor Study. Draft report submitted by Chester Environmental to the Allegheny County Sanitary Authority, Pittsburgh, PA, 1991.

CHAPTER 14

GIS for Oil Spill Response: Database Needs, Uses, and Case Studies from Florida and Texas

Michael Garrett

ABSTRACT

Geographic Information Systems (GIS) are gaining rapid acceptance as a means to manage and coordinate a maritime oil spill response. In this paper the essential components of a GIS dedicated to oil spill response are described and a recent spill in Tampa Bay, FL is used to illustrate successful application of GIS and Global Positioning Systems (GPS) technology for response support. An overview of the Texas General Land Office's (TGLO) Oil Spill Prevention and Response Program (OSPRP), which includes a GIS for management and coordination of a spill response, is undertaken.

INTRODUCTION

In December 1992, the tanker *Aegean Sea* ran aground and split up off the northwest coast of Spain. This spill impacted 124 miles of coastline. One month later, the tanker *Braer* was blown onto rocks off the coast of the Shetland Islands northeast of Scotland. Each vessel spilled in excess of 24 million gallons of crude oil onto economically important coastlines, and each spill was over twice the volume of the *Exxon Valdez* disaster. In January 1994, the barge *Morris J. Berman,* grounded on rocks off the coast of Puerto Rico and spilled three quarters of a million gallons of crude oil onto several miles of beach heavily frequented by tourists. These spills clearly demonstrate that despite the many precautions taken to prevent catastrophic oil spills, it has not been possible to completely eliminate the threat that spills pose to the natural and economic environments in which they occur. If a spill does occur, however, it is possible to ameliorate the potentially devastating effects through adequate contingency planning.

GIS technology recently has gained widespread acceptance as a means to map and document oil spills such as the *Exxon Valdez* (in Alaska, 1989), *World Prodigy* (in Rhode Island, 1989), *American Trader* (off California, 1990), and the barge *Ocean 255* (in Florida, 1993). These four spills illustrate a gradual increase in sophistication in use of GIS for spill response. GIS use during the *Exxon Valdez* and *American Trader* spills comprised basic mapping and documentation of the response efforts. The *World Prodigy* spill used GIS to both document and map the spill response and also to assess the impact of the spill on the natural environment (August et al., 1990). This use represents reaction to the spill rather than use of GIS technology to help manage, coordinate, and support the spill response effort. In order to use this technology for real-time management of a spill event it is necessary to prepare a GIS specifically for emergency response (August et al., 1990). Two GIS applications dedicated to emergency spill planning and response, the Marine Resources Geographic Information System (MRGIS) and the Florida Marine Spill Analysis System (FMSAS), have been under development since early 1992 by the Florida Department of Environmental Protection (DEP). When the barge *Ocean 255* collided with two other vessels near St. Petersburg, FL, in August 1993, the MRGIS provided real-time management and coordination of response assets and tactical maps of the spill. These response maps included environmental sensitivity indexing (ESI) of the shoreline, bathymetry, and locations of other culturally and ecologically sensitive areas (Friel et al., 1993).

Many coastal states are preparing emergency response databases similar to MRGIS and FMSAS including Maine, Texas, California, New Jersey, and Alaska. This paper describes the contribution which GIS can make to coordination and management of a spill response effort. It is not the intention of this paper to describe the specifics of each GIS component, however. Brief descriptions of the GIS database infrastructure which should be in place prior to a spill event, and other data prerequisites are given. An overview of the TGLO's Oil Spill Prevention and Response Program (OSPRP) is also given.

CARTOGRAPHIC OUTPUT

All oil spill responses make extensive use of cartographic products to disseminate information about the spill to response managers, the media, and general public. These maps typically show data such as current location of the spill, estimates of spill fate, and environmentally sensitive areas which should be protected. In the case of *Ocean 255*, these maps were deemed to be critical aids for Coast Guard command center which coordinated the response (Friel et al., 1993). Without the automated mapping capabilities provided by a GIS, these maps would have been hand drawn. Manual production would have significantly delayed dissemination of spill-related data and information. The first map of the *Ocean 255* spill was produced within hours of the spill (Friel et al., 1993). This rapidity clearly demonstrates the short cartographic response time which can be achieved using the automated mapping capabilities of GIS technology.

Rapid availability of data relating to a spill can affect decisions about which response options should be employed. Rome (1988) describes the concept of "windows of opportunity" during a spill response when different response options may be employed on a spill. Some windows of opportunity occur within a short time of a spill, and rapid evaluation of spill status is necessary for these opportunities to be exploited. The period during which *in situ* burning may be employed in a response is a good example of an option whose window of opportunity occurs immediately after a spill has taken place and may have a relatively short duration. A comprehensive GIS dedicated to spill response can provide an information and data infrastructure which can help identify and evaluate viable response options in a timely manner.

Spill response efforts often change geographic extents (or "scales") dramatically depending on spill location. For example, a spill may occur offshore and gradually drift toward the coast. Initially, maps produced at small scales are necessary to gain an understanding of the dynamics of the spill in its entirety and to evaluate the total area threatened. If a spill approaches the coast, larger scale maps which deal with individual bays and estuaries are needed for specific clean-up crews and strategies. The mapping capabilities of a GIS permit display and output of spatial data in an unlimited variety of scales and geographic extents while reusing the same spatial and attribute databases. This ability to change both scale and geographic extents permits many specific maps to be produced quickly depending on the needs of the response managers and also ensures that identical data sets are the basis for all maps.

The ability of GIS to incorporate and register collateral data from sources such as digital orthophotos, scanned nautical charts, and satellite images permit these data to be included in spill response maps. In the case of *Ocean 255*, response officials used scanned nautical charts as a background to the response maps (Friel et al., 1993). These charts provided data on the location of islands, aids to navigation, and marine hazards, and their familiar format made them easily understood by clean-up crews (Friel, 1994).

During a spill, data arrives at the command center from many different sources. For example, oil slick extents may be mapped using a variety of methods including hand-drawn sketches, Global Positioning System (GPS), radar, and aerial photography. The data paths and potential sources of data should be identified and provisions made to assimilate these data prior to a spill rather than during the response (Garrett and Jeffress, 1992). In the case of *Ocean 255*, a helicopter was used to fly a GPS receiver around the perimeter of the spill to record the spill location (Friel et al., 1993). GPS mapping of the spill was deemed essential to the response effort for two primary reasons:

1. The locations provided by the GPS were more accurate than hand-drawn sketches of the oil slick produced by response personnel (Friel, 1994).
2. The automation of the mapping task permitted response personnel to focus attention on oil characteristics rather than slick location (Friel, 1994).

Public support of a spill response effort is also a major concern during and after an oil spill. The success of clean-up efforts is often judged by the perception of success as reported by the media rather that the number of gallons of oil recovered or natural resources protected (Jenson, 1990). In the case of *Ocean 255*, the press conferences relating to the spill effectively became scheduled around when the next GIS map would be available (Friel, 1994). High-quality maps showing the spill location relative to threatened resources can dramatically increase the perceived competence of response managers and keep the public well informed about the extent and status of the response effort. A public relations effort which uses the latest technology and produces high-quality information and maps can help muster public support for the overall response effort.

NATURAL RESOURCE INVENTORY

Most spill response efforts attempt to minimize effects of the spill on the natural resources of the environment. In order for these resources to be protected, their location must be identified and included in the spatial and attribute databases of the GIS. The usefulness of natural resource inventories during spill response has been well demonstrated in both the *Ocean 255* and *World Prodigy* spills. In both spills, data sets documenting the location and size of different marine habitats produced prior to the spill enabled response managers to make more informed decisions regarding the deployment of response assets (Friel et al., 1993; August et al., 1990).

Two types of natural resource need to be included in the GIS database: (1) biological resources which include the habitats of various organisms prevalent in a geographic area, and (2) an inventory of the geomorphological characteristics of the shoreline. A biological resources inventory can help estimate the potential kill of different plant and animals species in an inundated area and can be used to generate "what if" scenarios in the GIS. A shoreline characteristics inventory can record data about what type of shoreline is prevalent along different parts of a coastline. The effort and cost to remove oil from an inundated shore is directly related to its geomorphological characteristics. Coastlines can be rated relative to each other in how easily oil can be removed from shoreline material. For example, a mangrove swamp or salt marsh is generally much more difficult to clean than a rocky coastline. A method of rating different shoreline types (i.e., Environmental Sensitivity Indexing) is described later.

Areas of economic and cultural sensitivity should also be located to provide response managers with additional information to aid deployment of response assets. For example, a tourist beach may be given higher priority if tourism is the mainstay of a particular coastal community.

ENVIRONMENTAL SENSITIVITY INDEXING OF SHORELINES

The concept of an environmental sensitivity index (ESI) for a particular segment of coastline is often used to rate the susceptibility of a particular shoreline type to oil inundation. The ESI concept permits a cartographic indication of the vulnerability of specific shorelines to oil spills (Friel et al., 1993). A visual representation of susceptibility of different shoreline segments to oil inundation provides response managers with the ability to quickly evaluate the potential effects of an oil spill on one coastline relative to another and place response assets accordingly.

There are several advantages to using a standardized method for evaluation and symbolization of ESI's for shoreline segments. Spill response crews generally operate in a national and international marketplace and a common form of cartographic representation for the ESI of a coastline will dramatically decrease the orientation time for a crew not local to the spill site. Standardization of ESI methodology also facilitates exchange of these data with other federal, state and local agencies involved in the clean-up efforts. Agencies in several states (Florida, California, Michigan, Texas, and Maine) already use a similar methodology and are making efforts to adopt a uniform standard among these coastal states (Martin, 1994; Friel, 1994). This methodology is based on a system developed by Research Planning Inc. (RPI)[1] for shoreline evaluation and standardization of cartographic symbols (Michel and Dahlin, 1992).

CULTURAL AND HUMAN USE INVENTORIES

The location of all culturally sensitive areas in a region should be included in the ESI mapping of the coast. Such features do not necessarily fall within the standard ESI geomorphological and biological criteria but are important in determining where spill response assets should be located. Typical examples of features which should be included are: archaeological sites on or near the coast, water intakes for cooling of power plants or desalination plants, tourist beaches, sewage outfalls, and submerged cables.

Although the foregoing inventories (geomorphological, biological, and human use) classify shoreline segments relative to each other, it is not possible to create a composite rating of the overall sensitivity of one section of coastline. A composite rating is not possible because the environmental sensitivity of these three inventories is not always constant. For example, a tourist beach in the middle of winter may be of less significance than in the height of summer. Annotation of each of these inventories as they exist on the map permits response managers to make decisions based on the prevailing environmental, economic, social, and political conditions.

GIS AS AN HISTORICAL DATABASE

Postresponse analysis of a spill will ultimately include assessment of the response effort and damage assessment of the environment. This analysis requires a well-documented chronological database of the spill response effort.

---

[1] Based in Columbia, SC.

A GIS can be a repository for such data because it permits the inclusion of both attribute and spatial data which can be time stamped if adequate preparations for time stamping and data accounting are made ahead of time. For example, all GIS maps created during the response could contain a standard header file which lists the data source, estimates of data accuracy, date and time it was created, and the user who created it. If GIS is used to record the geographical location of response equipment at different times during the spill, it would also be relatively easy to associate a map feature (which may represent a piece of spill response equipment) with a document which catalogs the reason for placing the equipment in a particular location. The placement of response assets and location of the spill on the GIS map can also be recorded and time stamped and thus provide "snapshots" of the response at different times during the response. The spatial query and polygon manipulation/calculation capabilities of GIS can produce estimates of the linear footage of coastline affected and number of square miles of various habitats inundated. In the *World Prodigy* spill, the Rhode Island Geographic Information System (RIGIS) was used to calculate the following: (1) the maximum and cumulative exposure of shorelines to oil, (2) the linear footage of swimming beaches closed each day, and (3) the area of shellfish grounds closed as a result of the spill (August et al., 1990).

In postresponse monitoring of a site, GIS can play a vital role as an archive for all data relating to the spill. Such a database permits temporal analysis of the effectiveness of the response and provides a database infrastructure for long-term monitoring of the site. A long-term database allows "before, during, and after" scenarios to be developed, contrasted, and replayed. Sophisticated analysis functions of a GIS permit visualization of environmental changes using functions which generate three-dimensional surfaces across geographical maps using attribute data as the elevation coordinate. For example, changes in chemical composition along a coastline could be viewed in this way.

In the litigation and financial settlement of a spill, an historical spatial database is vitally important as the main form of documentation for the spill response. It is important that actions taken during the response phase of the spill, the personnel involved, and the quality of GIS data be carefully recorded (August, 1990). Without this documentation, the integrity of the entire response data set may be brought into question in the postresponse analysis and litigation of the spill.

Many state agencies currently collect baseline chemical data for coastlines and coastal ecosystems. For example, the Texas Natural Resource Conservation Commission routinely analyzes sediments at different locations along the Texas coast to determine the content of elements and compounds including arsenic, cadmium, chromium, lead, copper, mercury, pesticides, oil and grease, phosphorous, Ph, dissolved oxygen, and total dissolved solids. These data can be imported into the GIS and used to quantitatively determine the net effect that a spill event has had on a particular ecosystem in relation to specific chemicals and compounds. Such measurements could help to quantitatively assist damage assessment and to form a basis for the financial settlement of a spill.

As can be seen, the use of GIS as an historical database has many benefits beyond oil spill response. Certainly, much of the data outlined previously is of great benefit to any coastal management program. A GIS which is prepared for responding to an oil spill can be considered a subset of a more generalized GIS for coastal zone management with a spill response capability as a major function.

LOCATION OF OIL SPILL RESPONSE EQUIPMENT AND PERSONNEL

Location and availability of spill response equipment is vitally important to response managers. In addition, quantitative and qualitative data should also be gathered on boat ramps, jetties, equipment staging areas, and any other information necessary to support a spill response effort such as public showers, rest rooms, and restaurant locations.

An inventory of available spill response equipment and their location(s) will help determine mobilization times and identify if equipment must be brought in from elsewhere. It is preferable, therefore, that an equipment inventory be implemented on a statewide basis rather than on a regional basis. For coastal areas along state borders, some form of interstate response plan should be developed to share response equipment and to ensure that jurisdictional problems are resolved prior to a spill rather than during the response. Maps produced by the GIS will be vitally important in calculating the quantities of spill equipment needed. Once a map of the spill has been produced, it is a relatively trivial exercise to measure distances and calculate areas and perimeters. An equipment inventory should also include contact names and addresses of key response personnel. The inclusion of the contact data will facilitate rapid notification of these personnel if a spill occurs.

AN OVERVIEW OF THE TEXAS GENERAL LAND OFFICE'S OIL SPILL PREVENTION AND RESPONSE PROGRAM (OSPRP)

In 1991, the Texas legislature passed the Oil Spill Prevention and Response Act (OSPRA) which designated the Texas General Land Office (TGLO) as the lead state agency in oil spill planning and response. The funding approved by OSPRA created the Oil Spill Prevention and Response Program (OSPRP). The primary goal of this program is to monitor the integrity of oil transport through Texas' coastal waters and to provide timely and efficient response to any spill. The OSPRP includes an information support system called the Oil Spill Planning and Response Interface (OSPRI). OSPRI is comprised of several components including: trajectory modeling software (a GIS database which contains ESI maps of bays and estuaries along the Texas coast and location spill of response equipment and contractors), and real-time data input from a network of tide gauges and meteorological platforms. These components are now briefly described.

GIS and the OSPRI

The GIS has been developed using ARC/INFO®[2] by Environmental Systems Research Institute, Redlands, CA. OSPRI is an interface which has been developed in Arc Macro Language®[3] to access the various GIS databases. OSPRI permits the user to display and query any combination of data layer and also produce summary reports of threatened areas. The interface includes the ability to store and recall predefined contingency plans for specific areas along the coast. Predefinition of response plans enables generalized contingency plans to be devised ahead of time and then customized during an actual spill response. OSPRI includes many routines to quickly assimilate data from external sources such as digitized aerial photography, hand-drawn sketches, and also from other GIS coverages. While the mechanics of these data conversion utilities have existed for some time, the OSPRI approach is unique in streamlining the set-up and digitizing procedures. The interface guides the user through the entire process based on the type of source document the user has specified. In a similar way, cartographic output can quickly be generated using predefined legends, title blocks, and layouts. The rapid production capability of OSPRI has been developed through identification of the types of data that may come available and the type of cartographic support needed during a spill response. GIS capability continues to be developed through response drills held in conjunction with Coast Guard and NOAA officials, and private industry.

ESI Response Maps

TGLO has map data from a variety of sources, both government and private, which catalog environmental sensitivity of parts of the Texas coast at varying scales. In order to update these maps and standardize scale and method of categorization, the OSPRP has embarked on a program to remap the Texas coast to show ecologically sensitive areas and other economic and cultural features which are relevant to a spill response. The ESI base map has been derived from United States Geological Survey (USGS) digital line graph data (DLG) originally produced at a scale of 1:24,000. The ESI maps currently being created are comprised of three components: biological/ecological, geomorphological, and human use. The creation of these three map components is now discussed.

Generation of biological/ecological inventory maps has been a cooperative effort between TGLO, Texas Parks and Wildlife Department (TPWD), and interested parties from the scientific and environmental communities along the Texas Gulf coast. Each TPWD region has compiled biological/ecological sensitivety maps at 1:24,000 using USGS Quadrilateral (Quad.) topographic sheets as the base map after first conferring with local scientific and environmental experts. After completion of the inventories, these maps were then passed to the

---

[2] Registered Trademark of Environmental Systems Research Institute, Redlands, CA.
[3] Registered Trademark of Environmental Systems Research Institute, Redlands, CA.

GIS department of TGLO for digitizing and editing. The first draft of these maps spans the length of the Texas coast and is currently being converted to ARC/INFO coverages. The most sensitive areas will be subject to a more detailed study over the next year. These areas extend from the southwest end of the Galveston Bay system to the Texas-Louisiana border.

The geomorphological component of the ESI maps has been developed in conjunction with the Bureau of Economic Geology (BEG) and RPI. These data sets were created by a helicopter field survey of the Texas coast. Each coastal segment was assessed for shoreline geomorphology, wave intensity, and wave energy. This survey used airborne videography, aerial photography, and site visits to both aid and document geomorphological classification.

Human-use data have been derived from several sources including DLG data, USGS Quad sheets, and data provided by the Maritime Spill Response Corporation (MSRC). MSRC has performed a GPS survey of jetties and boat ramps along the Texas coast and has supplied these data to TGLO for inclusion into the GIS. Street data were originally derived from USGS digital line graph (DLG) data. DLG road data are currently being replaced with data from the Texas Department of Transportation (TxDOT) which provides greater transportation and road nomenclature detail. Private contractors who specialize in the clean-up of hazardous spills [(Discharge Cleanup Organizations (DCO's)] are required to provide TGLO with an inventory of spill response equipment available. These data also reside in a database which can be queried; however, these data are not presently part of the GIS database.

TGLO is currently researching the use of color-infared Digital Orthophoto Quarter Quads (DOQQ) with a 1-m resolution for base maps. These maps will reside in the GIS database and provide a photographic backdrop to complement the existing spatial database. The ultimate goal of the ESI mapping program is to compile a series of atlases which will be produced and prepositioned at the regional spill response centers. The GIS database is comprised of over 50 thematic layers and research is presently underway to determine the most appropriate scales, extents, and data layer combinations.

Oil Spill Trajectory Modeling

TGLO has selected an oil spill trajectory model, Spillsim®, developed by Seaconsult Marine Research Ltd, Vancouver, BC. This software models the movement of an oil spill and relies on data regarding shoreline properties, currents, and meteorological data. At present, the trajectory modeling software and OSPRI are not integrated. Each system has the ability to import data from external sources which provides a data path for predicted spill movement from SpillSim to OSPRI. Development of the trajectory modeling has been performed in cooperation with the Texas Water Development Board (TWBD). Using tidal, bathymetric, river inflow, and meteorological data, modelers at TWDB can process hydrodynamic models for specific Texas bay systems to generate flow fields for use in the trajectory model.

The Conrad Blucher Institute for Surveying and Science (CBI) at Texas A&M University, Corpus Christi operates a series of tide gauges and meteorological platforms along the Texas coast as part of the Texas Coastal Ocean Observation Network (TCOON) (Garrett and Jeffress, 1992). These data are transmitted in near real-time to CBI and then retransmitted over Internet to TGLO. Once in TGLO databases, these data can be graphed to assess current water level and environmental conditions at the spill site. These data can also be imported into the TWDB hydrodynamic models which in turn provide input to the trajectory model.

An area of future development for TGLO is inclusion of real-time positioning data from equipment such as GPS, surface-current radar, airborne scanner imagery and floating buoys. The significant role GPS played in the *Ocean 255* spill supports the notion that GPS should be included as a mapping component for a spill response.

CONCLUSIONS

This paper has described the components necessary to use GIS as a tool to manage and coordinate a spill response. However, the GIS data layers mentioned in the first section of the chapter are not sufficient to achieve this goal on their own. The need for adequate planning for sources of collateral data and "ready to run" applications which generate meaningful and timely results is essential to successful use of GIS for disaster management. This type of preparation is generally an iterative exercise which should be tested with a series of drills (Friel et al., 1993; Garrett and Jeffress, 1992).

To date, the *Ocean 255* spill represents the most sophisticated use of GIS in a spill response. The maps produced by the FMSAS during the *Ocean 255* spill were deemed essential by the response managers (Friel, 1994). In this sense, the GIS is being used to manage and coordinate the spill response, but not interactively. The possibility of a team of high-level response managers gathered around a computer screen making decisions based on displayed data appears somewhat remote for the immediate future. However, there appears little doubt that GIS will continue to play a major role in emergency spill response.

The TGLO's contingency plans are a good example of interagency cooperation. It has relied on the expertise of other government agencies and the general public in the compilation of its spatial and attribute databases. The development of predefined applications to automate data input and output are good examples of the way input of collateral data has been accommodated.

The antidote to risk is information. A GIS dedicated to spill response will provide an invaluable asset to a response effort by providing an information and data infrastructure to support response decisions. In many cases data used are not only important for the response effort but also for litigation of the spill and for many years after the response in the long-term monitoring of the site.

ACKNOWLEDGMENTS

The author would like to acknowledge the comments and support of Chris Friel of the Florida Marine Research Institute (FMRI), Jeffery Dahlin of Research Planning, Inc., Mehrdad Moosavi, Dr. Robert Martin, Lee Smith of the Texas General Land Office (TGLO) and the Conrad Blucher Institute for Surveying and Science at Texas A&M University, Corpus Christi.

REFERENCES

August, P., Hale, S., Bishop, E., and Sheffer, E., GIS and environmental disaster management: *World Prodigy* oil spill, Proceedings: *Oil Spills: Management and Legislative Implications*, Spaulding M. L. and Reed, M., Eds., American Society of Civil Engineers, New York, 1990.

Butler, H. L., Chapman, R. S., Johnson, B. H., and Lower, L. J., Spill management strategy for the Chesapeake Bay, Proceedings: *Oil Spills: Management and Legislative Implications*, Spaulding M. L. and Reed, M., Eds., American Society of Civil Engineers, New York, 1990.

Dahlin, J. A., Senior Scientist, Research Planning Inc., Columbia, SC, personal communication, 1994.

Friel, C., Leary, T., Norris, H., Warford, R., and Sargent, B., GIS tackles oil spill in Tampa Bay, *GIS World*, 6 (II) 30, 1993.

Friel, C., Research Administrator/GIS Coordinator for Coastal and Marine Resource Assessment Group, Florida Marine Research Institute, personal communication, 1994.

Garrett M. and Jeffress G. A., Integration of real time environmental data into a GIS for oil spill management and control, Proceedings: GIS/LIS '92, San Jose, CA, 1992, 247–255.

Genamap helps battle oil spill, *GenaNews*, 1 (1) 3, 1990.

Jayko, K., Predicting the movement of the *World Prodigy* oil spill, Proceedings: *Oil Spills*: Management and Legislative Implications, Spaulding M. L. and Reed, M., Eds., American Society of Civil Engineers, New York, 1990.

Jenson, D. S., Coast Guard oil spill response research and development, *Mar. Technol. Soc. J.*, 24(4), 1990.

Martin, R. D., Chief of Research/State Scientific Support Coordinator, Oil Spill Prevention and Response Program, Texas General Land Office, personal communication, 1994.

Michel, J. and Dahlin, J., *Guidelines for Developing Digital Environmental Sensitivity Index Atlases and Databases*, Report to Hazardous Materials Response and Assessment Division, National Oceanic and Atmospheric Administration by Research Planning Incorporated, Columbia, SC, 1992.

Moosavi, M., Martin, R. A., and Smith, L. A., *Perspective of the Geographic Information Systems Division of the Texas General Land Office*, Texas General Land Office, unpublished, 1993–1994.

Rome, D. D., Readiness planning for Arctic offshore oil spill response, Workshop Proceedings: *Alaska Arctic Offshore Oil Spill Response Technology*, Jason N. H., Ed., National Institute of Standards and Technology, Washington, D.C., 1988.

CHAPTER 15

DEM Aggregation and Smoothing Effects on Surface Runoff Modeling

Baxter E. Vieux

ABSTRACT

A digital elevation model (DEM) may be used to model watershed-scale hydrologic processes. The raster data structure of the DEM is a discrete approximation of the continuous land surface. Depending on the hydrologic process, various model parameters may be extracted from the DEM to simulate the distributed effect of topography. Spatial variability of the topography affects the apparent slope and flow path length extracted from DEMs. Smoothing of the digital elevation data is a common procedure for reducing or eliminating peaks or pits prior to watershed delineation using a DEM. Cell resolution must be chosen such that the spatial variability is captured. The simplest form of aggregation is resampling the data at larger cell sizes. Error is propagated in simulations of direct surface runoff if the apparent slope is flattened by smoothing or if flow path length is shortened due to aggregation. Model error due to aggregation and smoothing is presented. A measure of the spatial variability is informational entropy and is related to hydrograph response. The magnitude of error propagation is measured using a finite element solution of direct surface runoff to generate the hydrographs. The log of the hydrograph error scales linearly with the log of the relative entropy loss. Further, low rainfall intensities produce larger errors than at higher intensities. This method of error analysis provides an *a priori* means of assessing the magnitude of the consequent error due to aggregation and smoothing.

INTRODUCTION

Runoff originating from precipitation intensities exceeding the infiltration capacity of the soil surface is known as Hortonian runoff (Horton, 1932). This process results in lateral inflow to streams and rivers and can cause downstream flooding and transport of contaminants to surface water and groundwater. Overland flow coupled with infiltration is an important hydrologic process at the watershed scale. Although other types of runoff processes such as subsurface runoff are important, types other than Hortonian or direct surface runoff are not considered herein. Because DEMs are increasingly used to model hydrologic processes, error propagated due to smoothing and the choice of raster cell sizes (aggregation) is investigated herein. A background of approaches using digital terrain models to simulate hydrologic processes is also presented.

Moore et al. (1991) reviewed applications of digital terrain analysis techniques in modeling hydrologic processes. They found the three major data structures and methods for digital representation of the topography to be grid cell (DEM), triangular irregular network (TIN), and contour-based methods. Vieux (1988, 1991) presented a method using a TIN network to supply land surface slope for a superimposed finite element mesh. This approach does not utilize the TIN facet directly, but rather the slope is sampled from the TIN at finite element nodal locations. This approach was used to simulate direct surface runoff from a small watershed basin.

Quinn et al. (1991) present the application of TOPMODEL which models subsurface flow at the hillslope scale. Model sensitivity to flow path direction derived from a DEM was investigated. This application used a 50-m grid cell resolution which is the default value of the United Kingdom database. Resampling at larger grid cell resolutions was found to significantly affect soil moisture modeling due to aggregation.

Tarboton et al. (1991) investigated stream network extraction from DEMs at various scale resolutions. They found the drainage network density and configuration to be highly dependent on smoothing of elevations during the pit removal stage of network extraction. In fact, if smoothing was not applied to the DEM prior to extraction of the stream channel network, nothing resulted that resembled a network. However, smoothing has deleterious effects viz., the flattening of the apparent slope between grid cells, and consequently, the ensemble of slope values in the watershed basin. This affects the slope of the watershed basin area derived from automatic delineation.

This investigation focuses on the effects on hydrologic modeling produced by two types of filters: smoothing and cell aggregation. Cell size selection is important for capturing the spatial variability of the DEM. Smoothing is often necessary before automatic delineation of the watershed and stream network. The relationship between information content loss due to filtering and the resulting error in the finite element output is presented as an *a priori* means of assessing the hydrologic model error due to filtering of the DEM.

METHODOLOGY

The data used in this study were taken from digital elevation data for the Spearfish South Dakota quadrangles. Thirty meter resolution DEMs are available for limited areas of the United States from the U.S. Geological Survey (USGS). Composed of two USGS 7.5-min topographic quadrangles, the DEMs used in this study are 1:24,000 scale and at 30-m intervals. Automatic watershed delineation is accomplished by successively examining each cell to identify the drainage direction. Beginning with the watershed basin outlet cell, neighboring cells are examined to determine which cell(s) drain into it. The four point or rook's move and the eight cell or queen's move are the most common algorithms for searching neighboring cells to determine the drainage network for the contributing drainage area. Successive consideration of each cell determines the drainage network and divides.

Two types of filters are investigated herein: aggregation and smoothing. The effects of smoothing are investigated by applying 3×3, 5×5, or 7×7 roving windows that replace the center cell with the arithmetic average of all the cells in the roving window. The effects of aggregation are investigated by resampling the 30-m original resolution at the center of each cell for sizes of 90, 150, and 210 m. The smoothing and aggregation are applied to the elevation data which is then used to derive slope and aspect maps. The automatic watershed delineation was performed using the original 30-m data so that the watershed basin area is invariant for the subsequent smoothing and aggregation algorithms. Smoothing and aggregation, if taken to the limit, produce a single lumped value of elevation for the watershed basin. Filtering has effects on watershed slope and flow path length. These effects cause errors in the prediction of direct surface runoff. The departures from the hydrologic model output for the original 30-m DEM are considered errors.

The Geographic Resources Analysis Support System (GRASS) developed by the U.S. Army Corps of Engineers at the Construction Engineering Research Laboratory, version 3.1 (Westervelt et al., 1988), and shell scripts developed by the author were used in this investigation to sample the smoothed and aggregated elevations, and to compute the information content loss due to filtering the DEMs. GRASS is a raster (grid cell) GIS that provides map analysis and overlay capabilities, an automatic watershed delineation program, and slope and aspect analyses.

To gauge the propagation of error in hydrologic model output resulting from smoothing and aggregation, the GRASS model, **r.water.fea,** is used to simulate the watershed basin produced by the automatic watershed delineation. The outflow hydrograph for each of the smoothed and aggregated data sets is compared to the hydrograph produced using the original 30-m data. The difference in hydrograph discharge values caused by smoothing and aggregation is compared to the information content loss. A relationship for predicting consequent errors in the hydrologic model output due to information content loss is presented.

Smoothing

The original 30-m resolution DEM is smoothed to generate three additional DEMs using the GRASS *r.mfilter* program. Smoothing applies a 3×3, 5×5 or 7×7 window, respectively, over the original 30-m resolution DEM and assigns the average elevation value within the window to the central cell of the window. After smoothing, resolutions remain at 30 m. Smoothing reduces possible random errors that occur due to systematic measurement and filters out

small pits or peaks in the original DEM. As smoothing progresses, the topographic data tends toward a data set with a constant elevation (the mean). This
reduces spatial variability in the elevations which propagates error in hydrologic
simulations. Smoothing is a filter often used to improve automatic delineation
of watersheds and stream networks by removing pits and spikes from the DEM.
These spurious elevations may result from the regular sampling interval over an
irregular surface. Aggregation in its simplest form is performed by enlarging the
cell resolution. Cell resolution must be chosen prior to development of a DEM.
The choice of grid cell resolution also affects the topographic features such as
slope and spatial variability.

Aggregation

The original 30-m resolution DEM is resampled into three more DEMs with
90, 150, and 210-m resolution, respectively, using the GRASS *g.region* program.
The GRASS *g.region* resamples the 30-m resolution DEMs into larger resolution DEMs by assigning the elevation value closest to the center of the larger
cell. As aggregation progresses, the topographic data becomes a data set of larger
and larger cells. As with smoothing, if taken to the limit, a lumped value results
with zero entropy. Aggregation is done by resampling at larger raster grid cell
sizes. Thus, aggregation is a means of investigating the effects of using larger
cell sizes in hydrologic modeling. Capturing the spatial variability with the appropriate cell size is necessary in order for the hydrologic model to accurately
represent the physical characteristics of the watershed.

DEM Information Content

Smoothing and aggregation reduce the spatial variability of elevation and the
derived slope data. The GRASS programs *g.region* and *r.mfilter* are both data
filters that reduce information content. The application of information theory and
specifically the concept of entropy, first introduced in the landmark theory by
Shannon and Weaver (1964), to hydrologic modeling is useful in understanding
the effects of data filters on the data used to extract hydrologic model parameters. Entropy, I, in this context is defined as

$$I = -\sum_{i=1}^{B} P_i \log(P_i) \tag{1}$$

where B is the number of bins or discrete intervals of the variate; and P_i is the
probability of the variate occurring within the discrete interval. A negative sign
in front of the summation is by convention such that increasing information content results in positively increasing entropy. (For a complete description of entropy in information theory context refer to Papoulis, 1984.) Base 10 is used in
this investigation yielding information in units of Hartelys. Information theory
commonly uses Base 2 when operating on bits of information yielding binits of

information. Entropy becomes a measure of spatial variability when applied to topographic surfaces defined by a raster DEM. As the variance increases so does entropy; conversely, as variance decreases, so does entropy. In the limit, if the topographic surface is a plane with a constant elevation, the probability is 1.0 resulting in zero entropy, zero uncertainty, and zero information content. Maximum entropy occurs when all elevations are equally probable which may be either a surface with highly variable elevations or one with uniform slope.

Measuring entropy at increasing resolutions provides an estimate of the rate of information loss due to aggregation to larger cell sizes. The rate of loss with respect to cell size can be put into terms of a noninteger or fractal scaling law. The fractal dimension is used in this study to indicate the effects of aggregation on hydrologic model output.

Fractal Dimension

Geographers have struggled with the notion of how long is a coastline. As smaller sized dividers are used, the length increases without bound. Length approaching infinity as the resolution of the measurement approaches zero is characterized by the fractal dimension. The application of this idea to topographic surfaces was made by Goodchild (1982) in the use of fractional Brownian processes in terrain simulation. Mark and Aronson (1985) measured the fractal dimension of several digital terrain models at 30-m resolution. Using a semivariogram technique, they found that the fractal dimension of the topography varied between 2.2 to 2.3 over scales of 0.6 km but that the fractal model was not correct over all scales due to periodicity and multiple fractal dimensions at differing scales. From a purely geographic or cartographic view, error can only be measured as error in location or attribute but not in terms of hydrologic simulation error. As shown by Goodchild and Mark (1987), error in estimating an area depends on the fractal dimension of the boundary and the size of the pixel representing the area. However, this offers little insight into the consequences to hydrologic modeling using a raster representation of elevations (DEMs).

The implication to the investigation presented herein is that as the sampling interval increases with increasing cell size, the information loss is greater for surfaces with higher fractal dimension. In turn, the information loss propagates errors in the hydrologic model output. Thus, the fractal dimension is an indication of how much information loss may occur because of filtering by either smoothing or aggregation. The rate of information loss would vary in any natural topography. High variance at a local scale demands smaller cell sizes to capture the information content. When variance is low (a surface with constant elevation), any number of cells is sufficient to capture the information content including just one cell. For example, a plane surface will have a fractal dimension equal to the Euclidian dimension of 2. Thus, smoothing and cell aggregation will not produce information loss. On the contrary, topographic surfaces possessing high variability with a fractal dimension greater than 2.0 will experience a higher degree of information loss as smoothing and aggregation progress.

The information dimension is the fractal dimension computed by determining the information content as measured by entropy at differing scales. Several definitions of dimension, including the information dimension, for dynamic systems exhibiting chaotic attractors are presented by Farmer et al. (1983). The fractal dimension is a measure common to both fields of application.

Self-Affinity

Information content measured at different scales is not a self-similar fractal because a self-similar fractal exhibits identical scaling in each dimension. In terms of a box dimension, a rectangular coordinate system may be scaled down to smaller grid cells of side ϵ. Following the arguments of Mandelbrot (1985), a self-affine function (such as entropy) follows the nonuniform scaling law where the number of grid cells is scaled by ϵ^H. Writing the scaling law in terms of a proportionality, the number of grid cells, $N(\epsilon)$, of side ϵ is

$$N(\epsilon) \propto \epsilon^H \qquad (2)$$

where ϵ is the grid dimension; and H is the Hurst coefficient. Note that the fractal dimension, d_I must exceed the Euclidian dimension and that $0 \leq H \leq 1$, therefore,

$$d_I = E + 1 - H \qquad (3)$$

H has special significance (Saupe, 1988); at 0.5, the surface is analogous to ordinary Brownian motion. If $H < 0.5$ there is negative correlation between the scaled functions, and if $H > 0.5$ there is positive correlation between the scaled functions.

The information dimension described by Farmer et al. (1983) is based on the probability of the variate within each cube in three dimensions or within a grid cell in two dimensions covering the set of data. The set of grid cells representing the watershed basin is a DEM which is a two-dimensional raster data structure. As a test case, we examined the dimension of a plane. The concept was extended to the more general topographic surface; the study watershed basin in following sections. The plane surface was subdivided into cells of ϵ = 1/5th, 1/125th, and 1/625th. The plane surface is of uniform slope such that each elevation has equal probability and $I(\epsilon) = \log(N(\epsilon))$. If a plane surface of uniform slope in the direction of one side of the cells is divided into a 5 × 5 set of cells, it will have five unique rows of equally probable elevations. Each bin will have a probability of one fifth and an entropy of 0.69897 which is simply $\log(B)$. If the plane is divided into successively smaller grid cells, the rate at which the entropy changes should be equal to one. To see this, Table 1 presents the calculations for the plane of uniform slope and one cell wide. The number of bins, B, equals the number of cells, N, and have equal probability.

The Hurst coefficient, H, is the proportionality factor relating the number of grid cells to the size of the cells. We expect that entropy scales with the size of

Table 1   Entropy of a Plane Surface with Uniform Slope

$1/\epsilon$	$N = B$	$I(\epsilon)$ (Hartleys)
(1)	(2)	(3)
625	625	2.7958
125	125	2.0969
25	25	1.3979
5	5	0.69897

From Vieux, B.E., DEM aggregation and smoothing effects on surface runoff modeling, *ASCE J. Computing Civil Eng.* 7(3) 310, 1993.
With permission.

the cell of side ϵ covering the set according to Equation 2. Taking logarithms of both sides of Equation 2, replacing the proportionality with a difference relation and recognizing that for equal probabilities, $\log(N(\epsilon)) = I(\epsilon)$, we find that H is the proportionality constant that relates the rate at which entropy changes with grid cell size. H is related to the fractal dimension, d_I by the following for a two-dimensional Euclidian space, $E = 2$.

$$H = \frac{\Delta I(\epsilon)}{\Delta \log(1/\epsilon)} \qquad (4)$$

where $I(\epsilon)$ is the information content computed for grid cells of side ϵ. Applying Equation 4 to Table 1, we find that $H = (2.79588 - 2.0969)/(\log(625) - \log(125)) = 1.0000$. Thus, for a two-dimensional Euclidian surface $E = 2$ and by Equation 3, $d_I = 2$ as would be expected for a plane surface. Furthermore, $H = 1.0$ (>0.5), and thus, the elevations are positively correlated.

The rate at which information is lost and error produced in the estimates of slope is measured by the information dimension. The hydrologic model output of the finite element simulations are based on the slope and flow path length parameters extracted from the smoothed and aggregated DEMs.

Finite Element Analysis

Finite element analysis allows simulation of the effects of the filtered data on the direct runoff hydrograph. Vieux (1988) developed the solution algorithm for overland flow without channel routing using the Galerkin formulation of the finite element method for solving the kinematic wave equations using one- and two-dimensional finite elements. Vieux et al. (1990) presents the one-dimensional solution for a simple two-plane watershed. This solution uses nodal values of slope which allow simulation of direct runoff for complex topographic surfaces with spatially variable slope and hydraulic roughness. Vieux (1988, 1991) presents the finite element solution linked with a vector-based GIS. In this solution, the complex topography of a small watershed basin is represented using a Triangular Irregular Network (TIN). A two-dimensional finite element grid is superimposed over the TIN. Land surface slope is sampled at each finite element node from the unique slopes of each facet. The study presented herein uses a similar technique except that a set of raster grid cells are aggregated to the basin level. The mean value of slope for the basin becomes the slope at the downstream node of finite element representing the basin.

Infiltration is not considered due to the inherent difficulties in prediction. Rather, several rainfall excess intensities are assumed to continue until the basin reaches equilibrium. Gauged events are not available for this basin nor would they be particularly relevant because the effects of infiltration uncertainties could potentially mask the effects that are the subject of this investigation.

Derivation of the finite element solution will not be presented in detail but the form of the equations for the Galerkin Formulation of the kinematic wave equations will be stated. The following sections present the kinematic wave equations, Galerkin Formulation, and finally the system of equations that simulate the hydrologic output of the watershed basin.

Kinematic Wave Equations

Conservation of mass accounting for overland flow and rainfall excess (rainfall minus infiltration) together with a simplified form of the conservation of momentum equation comprises the kinematic wave equations. Conditions where the full momentum equation may be simplified using the kinematic analogy are presented in detail by Woolhiser and Liggett (1967), Morris and Woolhiser (1980) and many others. The kinematic analogy requires that the overland flow be in the direction of the land surface slope implying that the friction gradient is parallel in direction and equal in magnitude to the land surface slope. This in turn leads to the use of the Chezy or Manning formula for open channel flow. The wide channel assumption results in the following form of the Manning formula for unit width discharge, q in the principal direction of slope,

$$q = \frac{c}{n} S^{1/2} h^{5/3} \tag{5}$$

where n is the Manning roughness coefficient, S is the land surface slope, h is the flow depth, and c is 1.0 for metric and 1.486 for English units.

Conservation of mass for open channel flow with distributed rainfall excess is written as

$$\frac{\partial h}{\partial t} + \frac{\partial q}{\partial x} - (r - i) = 0 \tag{6}$$

where $r - i$ is the rainfall minus infiltration or rainfall excess intensity in consistent units; and the other variables are as defined above. Equations 5 and 6 constitute the kinematic wave equations for overland flow.

Application of the Galerkin Formulation requires minimization of the residual over the system of elements. The residual is formed by substituting shape functions $N^{(e)}$ for the continuous variables and using the same shape functions for the weighting functions as required by the Galerkin Formulation. (For more details, refer to Vieux, 1988 and Vieux et al., 1990.)

The residual equation is expressed for an element (e) as

$$R^{(e)} = \int N^T\, N \partial x\, \dot{h} + \int N^T \left[\frac{\partial N_i}{\partial x}\ \ \frac{\partial N_j}{\partial x} \right] \partial x\, Q - \int N^T\, \partial x \partial y (r - i) \qquad (7)$$

where $\dot{h}$ is the time rate of change of the flow depth, and Q is the unit width flow rate in Equation 5 multiplied by the width of the watershed basin at each node. The matrix equation that results from the integration of Equation 7 is

$$R^{(e)} = C\dot{h} + BQ - F \qquad (8)$$

where

$$C = \frac{L}{6} \begin{bmatrix} 2 & 1 \\ 1 & 2 \end{bmatrix}$$

$$B = \frac{1}{2} \begin{bmatrix} -1 & 1 \\ -1 & 1 \end{bmatrix}$$

$$F = \frac{L(r - i)}{2} \begin{Bmatrix} 1 \\ 1 \end{Bmatrix}$$

A finite difference solution is employed to solve for the time-dependent solution. Each value of the flow rate vector, Q, computed at each time step results in hydrographs for each nodal location. The finite difference solution of Equation 8 in the time domain is

$$C\dot{h}_{new} = C\dot{h}_{old} - \Delta t B[(1 - \omega)Q_{old} + \omega Q_{new}] + \Delta t[(1 - \omega)F_{old} + \omega F_{new} \qquad (9)$$

where ω is one-half for a central difference solution, and Δt is the time step. At each time step a new value of the flow depth is computed. Because new values of the flow rate, Q, also occur on the right-hand side of Equation 9, a nonlinear system of equations results requiring an iterative solution at each time step.

The finite element used to represent the geometry of the watershed basin is a one-dimensional element with variable width at the nodes. The following sections present the results of the automatic watershed delineation; smoothing and aggregation of the DEM and resulting slope maps; the outflow hydrographs produced from the finite element simulations; and finally, relationships between the propagated error and relative entropy loss caused by smoothing the DEM. The finite element program presented by Vieux (1993) utilized a single element to represent the watershed. The simulations that follow utilize a finite element solution that is a part of GRASS called **r.water.fea** developed by Vieux and Gaur (1994). The solution using **r.water.fea** connects each grid cell together with a finite element. Each grid cell slope value is then represented in the solution.

RESULTS

Watershed basin delineation is performed by selecting an outlet and then finding all the cells that drain to it. The GRASS program *r.watershed,* which uses an 8-cell search, is applied to the original 30-m DEM for the automatic watershed delineation. Slope and aspect maps are then derived from the elevations. For each of the smoothed and aggregated DEMs, the slope map is produced using the GRASS "Slope.aspect" program. Delineation was done only once using the original 30-m data so that watershed basins of different configurations would not result. The hydrologic model parameters were extracted from the slope maps produced from the smoothing and aggregation filters.

Beginning with the outlet cell, an 8-cell search is made for each cell to determine the drainage network. Once the drainage network is completed, the basin becomes the set of those cells that drain to that outlet. The boundary enclosing these cells is the watershed. Figure 1 is a three-dimensional view of the original 30-m DEM used to delineate the watershed basin. The aspect map is draped on top of the topography. Figure 2 shows the resulting drainage network and watershed basin delineated by GRASS. The three basins shown in the inset in Figure 2 are simulated using **r.water.fea**.

The stream network is derived by considering whether a minimum number of cells drain to a particular cell. The minimum number is called an accumulation threshold. In this case, a threshold of 300 was used which indicates that the upper most cells in the stream network have at least 300 cells contributing drainage to them. Smaller accumulation threshold values result in a more space-filling stream network.

Figure 3 shows the original and smoothed DEMs at 30-m resolution. There is a high degree of spatial variability and therefore entropy in slope. Flatter slopes occur in the streams. This variability decreases if the DEM is first smoothed, and then the slope is derived from the smoothed DEM. Because water organizes itself according to the drainage network and the enclosing watershed boundary, it is important to consider the spatial variability across the watershed basin being analyzed. Since variability will change at different scales and across different regions, variability measures such as entropy are more meaningful when applied at the watershed basin to gauge the effects of smoothing or aggregation on hydrologic processes affected by topography. The hydrologic model **r.water.fea** is used to measure propagated error. Parameters are extracted and simulation effects are compared at the watershed basin outlet.

Smoothing Effects

Figures 3 and 4 show the effects of applying the 3×3, 5×5, and 7×7 roving windows, respectively, to the original 30-m DEM. Although the mean elevation is unchanged, the spatial variability is reduced. Entropy decreases as more smoothing takes place with the application of larger and larger roving windows. Smoothing of the original 30-m DEM using a 7×7 roving window causes lost information content of $2.090 - 1.404 = 0.686$ Hartleys.

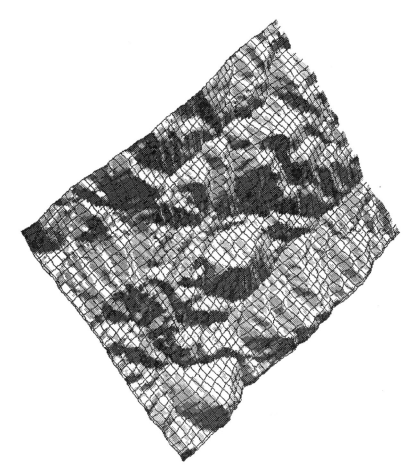

Figure 1   Three-dimensional view of the original 30-m DEM used to delineate the watershed basin. (From Vieux, B. E., *ASCE J. Computing Civil Eng.*, 7(3) 310–338,
1993. With permission.)

Entropy is also computed for the slope maps derived from the smoothed
DEMs. The effects of the roving windows applied to the original DEM are measured by observing the change in mean slope and entropy. It is important to note
that the slope maps are derived from the smoothed DEMs. To see how smoothing may affect the apparent slope of the watershed, a frequency of occurrence
is calculated for slope intervals of one degree. Entropy is calculated as the negative sum of the discrete values using Equation 1. The discrete intervals were
chosen as increments of one degree since the slope is measured in integer degrees. Besides getting flatter, variability is reduced as more and more smoothing occurs. If continued with more passes of the smoothing window, the basin
would tend toward a uniform slope with zero entropy and no spatial variability.

Figure 5 shows the histograms of slope magnitude derived from the smoothed
DEMs. Smoothing of the elevations results in a shift of the mode and mean of

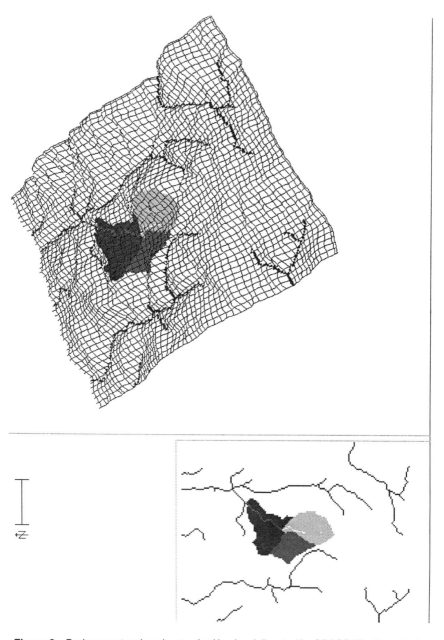

Figure 2　Drainage network and watershed basins delineated by GRASS. The three basins shown in the inset are simulated using **r.water.fea.** (From Vieux, B. E., *ASCE J. Computing Civil Eng.*, 7(3) 310–338, 1993. With permission.)

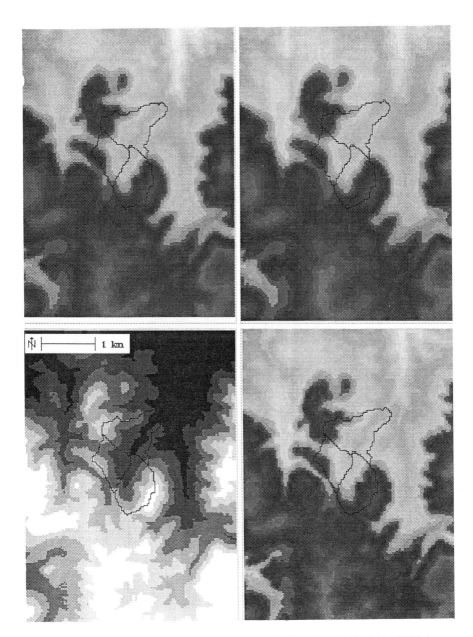

Figure 3 Original and smoothed DEMs at 30-m resolution. *Lower left*, original DEM; *lower right*, 3 × 3 smoothed DEM; *upper left*, 5 × 5 smoothed DEM; *upper right*, 7 × 7 smoother DEM. (From Vieux, B. E., *ASCE J. Computing Civil Eng.*, 7(3) 310–338, 1993. With permission.)

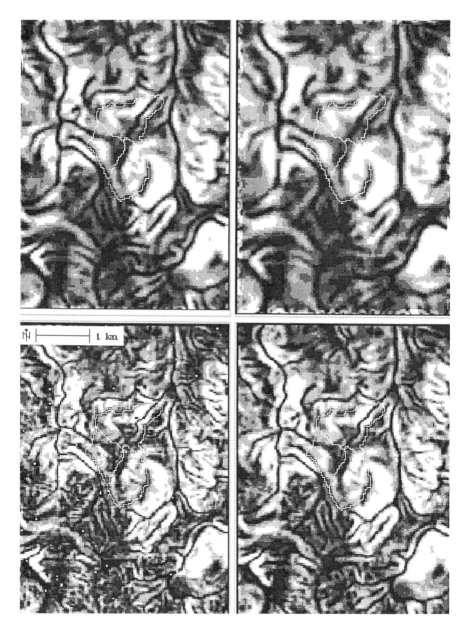

Figure 4   Slope maps produced from the original and smoothed DEMs. *Lower left*, slope produced from original DEM; *lower right*, slope produced from the 3 × 3 smoothed DEM; *upper left*, slope produced from 5 × 5 smoothed DEM; upper right, slope produced from 7 × 7 smoother DEM. (From Vieux, B. E., *ASCE J. Computing Civil Eng.*, 7(3) 310–338, 1993. With permission.)

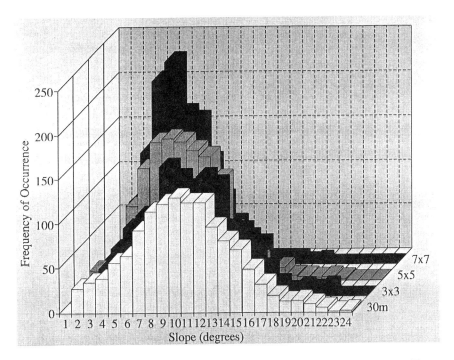

Figure 5 The histograms of slope magnitude derived from the smoothed DEMs. (From Vieux, B. E., *ASCE J. Computing Civil Eng.*, 7(3) 310–338, 1993. With permission.)

the slope. Table 2 summarizes the effects of smoothing on elevation. Table 3 summarizes the effects on the derived slope magnitudes. As a consequence of the averaging in the roving window, the mean elevation remains the same as the original 30-m data. Thus, entropy decreases with smoothing even though the mean is invariant. Entropy measures the effects of smoothing indicating that smoothing is a lumping process.

As smoothing progresses, the arithmetic mean slope decreases from 12.237° in the original data to 8.044° after 7×7 smoothing. The mode and entropy follow similar trends. Filtering by smoothing elevations before calculating slope alters the statistics and distribution of the slope data and produces significant errors in slope magnitude. The propagation of this error in the hydrologic model output will be measured by using this slope magnitude in a finite element simulation of direct runoff.

Aggregation Effects

The effect of aggregating the DEMs is shown in Figure 6. Decreased variation in elevation is evident as cell size increases. The overall watershed becomes a lower elevation as higher elevations drop out of the ensemble of values. Figure 7 shows the slope maps derived from the aggregated DEMs. The steeper slope

Table 2 Effects of Smoothing on Elevation

Smoothing window (1)	Mean (m) (2)	Entropy (Hartleys) (4)
Original	1563.89	2.090
3 × 3	1563.93	1.877
5 × 5	1563.93	1.594
7 × 7	1563.93	1.404

From Vieux, B.E., DEM aggregation and smoothing effects on surface runoff modeling, *ASCE J. Computing Civil Eng.* 7(3), 310, 1993. With permission.

Table 3 Slope Derived From Smoothed DEMs

Smoothing window (1)	Mean (degrees) (2)	Mode (degrees) (3)	Entropy (Hartleys) (4)
Original	12.237	10	1.266
3 × 3	10.738	8	1.193
5 × 5	9.210	7	1.123
7 × 7	8.044	6	1.055

From Vieux, B.E., DEM aggregation and smoothing effects on surface runoff modeling, *ASCE J. Computing Civil Eng.* 7(3), 310, 1993. With permission.

areas in many parts of the watershed are flatter than the original 30-m slope map. As the watershed becomes flatter, the variability of the slopes also decreases. Cell size aggregation reduces information content. As the cell size increases, entropy decreases because variability is decreased.

Entropy and information content decrease due to aggregation but not without a loss in accuracy in the slope derived from the smoothed DEMs. Filtering by aggregating elevations before calculating slope alters the statistics and distribution of the slope data in a manner similar to smoothing. Table 4 summarizes the following: (1) effects of aggregation on mean elevation, (2) entropy, and (3) number of cells representing the watershed area. Table 5 shows similar effects on slope magnitude. The histograms in Figure 8 show the mode shifting toward the flatter end of the distribution.

Summary

Smoothing of the DEM does not cause the watershed area or boundary to vary unless, after each smoothing operation, another delineation is performed. In order to make comparisons, delineation is done only once. As cell sizes increase through aggregation, meanders present in the original data are short circuited causing a reduction in the flow path length from the outlet where delineation began to the upper reaches of the watershed basin. The flow path length and slope vary as cell sizes increase. Because the watershed area is approximated by the number of grid cells multiplied times the area of an individual grid

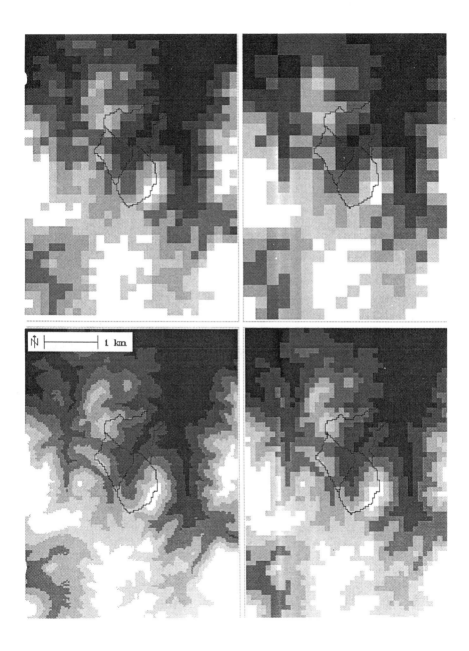

Figure 6 Original and aggregated DEMs. *Lower left*, original DEM; *lower right*, 90-m DEM; *upper left*, 150-m DEM; upper right, 210-m DEM. (From Vieux, B. E., *ASCE J. Computing Civil Eng.*, 7(3) 310–338, 1993. With permission.)

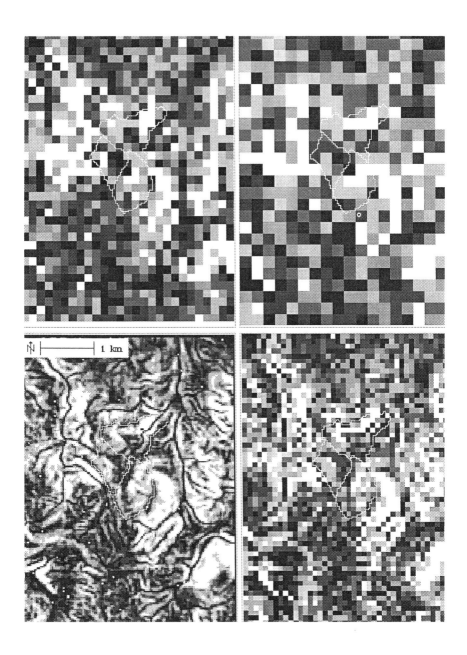

Figure 7　Slope maps derived from the aggregated DEMs. *Lower left*, slope produced from original DEM; *lower right*, slope produced from 90-m DEM; *upper left*, slope produced from 150-m DEM; upper right, slope produced from 210-meter DEM. (From Vieux, B. E., *ASCE J. Computing Civil Eng.*, 7(3) 310–338, 1993. With permission.)

Table 4   Effects of Aggregation on Elevation

Cell size (m) (1)	Mean (m) (2)	Entropy (Hartleys) (3)	No. of cells (4)
30	1563.89	2.090	1306
90	1554.05	1.877	141
150	1535.93	1.594	54
210	1510.76	1.404	28

From Vieux, B.E., DEM aggregation and smoothing effects on surface runoff modeling, *ASCE J. Computing Civil Eng.* 7(3), 310, 1993. With permission.

Table 5   Effects of Aggregation on Slope

Cell size (m) (1)	Mean (degrees) (2)	Mode (degrees) (3)	Entropy (Hartleys) (4)
30	12.237	10	1.266
90	12.141	11	1.229
150	11.746	11	1.180
210	11.138	12	1.006

From Vieux, B.E., DEM aggregation and smoothing effects on surface runoff modeling, *ASCE J. Computing Civil Eng.* 7(3) 310, 1993. with permission.

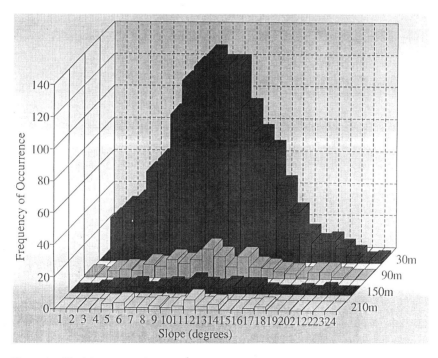

Figure 8   The histograms of slope magnitude derived from the aggregated DEMs. (From Vieux, B. E., *ASCE J. Computing Civil Eng.*, 7(3) 310–338, 1993. With permission.)

cell, the area varies slightly due to edge effects. More details on the effect of aggregation on flow path length may be found in Vieux (1993).

Finite Element Simulations

Simulation of the effects of smoothing and aggregation performed by Vieux (1993) used one finite element the length of the main channel flow path with nodal slopes equal to the mean slope for each case. The hydrograph simulations that follow use the GRASS program **r.water.fea**. Simulations are conducted for 9000 s which is sufficient to reach equilibrium for the rainfall excesses tested. Simulations for the smoothed data sets are shown below for rainfall excess intensities of 1.1, 2.2, and 4.4 cm/h. At equilibrium, the inflow of rainfall excess over the watershed basin equals the outflow at the outlet, and therefore the effects of slope and length are not evident. Because the propagated error is manifest in the rising limb but not during equilibrium, our attention is focused on the rising limb. Even though the falling limb would be equivalent, the error during the rising limb is sufficient to establish a relationship of error vs. information content loss. Further, in the case of partial equilibrium, the rising limb of the hydrograph determines the peak discharge rate which is often the most important information needed in the design of hydraulic structures.

Error Propagated by Smoothing

Four rainfall excess intensities were tested. The simulated hydrographs in Figure 9 show the effects of smoothing the DEM using 3×3, 5×5, and 7×7 roving windows. The original 30-m data set is used as the standard for comparison. Departures from this hydrograph are considered an error. The measure of the error is the *L2* norm defined as

$$L2 = \sum_{t=1}^{T} (\overline{Q}_{30_t} - \overline{Q}_{\beta_t})^2 \qquad (10)$$

where $\overline{Q}_{30_t}$ is the discharge at each time step normalized by the equilibrium discharge for the respective rainfall excess intensities, Q_{β_t} is the hydrograph discharge values for each $\beta = 3 \times 3$, 5×5, 7×7 smoothing windows. Thus, the normalized hydrograph values vary between zero and one making them comparable to each other.

Figure 9 shows that as the mean slope of the watershed basin is flattened, the response of the hydrograph is delayed. Thus, any hydrologic model, including *r.water.fea,* will have a delayed response due to an apparent watershed slope that is flattened due to filtering. Similar effects result due to aggregation except that the response is accelerated because of shortened flow path lengths (Vieux, 1993).

The error computed using the L2 norm in Equation 11 can now be related to the amount of entropy loss. Because the L2 norm measures the departure from the hydrograph derived using the original 30-m DEM, the entropy loss associ-

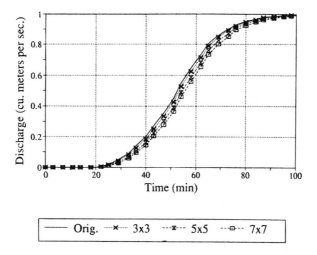

Figure 9 Hydrograph response using slopes derived from smoothed DEMs. (From Vieux, B. E., *ASCE J. Computing Civil Eng.*, 7(3) 310–338, 1993. With permission.)

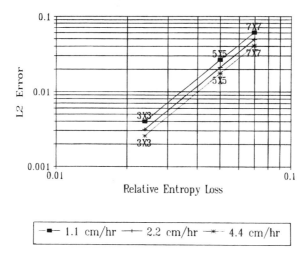

Figure 10 L2 norm vs. relative entropy loss. (From Vieux, B. E., *ASCE J. Computing Civil Eng.*, 7(3) 310–338, 1993. With permission.)

ated with slope is expressed relative to the original 30-m slope entropy. The finite element simulations are run for three intensities, 1.1, 2.2, and 4.4 cm/h, with the results shown in Figure 10.

Since entropy is expected to follow a power scaling law according to Equation 2, the propagated error measured by the L2 norm of the normalized discharges is plotted against the relative entropy loss. The relative entropy loss is computed by dividing the loss of entropy caused by smoothing or aggregation by the entropy of the original slope map derived from the original 30-m DEM.

Figure 10 shows that the logarithm of the L2 norm is linearly related to the logarithm of the relative entropy loss. The intercepts are inversely proportional to the rainfall excess intensities. Thus, the topographic variability manifests itself in terms of the rate at which error is propagated vs. the relative entropy loss; whereas, the rainfall excess manifests itself as a scaling parameter that linearly increases the amount of propagated error. This is significant because now we have an *a priori* means of assessing the impact of filtering (smoothing and aggregation) on propagated error in a finite element simulation of direct runoff. Further, the rainfall excess is independent of topography and, as such, is an independent factor that linearly decreases the magnitude of the error introduced by smoothing and aggregation.

The form of the error relationship in Figure 10 is

$$L2 = \alpha(I_{rel})^m \tag{11}$$

where α is the intercept, I_{rel} is the difference between the original entropy and the smoothed or aggregated entropy divided by the original entropy, and m is the slope on the log-log plot. The slope of the relationships for smoothing in Figure 10 are constant with the effects of rainfall (error intercepts) separated from the effects of topographic smoothing (rate of error increase).

Fractal Dimension

The fractal dimension and the Hurst coefficient provide a convenient means of assessing the rate at which entropy is lost due to smoothing or aggregation. Figure 11 shows a plot of the entropy of the elevations at 30-, 90-, 150-, and

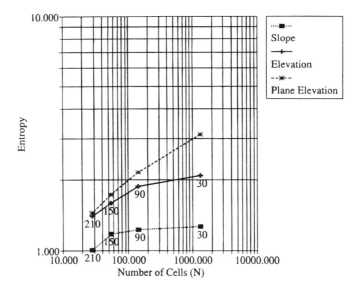

Figure 11   Entropy vs. number of cells for equal probability plane, elevation, and slope maps. (From Vieux, B. E., *ASCE J. Computing Civil Eng.*, 7(3) 310–338, 1993. With permission.)

210-m resolutions including the plane surface entropy where all elevations are equally probable and entropy is log(N). As aggregation proceeds, the elevation and slope entropy converges towards the line representing an equal probability surface.

In Table 6, Column 4 is the entropy of the plane and the slope of this line on the log–log plot is one. Thus, the Hurst coefficient in Column 5 is 1.0 and the fractal dimension is 2.0. The Hurst coefficient in Column 6 is that of the entropy of the elevations. Since it is smaller at smaller resolutions, the fractal dimension is not constant. This is consistent with the findings by Mark and Aronson (1985) who found multiple fractal dimensions of 2.2 and 2.3 over scales of 0.6 km.

CONCLUSIONS

The functional form between error propagated in a hydrologic model due to smoothing or aggregation of a DEM is linear on log–log plots of L2 norm error of normalized discharges vs. relative entropy loss. This provides a convenient method of assessing the rate of the error propagated due to entropy loss caused by filtering (smoothing and aggregation). Further, the rainfall excess intensity was found to magnify linearly the amount of error propagated. Thus, the effects of topographic variability and rainfall excess causing direct surface runoff are separated in the error analyses. By measuring entropy losses during filtering and preprocessing of DEMs for automatic watershed delineation, an estimate of the error propagation in subsequent direct surface runoff simulations can be developed.

The L2 vs. I_{rel} plot is linear on log-log paper and provides a convenient method of estimating error propagation. During database development, choice of cell size can now be related to consequent error in the hydrologic model thus answering the questions:

1. What cell size should be used in the database and analyses using digital elevations? and
2. What will be the consequences of smoothing DEMs on the hydrologic simulation?

Table 6 Entropy, Hurst Coefficient, and Fractal Dimension of the Aggregated DEM and a Plane Surface

DEM cell size (1)	No. of cells (2)	DEM entropy (3)	Plane entropy (4)	Plane H (5)	DEM H (6)	Fractal d = 3-H (7)
30	1306	2.090	3.116	—	—	—
90	141	1.877	2.149	1.00	0.22	2.78
150	54	1.594	1.732	1.00	0.68	2.32
210	28	1.404	1.447	1.00	0.67	2.33

From Vieux, B.E., DEM aggregation and smoothing effects on surface runoff modeling, *ASCE J. Computing Civil Eng.* 7(3), 310, 1993. With permission.

The method of estimating propagated error from relative entropy loss has potential applications whenever spatially distributed processes are simulated using spatially variable input data. The exact relation between the rate of error propagation and the Hurst coefficient requires further analysis. The method is envisioned to be generally applicable to simulation of surface or subsurface distributed hydrologic processes when DEMs or other spatially distributed data are used as model parameters.

REFERENCES

Farmer, J. D., Ott, E., and Yorke, J. A., The dimension of chaotic attractors, *Phys. D; Nonlinear Phenomena*, 7, 153, 1983.

Goodchild, M. F., The fractional Brownian process as a terrain simulation model, Proceedings of the Thirteenth Annual Pittsburg Conference on Modeling and Simulation, 13, 1133–1137, 1982.

Goodchild, M. F. and Mark, D. M., The fractal nature of geographic phenomena, *Ann AAG*, 77(2), 265–278, 1987.

Horton, R. E., Drainage-basin characteristics. *Trans. AGU*, 13, 350–361, 1932.

Mandelbrot, B. B., Self-affine fractals and fractal dimension, *Phys. Scr.*, 32, 257–260, 1985.

Mark, D. M. and Aronson, P. B., Scale-dependent fractal dimensions of topographic surfaces: an empirical investigation, with applications in geomorphology and computer mapping, *Math. Geol.*, 16(7), 671–683, 1985.

Moore, I. D., Grayson, R. B., and Ladson, A. R., Digital terrain modeling: a review of hydrological, geomorphological and biological applications, *J. Hydrol. Process*, 5, 3–30, 1991.

Morris, E. M. and Woolhiser, D. A., Unsteady one-dimensional flow over a plane: partial equilibrium and recession hydrographs, *J. Water Resour. Res.* 16(2), 355–360, 1980.

Papoulis, A., *Probability, Random Variables, and Stochastic Processes*, 2nd ed., McGraw-Hill, New York, 1984, 500–567.

Quinn, P., Beven, K., Chevallier, P., and Planchon, O., The prediction of hillslope flow paths for distributed hydrological modeling using digital terrain models, *J. Hydrol. Process*, 5, 59–79, 1991.

Saupe, D., *The Science of Fractal Images*, Peitgen, H.-O. and Saupe, D., Eds., Springer-Verlag, New York, 1988, 82–84.

Shannon, C. E. and Weaver, W., *The Mathematical Theory of Communication*, University of Illinois Press, Urbana, IL, 1964.

Tarboton, D. G., Bras, R. L., and Rodriguez-Iturbe, I., On the extraction of channel networks from digital elevation data, *J. Hydrol. Process*, 5, 81–100, 1991.

Vieux, B. E., *Finite Element Analysis of Hydrologic Response Areas Using Geographic Information Systems*, Ph.D. Dissertation, Department of Agricultural Engineering, Michigan State University, MI, 1988.

Vieux, B. E., Bralts, V. F., Segerlind, L. J., and Wallace, R. B., Finite element watershed modeling: one-dimensional elements, *J. Water Resour. Manage. Planning*, 116(6), 803–819, 1990.

Vieux B. E., Geographic information systems and non-point source water quality and quantity modeling, *J. Hydrol. Process*, 5, 101–113, 1991.

Vieux, B. E., DEM aggregation and smoothing effects on surface runoff modeling, *ASCE J. Computing Civil Eng.*, 7(3), 310, 1993.

Vieux, B. E. and Gaur, N., Finite element modeling of storm water runoff using GRASS GIS, *Microcomputers in Civil Engineering*, Elsevier, 1994, in press.

Westervelt, J., Shapiro, M., and Goran, W. D., *GRASS User's Reference Manual*, ADP Report N-87/22, U.S. Army Corps of Engineers, Construction Engineering Research Laboratory (USACERL), Champaign, IL, 1988.

Woolhiser, D. A. and Liggett, J. A., Unsteady, one-dimensional flow over a plane—the rising hydrograph, *Water Resour. Res.*, 3(3), 753–771, 1967.

CHAPTER 16

Incorporation of Real-Time Environmental Data Into a GIS For Oil Spill Management and Control

Michael Garrett, Gary A. Jeffress, and Donald A. Waechter

ABSTRACT

The ability of geographic information systems (GIS) to manipulate and display complex spatial attributes and relationships rapidly makes them ideal for the management of incidents where time is a critical factor. An oil spill is an example of such an incident. This paper outlines development of a pilot system to incorporate wind speed, wind direction, water temperature, air temperature, and other environmental data into a graphical display within ARC/INFO for a potential oil spill site. These data are presently collected automatically and in near real-time as part of the Texas Coastal Ocean Observation Network (TCOON). This paper explains the method whereby data are collected in the real-time network (RTNET) and transferred to the GIS database. Implementation of the graphical user interface for ARC/INFO is briefly described as is the role of GIS in management of an oil spill response.

INTRODUCTION

TCOON presently comprises 45 Data Collection Platforms (DCPs) collecting a variety of environmental data in bays, estuaries, and offshore along the Texas Gulf coast. This network was established in 1989 by the Conrad Blucher Institute for Surveying and Science (CBI) at Texas A&M University, Corpus Christi. The original network was comprised of three remote reading tide gauges along the shores of Corpus Christi Bay. These gauges were used primarily to provide real-time water level data for hurricane preparedness for the city of Corpus Christi.

As water level data became available from the original three gauges, other users of water level data became interested in accessing the network. This has

Based on Garrett, M. and Maynard, G., Managing oil spills and Texas GIS integrates near real-time environmental data, *Geo Info Systems*, Jan. 1993. With permission.

led to cooperative agreements between CBI and other parties to expand the network and the types of data collected. The focus of TCOON has now shifted from hurricane preparedness to water level measurement for use in determination of littoral boundaries and scientific research. Additionally, some DCPs collect data on wind speed, wind direction, water temperature, air temperature, and barometric pressure. Current users and sponsors of TCOON include the Texas General Land Office (TGLO), Texas Water Development Board (TWDB), the Environmental Protection Agency (EPA), and the National Oceanic and Atmospheric Administration's (NOAA) National Ocean Service (NOS). As mentioned, the primary objective of the network is to provide water level data and consequently all gauges have been established to NOS standards. This allows water level data to be measured reliably and addresses legal concerns regarding admissibility of data for determination of littoral boundaries (Jeffress, 1991). Figure 1 shows the location of DCPs which presently comprise TCOON.

This paper outlines development of a pilot application to incorporate wind speed, wind direction, water temperature, air temperature, and other environmental data into a graphical display within ARC/INFO™.[1] These data are presently collected automatically and in near real-time as part of TCOON. This paper explains how data are collected in the RTNET and transferred to the GIS database.

Implementation of the graphical user interface for ARC/INFO is briefly described. The role of GIS in the management of an oil spill response is discussed and conclusions drawn regarding the contribution that RTNET and TCOON will make to management of an oil spill.

INSTRUMENTATION

The equipment which constitutes TCOON are SUTRON 9000 Remote Terminal Units (RTU)™[2] and Vitel[3] data loggers controlling Aquatrak™[4] acoustic transducers for measurement of water level. The modular design of the SUTRON 9000 data logger allows inclusion of additional sensors to measure other environmental and meteorological data; some data loggers have been interfaced to wind speed and direction sensors, barometers, and thermistors. Currently, all DCPs within the network measure water level. Some DCPs have also been interfaced to Hydrolab H20™[5] multiparameter water quality sensing systems. These platforms provide information on water pH, salinity, dissolved oxygen, and specific conductivity. Table 1 lists stations currently comprising TCOON and shows what types of data are collected at each platform.

The reader is referred to Jeffress (1991) for a more detailed description of TCOON. For the purposes of this paper, it is sufficient to state that data arrive

[1] ARC/INFO is a trademark of Environmental Systems Research Institute, Redlands, CA.
[2] SUTRON 9000 RTU is a trademark of Sutron Corp., Herndon, VA.
[3] Vitel Corp., Chantilly, VA.
[4] Aquatrak is a trademark of Bartex, Inc., Annapolis, MD.
[5] H20 is a trademark of Hydrolab Corp., Austin, TX.

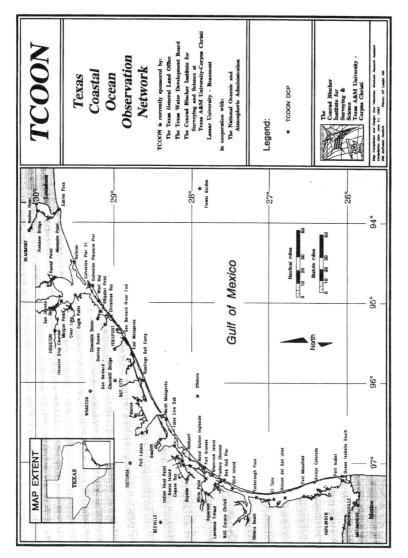

Figure 1   Texas Coastal Ocean Observation Network.

Table 1   TCOON Data Collection Platforms and Environmental Data Captured.

Station Name	Water Level	Water Temp	Cal. Air Temp.	Barometric Pressure	Wind Speed	Wind Direction	Hydrolab
1. PORT ARANSAS	X		X		X	X	
2. NAVAL STATION INGLESIDE	X		X				
3. WHITE POINT	X						
4. TEXAS STATE AQUARIUM	X	X	X	X	X	X	
5. LAWRENCE STREET T-HEAD	X	X	X	X	X	X	
6. CORPUS CHRISTI NAVAL AIR STATION	X		X				
7. PACKERY CHANNEL	X	X					
8. BIRD ISLAND CHANNEL	X		X		X	X	
9. RIVIERA BEACH	X	X					
10. YARBOROUGH PASS	X	X					
11. EL TORO	X	X	X	X	X	X	
12. RINCON DEL SAN JOSE	X	X	X				
13. CHRISTMAS BAY	X	X	X				
14. CLEAR LAKE	X						
15. OFFSHORE	X	X	X	X	X	X	
16. SABINE PASS	X	X	X				
17. ROCKPORT	X	X	X				
18. CORPUS CHRISTI BOB HALL	X	X	X				
19. PORT MANSFIELD	X	X	X				
20. PORT ISABEL	X		X				
21. NORTH MATAGORDA	X	X	X				
22. QUEEN ISABELLA	X		X	X	X	X	
23. FLOWER GARDEN	X		X				
24. BAYSIDE	X		X				
25. SEADRIFT	X		X				
26. EAST MATAGORDA BAY	X		X				
27. PORT LAVACA	X		X				
28. PALACIOS	X						
29. SHAMROCK, CORPUS CHRISTI BAY	X						
30. GOOSE ISLAND, ST. CHARLES BAY	X						
31. INDIAN POINT, ST. CHARLES BAY	X						
32. FALSE LIVE OAK, SAN ANTONIO BAY	X						
33. COPANO CAUSEWAY	X						
34. MORGANS POINT	X		X				
35. GALVESTON PIER 21	X	X	X				
36. GALVESTON PLEASURE PIER	X	X	X				
37. SAN BERNARD-CHURCHILL	X		X				
38. SAN BERNARD-RIVERS END	X		X				
39. COLORADO RIVER (RAWLINGS)	X		X				
40. ARROYO COLORADO	X		X				
41. RAINBOW BRIDGE	X		X				
42. MESQUITE POINT	X		X				
43. EAGLE POINT	X		X				
44. ALLIGATOR POINT	X		X				
45. ROUND POINT	X		X				

at CBI via at least one of three communication channels. Most DCPs in TCOON transmit data via the Geostationary Operational Environmental Satellite (GOES) to the National Environmental Satellite Data and Information Service (NESDIS), Wallops Island, VA. This mode is not real-time but is presently the most reliable link CBI has to its DCPs. Data from NESDIS are currently downloaded via telephone to CBI automatically, six times daily. Five gauges have the ability to measure wind speed and direction and initiate their own data transfers to CBI every 8 to 15 min. These five gauges comprise the real-time network which provides data for this pilot project. The data from all DCPs are stored in digital form in databases at CBI.

UPGRADING THE NETWORK

TCOON is currently being upgraded. New software which is currently in beta testing will "poll" the majority of DCPs in the network every 15 min via packet radio. The number of gauges polled and frequency at which they are polled can be varied and will permit a much more flexible approach to the data collection process. If a spill were to occur, gauges in the vicinity of the spill could be identified and their status in the polling hierarchy increased. These changes will significantly improve the real-time capability of the existing network. As stated above, only the five beaconing gauges can be presently (i.e., gauges which initiate their own data transfers every 8 to 15 min), considered real-time. CBI has been granted a total of three dedicated radio frequencies for its packet radio network and will receive an additional frequency in the future. This means that once the new polling software has been fully incorporated into the network, three radio channels will be available for simultaneous polling. Preliminary testing of the polling software indicates that data from a DCP can be downloaded to CBI in approximately 40 s.

WHAT IS REAL-TIME?

Does TCOON constitute real-time measurement of prevailing environmental conditions? What is real-time? The answer to these questions depends on the application for which data are being measured. For example, an electrical engineer or physicist may consider a real-time measurement as one that is available within a few nanoseconds after the event occurred. This may be necessary because the parameters being measured change rapidly. A real-time measurement, therefore, is a function of both the rate of change of the quantity being measured and how rapidly this information needs to be displayed and reacted on by those monitoring a particular incident.

Water level measurements within TCOON are made every 6 min in accordance with NOS standards. This 6-min measurement represents an arithmetic mean of many observations. Each 6-min water level measurement comprises 180 observations over a 3-min period. These observations are then averaged, outliers rejected, and mean recomputed. In a normal operational mode, data will transmitted to CBI approximately every 15 min. For the purposes of supplying these data to an oil spill trajectory model, and to managers remote from the site, the age of these data is probably acceptable. Although the data are perhaps not technically real time, under most circumstances, they are well within the rate of change of the parameters being measured.

INCLUSION OF THE DATA INTO CBI DATABASES

The gateway to CBI for data is through an 80486 PC running under a UNIX[TM6] operating system. This transfers data over the local area network to the primary host computer, a Hewlett-Packard[7] 9000 model 735 RISC workstation. Once data arrive at CBI, they are imported into an observation database developed on a relational database (Ingres[TM8]). This database contains data for the entire TCOON including the DCPs which comprise RTNET.

The Ingres TCOON database can be queried to extract readings from DCPs which comprise RTNET. The amount of historical data for each environmental sensor can be varied if necessary. For example, temperature data may only be extracted for 24 h and water level data for 96 h. For the RTNET pilot interface, data for all sensors are extracted for a 96-h period which permits generation of 4-day historical graphs of the various parameters. Graphs of historical data give the user an intuitive feeling for rates of change of different parameters and trends which may be occurring in the data. For example, historical water level data can provide an indication of whether water level is currently rising or falling, which may in turn dictate whether particular vessels can take part in the spill response.

The station identifier for each station to be included in RTNET, the identifier, and amount of historical data required for each sensor are stored in separate files. This allows addition or deletion of a station and/or sensor from RTNET as real-time capability and sensors are added to other DCPs in TCOON. In addition, a greater or lesser amount of historical data can be extracted for any particular sensor if a situation dictates. The file created by extraction of data from the TCOON database is transmitted automatically over the Internet to TGLO's Oil Spill Prevention and Response Unit in Austin, TX. This permits TGLO to always have the most recent data for the RTNET.

An advantage of the data path through the CBI network and TCOON database is that data become a subset of a large, well-maintained database of environmental data. August et al. (1990) notes the importance of maintaining historical records of all data layers in a GIS database used for oil spill response. The TCOON databases record prevailing conditions and data used in the decision-making process. Data from the CBI databases can be recalled during analy-

[6] UNIX is a trademark of UNIX System Laboratories, USA.
[7] Hewlett-Packard, Cupertino, CA.
[8] Ingres is a trademark of Ingres Corp., Alameda, CA.

sis of the spill response to determine whether actions taken were appropriate to prevailing conditions.

CONNECTIONS TO THE TGLO

The primary data connection between the CBI and TGLO is over the Internet. In addition, a back-up facility is available which permits connection via modem. This back-up facility would be initiated by TGLO in the event that regular connection over the Internet fails. Figure 2 shows the data path observation platform to TGLO's computers in Austin.

Presently, RTNET data are transmitted to TGLO every 15 min. However, these update times may be varied, if necessary, to occur more, or less, frequently. It is anticipated that under normal circumstances (i.e., there are presently no oil spills) average time delay from initial receipt of a packet of data from a TCOON platform until receipt of these data in Austin will be approximately 10 min. It will be possible to reduce this time to approximately 1 to 2 min if an urgent situation dictates.

The RTNET data file transferred to TGLO consists of a standard text file. From this file, data can be reformatted to allow import into a variety of other applications such as hydrodynamic models which calculate flow fields for oil spill trajectory models. Spill trajectory models rely on weather forecasts and on-site observations to model prevailing conditions over an area and predict spill fate. The Applied Science Associates[9] (ASA) spill trajectory model used in the 1989 *World Prodigy* spill off Rhode Island used 48 h predictions of wind speed and direction from NOAA's National Weather Service. These predictions were updated once or twice each day (Jayko, 1990). Jayko emphasizes the importance of accurate estimates of wind velocity over a spill site to trajectory model predictions of spill fate. Use of real-time TCOON observations from the spill site should aid accurate forecasting of prevailing winds for the immediate future (e.g., 4 to 8 h). The value of these observations in forecasting 48 h in advance is questionable. There is certainly some value in using these data to determine deviations from predicted conditions due to local geographical and meteorological anomalies. These deviations may help spill response managers customize a longer-term area forecast to the unique nature of a spill site.

THE GIS USER INTERFACE

The user interface for RTNET was developed by CBI using Arc Macro Language™ (AML).[10] This interface implemented a variety of graphical displays which permitted panning and zooming to different bay systems along the Texas coast, and also display and graphing of RTNET data.

[9] Applied Science Associates, Narragansett, RI.
[10] AML is a trademark of Environmental Systems Research Institute, Redlands, CA.

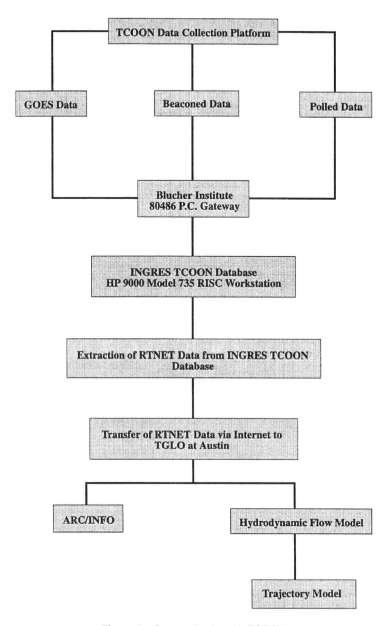

Figure 2 Data paths through TCOON.

Once the RTNET interface is initiated, the outline of the Texas coast is displayed in addition to locations of all RTNET platforms. During initialization of the interface, the current RTNET data file (received via the Internet) is also imported into INFO. The user may then zoom into the appropriate area on the map or select a geographic extent from a predefined list of bay and estuary systems.

Through a series of selection screens, the user is able to choose a particular RTNET platform to query. Once selected, the interface opens a subsidiary window which displays a graph of the preceding 96 h of historical data for each of sensors included on that particular platform. All displays are designed to give the user an intuitive feel for both the type of data displayed and also for navigation around the RTNET map coverages.

FUTURE DEVELOPMENT OF THE RTNET INTERFACE

There are several areas of future development for the RTNET interface. The present historical graphs of RTNET data rely on functions native to ARC/INFO. Development is currently underway to migrate graphing functionality to external X-Window software developed by CBI. This will eliminate use of the INFO database, decrease response time for data updates, and permit hardcopy output to laser printers. Development is also underway on software to predict the next high or low water based on recently collected data. This will help response managers determine what size vessels are able to participate in a clean-up and also for how long.

Development has commenced on a system which will "serve" data to TGLO computers following a request from a "client" RTNET application. An advantage of the client/server approach is that CBI computing resources will only be used when RTNET client applications request data compared with every 15 min as is presently the case. Another advantage of the client/server approach is that TCOON data will be available to other scientific users. These users can be provided with the same client software as the RTNET interface and can use this software to retrieve data from the TCOON database.

USE OF GIS FOR OIL SPILL MANAGEMENT AND RESPONSE

GIS has recently been used in response to oil spill disasters such as *Exxon Valdez* (1989), *World Prodigy* (1989), and *American Trader* (1990). This use has generally been confined to mapping and recordkeeping of data related to these spills (August et al., 1990a; GenaNews, 1990) and represents reaction to the spill, rather than use of GIS as a tool to manage and coordinate the overall spill response. To effectively use GIS for this task requires establishment of extensive databases prior to a spill event. The reader is referred to Chapter 14 for a more detailed discussion of the type of GIS preparation which should be performed.

It is an enormous task to prepare and maintain a large spatial database. The U.S. Coast Guard has considered the establishment of such a database (Jenson, 1990). Many local authorities are also undertaking compilation of such databases as part of resource mapping for coastal zone management. The preparation of comprehensive oil spill contingency plans which reside within a coastal zone management database are an essential part of all coastal zone management programs.

It is a mandatory undertaking if the full functionality of GIS to manage an oil spill response is to be realized.

CONCLUSION

Measurement of prevailing wind conditions and accurately predicting future conditions is important for oil spill trajectory modeling (Jayko, 1990). RTNET will provide valuable data to help predict spill fate in the short term and allow managers remote from the site to be kept fully informed of site conditions as they change.

It is anticipated that RTNET will provide the most up-to-date data for conditions along the Texas coast in the event of an oil spill. It is hoped that in conjunction with a comprehensive GIS based response strategy, RTNET will provide valuable and timely input regarding the management of an oil spill response and help reduce the impact of an oil spill on our fragile environment.

ACKNOWLEDGMENTS

Support for this paper has been provided by the Texas General Land Office through contract No. 92-170R and the Conrad Blucher Institute for Surveying and Science, Texas A&M University, Corpus Christi.

REFERENCES

August, P., Hale, S., Bishop, E., and Sheffer, E., GIS and environmental disaster management: *World Prodigy* oil spill, *Proceedings: Oil Spills: Management and Legislative Implications*, Spaulding M. L. and Reed, M., Eds., American Society of Civil Engineers, New York, 1990.

Butler, H. L., Chapman, R. S., Johnson, B. H., and Lower, L. J., Spill management strategy for the Chesapeake Bay, *Proceedings: Oil Spills: Management and Legislative Implications*, Spaulding M. L. and Reed, M., Eds., American Society of Civil Engineers, New York, 1990.

Genamap helps battle oil spill, GenaNews, Genasys II Inc., 1(1), 3, 1990.

Jayko, K., Predicting the movement of the *World Prodigy* spill, *Proceedings: Oil Spills: Management and Legislative Implications*, Spaulding M. L. and Reed, M., Eds, American Society of Civil Engineers, New York, 1990.

Jeffress, G. A., Next generation water level measurement for the Texas coast, *Proceedings: The Second Australasian Hydrographic Symposium*, The Hydrographic Society, 1991.

Jenson, D. S., Coast Guard oil spill response research and development, *Mar. Technol. Soc. J.* 24(4), 1990.

Liebermann, T. D. and Ciegler, J. C., Interactive display and query of near real-time data from a hydrologic alert network, Proceedings: 12th Annual ESRI User Conference, ESRI, 1992.

Section IV

Additional Applications and Background

CHAPTER 17

Introduction to Other
Issues and Applications

John G. Lyon and Jack McCarthy

In regional wetland analyses, satellite or high altitude aerial photographic data sets provide a valuable resource or focal point for data analyses. They represent the "base map" coverage, and a start for research efforts. Naturally, it is desirable to use a variety of other data sources in the analysis of remote sensor data. In particular, other remote sensor data of similar scale or of detail can supply important information. These sources should be identified and used in analyses as researchers see fit.

Several of the following chapters present and make use of remote sensor data. These remote sensing products or results, along with one or two characteristics measured in the field or laboratory, can supply good spatial information for modeling. The combination of different technologies can make the best input and data analysis for the evaluation of wetland resources.

Over the years, many types of remote sensor data sources have been identified as being beneficial in analyses of land and water resources, and have been shown to be compatible with other types of data products. For example, current and historical aerial photographs are available from archive. They can be used to evaluate land and water resource conditions, and to do so over space and over time. Also available are land cover maps such as Nation Wetland Inventory maps, which are available over large areas of the continental United States. Several of these data sources can supply a wealth of information. They are very amenable to GIS analyses and to the creation of layers in GIS databases. A number of these sources and their application are made in the chapter presented here. These contributions are discussed at the end of this introduction to this section.

Currently, there are a variety of data sources that can be used to support GIS and remote sensor database production and analyses. Most of the variable contributions are discussed below. The objective is to alert the reader to opportunities so that they can use these sources of data for their work. Later chapters in this section present actual uses of a number of these data sources in GIS analyses.

NORTH AMERICAN LANDSCAPE CHARACTERIZATION (NALC) DATA

The NALC data products are valuable as additional or ancillary data products to support a variety of activities in an given project. It is necessary to use other data sets in evaluations for a variety of reasons. These satellite sensor data sets are essentially independent measurements and can be used to calibrate TM data sets or other data sets that are involved in the project. They may also be used to check the accuracy of TM results and to fulfill a data accuracy assessment or validation that is so important to the utility of model results on an given basis.

A valuable feature of NALC data sets are the three multispectral scanner (MSS) images. Each example was selected from the archive during a certain period and hopefully during the same or similar season. Data were acquired either recently for this project, or were taken from archives. The triplicate consists of an image from the 1990s, as well as from the 1980s and 1970s. This consistent production of coverage for the continental United States provides the data input for evaluations of change over time in the MRLC program.

The NALC products can provide data for local and regional evaluations of change. In this fashion, NALC can support the research goals of MRLC and EMAP, and cooperating groups. Where there is a need for change detection in land covers, or the addition of digital elevation model (DEM) data for analysis, NALC data sets will be of great service.

The NALC products or triplicates are available at nominal cost from user services at the U.S. Geological Survey (USGS), EROS Data Center, Sioux Falls, SD. Users can also learn about NALC by browsing the EDC program GLIS over internet. Written publications include: Lunetta, R., J. Sturdevant, and J. Lyon, 1993. North American Landscape Characterization (NALC), Research Brief. US Environmental Protection Agency EPA/600/S-93/0005, 8 p., or Lunetta, R., Lyon, J., Worthy, D., Sturdevant, R., Dwyer, J., Yuan, D., Elvidge C., and Fenstermaker, L., North American Landscape Characterization (NALC), Technical Plan. US Environmental Protection Agency EPA 600/R-93/135, EMSL, TS-AMD, Las Vegas, NV, 1993, 417 pp.

ADVANCED VERY HIGH-RESOLUTION RADIOMETER (AVHRR) DATA

AVHRR data products and enhanced products such as normalized vegetation index (NDVI) images can be very useful. AVHRR data are available from on-going USGS and cooperator programs of data processing, analyses, and distribution. Data are available in the United States on a weekly or bi-weekly basis, and can be supplied to users on CD-ROM and other standard media to greatly ease the input of these data for analyses.

AVHRR data can be valuable because it supplies additional data with some similar and some different characteristics as compared to Landsat sensor products. In the analysis of TM images, the AVHRR time series can identify the general trend of phenology or seasonality of vegetation for a given year. AVHRR data also can provide the link to help scale images and provide analyses going from the regional or TM "scale" to the continental and global scale.

AVHRR enhanced products are available from the National Oceanic and Atmospheric Administration (NOAA) or the USGS. Contact user services at the

USGS, EROS Data Center, Sioux Falls, SD. The metadata browse program GLIS can also be used to evaluate data sources and to determine whether they are available for a given time and location.

HIGH-ALTITUDE PHOTOGRAPHS

High-altitude photographs have been acquired under many governmental programs over many years. In particular, the USGS program and the NASA program of high altitude photography have supplied a large amount of photos of several different emulsion types. These photos can be very useful in the analysis of TM scenes because of their small scale. The photos are at scales from 1:40,000 to 1:65,000. The relative scale between the TM image products or images can be displayed on a monitor and in these photos is the same. The similar scales and display of earth features can be a great help in interpretation of land cover and landscape characteristics. High-altitude photographs can be obtained from the USGS at the address mentioned above.

OTHER PHOTOGRAPHIC PRODUCTS AND VIDEO

The Agricultural Stabilization and Conservation Service (ASCS) conducts a number of aerial photography programs. The most interesting is an ASCS program developed for evaluating compliance with crop production program. The program involves the acquisition of small format (35 mm) color aerial photographs of agricultural areas. The coverage is performed once during the growing season. This coverage is valuable in other programs such as TM analysis to help evaluate vegetation presence and condition. The photographs are archived at the county level and are available for use since the early 1980s.

Other photography programs provide a great deal of coverage each year and supply material for archives back to 1945. The ASCS also archives and distributes photographs from its center in Salt Lake City, UT. Among these archives are ASCS photos, Soil Conservation Service (SCS) photos, and other U.S. Department of Agriculture (USDA) agency photos including those from the U.S. Forest Service. To determine the available coverage for an area of interest, contact the Aerial Photography Field Office at 2222 West 2300 South, POB 30010, Salt Lake City, UT 84130-0010.

The other big governmental archive and source of photographs is the holdings of the USGS at the EROS Data Center in Sioux Falls, SD. Mapping photographs used by the USGS are archived there and copies are available for sale. Other photographs such as those of the USGS and NASA high altitude photography programs are available for purchase. The address is mentioned above.

Video products have proved very useful in a number of projects. These projects included the U.S. Fish and Wildlife Service GAP projects, U.S. Environmental Protection Agency Environmental Mapping and Assessment Program projects, NASA projects, and in several other research efforts. These data are gathered on a custom basis, and no national, organized program exists. Nonetheless, if data are available or can be acquired, they should be used in analyses.

DIGITAL ORTHOPHOTO QUADRANGLE PRODUCTS

A new, exciting program of the USGS is the creation of orthophoto quadrangles in digital form. Where available, these data can be used just like an aerial photograph. They can be interpreted like a photograph and have very good resolution like a photograph. An added advantage is that it can be enhanced digitally using a variety of electrical engineering technologies. One can utilize these products as a valuable, digital adjunct source of data. These products will be available from private aerial mapping companies, or from the USGS. Sources can be found at the EROS Data Center or call 1-800-USA-MAPS.

MAP PRODUCTS

A number of paper and digital map products exist. They naturally provide a uniform information base to obtain general information. They also may be out of date in some map themes, due to the growth of the nation and the pace of map revision. In particular, the use of digital elevation model (DEM) or digital line graph (DLG) can be valuable in analyses or in the creation of high quality, informative products. DEM products present point elevations. They can be enhanced with software to display slope and aspect characteristics. Likewise, shade relief products can be made to display relative topography. DLG products present planimetric or cultural details. Currently, many DLG products are available which display the transportation and drainage patterns of selected quadrangle areas in the United States.

DEM products are distributed for each NALC triplicate. The ground location of a MSS NALC image or a TM image is approximately the same. The DEM data from NALC products should be useful for TM analyses because of the same local coverage, and the individual DEM may be identified directly from the NALC triplicate product. These products are all available from the USGS.

RECORDS, LITERATURE, AND STATISTICS

The sharing of records and statistics between cooperating agencies is a goal the industry is striving for. Although the utility of such information can be variable, it is clear that the cost and ease of access are very good. The GAP program really demonstrates the value of records in the incorporation of animal-plant distributions from historical records. In the future we will see more advances in the use of historical, qualitative records in our evaluation of land covers and habitat questions.

Literature sources may provide useful information for various project activities. Certainly, the value of general and regional sources of data can seen in analyses of satellite data. Certainly, the use of USDA Soil Survey documents and the U.S. Fish and Wildlife Service National Wetland Inventory (NWI) maps can provide a vital source of ancillary data for analyses of TM scenes.

FIELD SAMPLING DATA SOURCES

The collection of high quality field sampling data is an important goal of any project or program. Field sampling is vital to analyses, yet is often the source of problems due to the expensive efforts in terms of personnel and costs.

Cooperations between groups can extend the utility of a given field work campaign. These cooperations, including field data collection work sheets, are discussed elsewhere in the document. Naturally, the user is concerned with supplying field data that will support the overall efforts. Several projects have demonstrated how judicious collection of the appropriate field data is a great boost to any project.

GIS and remote sensor technologies supply part of the information that is usually required to meet the goals and objectives. Field sampling is the important effort that assists analysts in their evaluations of the data, and it also may supply the data necessary for conducting an accuracy assessment of the products.

WETLAND AND ENVIRONMENTAL APPLICATIONS

The following chapters further examine the use of GIS and the supporting role of other technologies. In particular, remote sensor data can be a vital data source for GIS. It may also supply independent measurements of resources that can provide valuable information. Chapter 18 demonstrates some of the value of remote sensor inputs to GIS.

GIS and remote sensor data can greatly facilitate modeling. The level of success depends on a number of factors. As mentioned, GIS information can be input directly to the model as primary data. GIS results can be used to estimate the value of model coefficient, thereby making the coefficient more realistic as compared to nature. GIS results can be compared with model results as their independent check or verification.

This section has several chapters that address the capabilities of GIS and related technologies. Chapters 19 and 21 both show how complex questions can be addressed by using a variety of protocols and sources of data. These applications provide a good idea as to the level of detail that can be provided when using a number of GIS and other technologies. A number of errors can be associated with the use of GIS. Chapter address many of these issues and provide background on their characteristics. A glossary is also presented to help with learning and to supply a reference for the professional.

CHAPTER 18

Wetlands Detection Methods

K. H. Lee and R. S. Lunetta

ABSTRACT

The purpose of this investigation was to research and document the application of remote sensing technology for wetlands detection. Various sensors and platforms are evaluated for: (1) suitability to monitor specific wetland systems; (2) effectiveness of detailing wetland extent and capability to monitor changes; and (3) relative cost-benefits of implementing and updating wetlands databases.

The environment to be monitored consists of physiographic and ecological wetland resources affected directly or indirectly by anthropogenic activity. Aircraft and satellite remote sensing can be used to record and assess the condition of these resources. Monitoring of environmental conditions is based on the observation and interpretation of certain landscape features. Although some forms of monitoring are continuous, resource monitoring from aircraft and satellite platforms is periodic in nature, with change being documented through a series of observations over a given span of time.

This report summarizes the findings of a bibliographic search on the methods used to inventory and/or detect changes in wetland environments. The bibliography contains numerous citations and is not intended to be all-inclusive. Books, major journal and symposium proceedings were examined. The findings documented will provide the potential user with a basic understanding of remote sensing technology as it is applied to wetland monitoring and trend analysis.

INTRODUCTION

There are two primary types of remote sensing projects relative to wetlands. The first, resource mapping, involves acquisition of baseline data on type, extent, and health of wetland communities. The second involves detection of change, either natural or anthropogenic, in those communities.

Most of the potential users of this data are responsible for the management and protection of the wetland resources within their respective states. To accomplish this, various types of data are required for decision making. In considering use of remote sensing sensors and systems to provide such data, sev-

An earlier version of this chapter was published as Lee, K., Wetland Detection Methods Investigation, U.S. Environmental Protection Agency, Report No. 600/4–91/014, Las Vegas, NV, 1991.

eral factors must be considered. What levels of precision and accuracy are required in identifying and measuring wetland resources? How much "leeway" will be tolerated in the correct identification of wetland communities? How "close" can the estimation of areal extent of specific wetland types be when compared to actual acreage? This factor of precision must be viewed from an economic standpoint, i.e., what is the relative cost of each system vs. its degree of precision?

Wetlands are dynamic ecosystems which are defined by federal regulating agencies as possessing three essential characteristics: (1) hydrophytic vegetation, (2) hydric soils, and (3) wetland hydrology, which are the driving force creating all wetlands (Federal Interagency Committee for Wetland Delineation, 1989). These ecosystems are difficult to quantify because of complex interactions between these parameters. Water level fluctuations and diverse geographic settings make the many types of wetlands difficult to characterize as a group. Changes occur in wetlands and along boundaries in response to variations in the hydrologic cycle in which seasonal, annual, and long-term fluctuations are the driving factors. Wetlands are also changed, altered, and encroached upon by human activity, such as urban and agricultural developments, stream channel alterations, and draining and damming activities.

Remote Sensing of Wetlands

When wetlands are to be inventoried several issues need to be addressed about the method in which the baseline or change data should be acquired, categorized, verified, integrated, stored, and distributed. Because wetlands generally have poor accessibility due to uneven and unstable terrain and frequently tall vegetation, any field work undertaken to inventory them is usually expensive, time consuming, and sometimes inaccurate in location. Reducing the amount of field work via remote sensing (aerial photography, aircraft or satellite multispectral scanners) is a viable solution. Baseline inventorying should consider climatological influences, and generally should have a multiseasonal approach to capture all the inherent variances of plant phenology. If wetland maps are based on persistence and extent of surface water alone, their boundaries will vary seasonally. Change detection and trend analysis requires that ecological information be collected under optimum conditions so valid comparisons can be made between several points in time. The availability of current and historical climatological and image data are relevant in the wetland inventory planning process. Roller (1977) lists the advantages to monitoring wetlands with remotely sensed data over conventional field surveys as: (1) economy, (2) timeliness, (3) favorable viewing perspective, (4) synoptic observation, and (5) permanent graphic records. It should be added that with the advent of Geographic Information Systems (GIS), land cover/land use data layers derived from aircraft or satellite-mounted multispectral sensors or digitally encoded aerial photographs can be incorporated fairly easily since their formats are digital. Furthermore, remotely sensed data collected for wetland change or monitoring activities may be utilized in future projects not under consideration at the time of acquisition. With more sophisticated digital image processing techniques available, imagery can be enhanced to answer other environmental questions. Limitations of monitoring wetlands with remotely sensed data are evident when detailed environmental information is required, such as a complete floristic make-up of a plant community, or if controversial regulatory decisions are to be made regarding boundary delineation. Generally there are requirements for field verification (reference data) in order to attach reliability or confidence estimates to the inventory. Logistic difficulties can arise because ideally ground reference data collection should coincide with the acquisition of the remotely sensed data.

When planning wetland inventories or change detection studies which will utilize remotely sensed data, a classification system must be selected or derived which will satisfy the users' requirements. Many times the resulting wetland information derived may not be directly compatible or comparable to previously used classification systems or available reference data. Therefore, the end-user must decide the inventory level of detail (e.g., vegetal components, minimum mapping unit, classification scheme, etc.) and whether multistage and/or multi-temporal sampling is required. Relationships that exist between the desired mapping parameters and pre-existing classifications may form the basis for the final classification scheme. If the classification system is hierarchical, it can be amended when information is attained at a higher level of detail. For instance, satellite data could be used to map an entire state and/or region, small scale photographs for regions and counties, large scale photographs for counties and townships, and hand-held oblique photos from light aircraft or helicopter, with ground transects or plot sampling for individual wetland systems (Roller, 1977).

Howland (1980) found that differentiation between vegetation signatures was best in the longer wavelengths (red through infrared) for the broad variety of wetlands occurring in northern New England. This phenomenon is attributed to the green biomass of vegetation which uniquely absorbs visible light and strongly reflects near infrared (IR). Chlorophyll absorption in the visible spectrum and multiple scattering in the near IR due to internal plant structure allows for the characterization of vegetative communities based on spectral patterns. In general, vegetation canopies have bidirectional reflectance, but patterns can be further influenced by the understory component, leaf structures, density of vegetation, etc. Furthermore, water absorbs a majority of the infrared energy which results in a sharp contrast between vegetation and water in the IR wavelengths.

Visible wavelengths are more sensitive to atmospheric haze and particle scattering, although the blue-green spectrum can penetrate water for short distances. When submergent vegetation is being targeted, collection of remotely sensed data in the visible wavelengths can provide superior results under the right atmospheric and hydraulic conditions.

Schloesser et al. (1985) found that low-altitude color aerial photography (1:6000 scale) with limited ground survey information, could economically identify beds of submerged macrophytes in the St. Clair-Detroit River system. Conditions of minimal cloud cover with acquisition time at approximately solar noon allowed maximum solar illumination and water penetration with minimal sun glint. Five macrophyte genera and a sand substrate category were delineated on the photos with a overall accuracy of 68% by six individual interpreters of varying experience. The authors suggest that improved accuracy could result from a more detailed dichotomous key, and greater use of biological experience in developing the key.

Classification Systems

There are a variety of classification systems used which are generally tailored to satisfy the needs of resource managers who must monitor and regulate wetlands under specific mandates. The result of different systems is maps and inventory products which are not mutually compatible, i.e., different scales, media formats, minimum mapping units, and classification schemes. Moreover, all classification systems are not designed to incorporate the results of digital remote sensing, thereby requiring modifications and/or ancillary data to be incorporated into the inventory to make them complete or compatible.

In 1979, the U.S. Fish and Wildlife Service (USFWS) published a nationwide classification system (Cowardin et al., 1979) for use in the national wetlands inventory (NWI). The Cowardin classification strategy utilizes a hierarchial approach where "Systems" form the highest level. Five systems are defined as marine, estuarine, riverine, lacustrine, and palustrine. Marine and estuarine each have two subsystems—subtidal and intertidal; the riverine has four subsystems—tidal, lower perennial, upper perennial, and intermittent; the lacustrine has two—littoral and limnetic; and the palustrine has no subsystems (Cowardin et al., 1979). Subsystems contain classes which are based on substrate material, water regime, or vegetative type, and can be further defined within subclass. Dominance type is a level subordinate to subclass and describes the dominant plant or animal species. Special modifiers can also be identified, such as water regime or water chemistry modifiers, and others which describe animal or human modifications to the wetland. This FWS classification is designed to allow: (1) ecological units which contain homogenous attributes to be described and grouped; (2) organization of these units to facilitate resource management decisions; (3) consistent inventory and mapping of wetlands on a national level; and (4) uniformity in concepts and terminology (Cowardin et al., 1979). The Cowardin system replaced the previous system employed by FWS in their 1955 nationwide wetland inventory which was designed to emphasize the value of wetlands for general wildlife and waterfowl habit, i.e., Circular 39 (Martin et al., 1953; Shaw and Fredine, 1956).

The accuracy of the NWI has never been tested for the nation as a whole, but a few assessments have been done which compare the NWI classification to localized wetland mapping results. Problems arise when comparisons are made between digitally processed land cover/land use data and a classification system like Cowardin which is designed for use with medium-to-large scale aerial photography. Only generalized classes can be compared when utilizing digital classifications of satellite or aircraft based scanners for comparisons with existing NWI data. An overall identification accuracy of 95% was determined by Swartwout et al. (1981) for the NWI in Massachusetts. Hardin (1985) found similar accuracies for the NWI in Delaware. Luman (1990) found good correspondence in Illinois when the NWI information was regrouped and then compared to satellite-based classifications. Pickus (1990) also found high correspondence (90%) when comparing NWI digital data which was re-grouped as wetland vs. nonwetland, to a satellite-based land cover classification in southeast Louisiana.

There is also a classification system developed by the U.S. Geological Survey (USGS) which is a multipurpose land use/land cover system designed for remotely sensed data (Anderson et al., 1976). This system is also hierarchial with a framework of nine general Level 1 categories that are further subdivided into 37 Level 2 categories. Higher detailed categories can be designed into the system to suit the user at the third and fourth levels. Wetland is one of nine Level 1 categories, with a subdivision at the Level 2 for forested vs. nonforested wetlands. The flexibility to either generalize or specify species at higher levels allows inventory compatibility when the remotely sensed or ancillary inventory data obtained is not described to the same level of detail. The higher levels (i.e., 3 and 4) are not usually attainable strictly from remotely sensed data, and normally require ancillary information.

AERIAL PHOTOGRAPHY APPLICATIONS

Photo-Interpretive Process

The interpretation of aerial photographs to accurately map specific wetland vegetation and boundary information is the most common method used by federal, state, and local agencies. Photo analysis involves the identification and delineation of specific features recognizable by their distinct "signatures" (combinations of image characteristics). These characteristics include: tone, texture, shape, size, shadow height, and spatial relationship. Additionally, examination of the aerial photos stereoscopically enables the interpreter to observe the vertical as well as the horizontal spatial relationships of the subject features. Due to the complexity of the interpretative process and the wealth of data within aerial photos, accurate photo interpretation requires considerable expertise. The accuracy of the interpretations depends on the quality of photography, the experience of the photo-analyst with specific wetland settings, and the amount of ground truth verification to be conducted.

Photographic systems acquire spectral information with films of various spectral sensitivities. In order to maximize the photo-interpretation result, it is important to select a film type which will provide maximum contrast between different plant communities. Choices available for camera systems are color, color infrared, and panchromatic (i.e., black/white) films. Table 1 lists the spectral region and season recommended by Roller (1977) to collect aerial photography

Table 1 Spectral Regions for Wetland Vegetation Discrimination

Vegetable type	Season of maximum contrast	Spectral region
Aquatic submergent		
shallow depths	Summer	Near infrared
deeper depths	Fall	Visible
Floating	Summer	Near infrared
Marsh emergent and meadow	Fall	Visible
Shrubs	Summer or Fall	Visible or near infrared
Trees	Fall	Visible

From Roller, N.E.G., Remote Sensing of Wetlands, Environmental Research Institute of Michigan (NASA-CR-153282), Ann Arbor, MI, 1977, 165 pp.

that would take advantage of phenological differences among common wetland vegetation.

Small-Scale Photography

The most notable resource mapping effort is that of the NWI conducted by the USFWS. Begun in 1977, it creates in map form a database of the nation's wetlands and deepwater habitats. Updates of specific areas are also conducted to document wetland gains or losses. NWI products have included detailed maps, wetland soils and plant lists, and reports on specific regions or states. Color-infrared photography (1:58,000 scale) was acquired nationwide from 1981 to 1984 by the USGS under the National High Altitude Aerial Photography Program (NHAP). Photo interpretations derived from NHAP photography for the NWI were categorized by dominant vegetation and hydrologic characteristics according to the USFWS classification system. The NWI identifies and maps wetlands to the smallest acreage visible on the photographs being used (Carter, 1982). The resulting NWI maps were produced using USGS 1:24,000 scale topographic maps as a base. These topographic quadrangles (when available) are used as base maps because they provide a geometrically accurate starting point on which to compile more recent photo-interpreted information. Standard USGS topographic maps depict vegetative categories by unbounded symbols which makes them useful only for gross generalizations. The smallest area that can be depicted on a 1:24,000 scale map is about 0.4 ha (0.9 acre). Starting in 1988, the USGS has initiated another nationwide acquisition of high flight CIR photography. The National Aerial Photography Program (NAPP) will collect 1:40,000 scale CIR, leaf-on, quarter quad-centered photography on a state-by-state basis.

Small-scale CIR photography other than NHAP and NAPP have been utilized to map wetlands. Carter et al. (1979) found that a 1:130,000 scale photograph had a recognizable or interpretable limit of approximately 0.5 ha (1.2 acre). This fact underscores the need to define the level of detail the wetland inventory will require based on its ultimate use. Wetland mapping projects are usually individually designed programs geared to satisfy either legal requirements for regulation, the needs of the mapping agency, or organizations conducting environmental assessments. For most regulatory permitting procedures and site specific planning activities, accurate boundary delineation is critical. The use of small scale photography (i.e., 130,000 scale or smaller) is primarily limited to the identification of plant communities with distinct spectral characteristics.

Medium- to Large-Scale Photography

Specific locations and vegetation types are easily identifiable on large-scale photography, even by photo users with minimal experience. Very-large-scale photographs are also useful in investigations of wetlands that are small, isolated, or narrow and linear. These characteristics, especially when combined, make these type wetlands virtually impossible to detect with small-scale photography.

Aerial Photo Previous Work

In a study reported by Howland (1980), color infrared (CIR) film was found to be superior to color and multiband black and white photography for vegetation discrimination in a diverse inland wetland in Vermont. Besides film choice, attention must also be given to the time of year and even the time of day when acquiring imagery. Grace (1985) found that for inland nonpersistent emergents in the southeastern United States, mid-August to mid-September was the best time to collect photography for annual comparisons since most species would be represented and in a mature state. A study by Carter et al. (1979) found that the time period between August to mid-October was optimum for identifying open water and marsh categories in Tennessee. In another study by Gammon and Carter (1979) of the Great Dismal Swamp, it was found that photographs obtained during vegetation dormancy allowed for better identification of wetland boundaries, areas covered by water, drainage patterns, separation of coniferous forest from deciduous, and classification of some understory components. In tidal environments, Brown (1978) found August to early October best for saline wetlands, late June to early September best for fresh/brackish wetlands, and that fresh water wetlands needed coverage twice, mid-to-late June and late August due to specific plant species being predominant at different times.

Wetland resource mapping projects have been performed for advanced wetlands identification projects at EPA's Environmental Monitoring Systems Laboratory (EMSL) in Las Vegas, and the Photo Interpretation Center (EPIC) located in Vint Hill Farms, VA (Norton, 1986a, b; Duggan, 1983; Williams, 1983; Mack, 1980). Hydrogeologic and ecologic evaluation methods are applied in field work to assist in advanced wetlands mapping efforts. Utilizing this type of information allows managers to take a "proactive" stance relative to jurisdictional determinations and suitability of land parcels for development, rather than obtaining data on a case by case basis. Other detailed resource mapping is conducted for EPA's Regional Offices for wetland areas of special or critical concern (Norton et al., 1985).

The following list of identifiable features is provided to acquaint the reader with data available from such projects. A nominal scale of 1:12,000 to 1:24,000 and use of color/CIR photos is assumed.

1. Wetland boundary delineation
2. Area
3. Edge, drainage length/densities
4. Shape of upland/wetland edge
5. Fetch/exposure
6. Vegetation growth form
7. Cover density and distribution
8. Species composition and health
9. Tidal flooding regime
10. Tidal conduits, inlets, outlets
11. Erosion

Change Detection with Aerial Photography

Change detection projects which utilize aerial photography involve the use of one to a number of historical sets of photographs to document changes either natural or anthropogenic (Williams, 1985, 1989; Grace, 1985; Niedzwiedz and Batie, 1984; Hardisky and Klemas, 1983). Listed below are features that are useful in change detection analyses which are interpretable from color/CIR photography with scales ranging from 1:10,000 to 1:24,000.

1. Predevelopment vegetation patterns
2. Natural vegetation removal
3. Surface drainage network
4. Sediment
5. Dredging, turbid water
6. Fill or spoil deposition
7. Erosion
8. Ground staining, scarring
9. Vegetation stress or damage
10. Contaminant sources

Although there is no "average" change detection project, a description follows of some requirements and products which are typical. Federal agencies charged with enforcement of Section 404 of the Clean Water Act, which authorizes permits for the discharge of dredge or fill materials into the waters of the United States, extensively utilize aerial photointerpretive data to document alleged illegal development of wetland properties. Many times chronological documentation must be performed so that the illegal filling of wetlands can be substantiated. Regulatory enforcement related projects typically involve numerous historical photographs with detailed comparative analysis, interpreted overlays, field verification of interpreted results, and very large scale graphic displays. Therefore, costs for in-depth wetland delineation are substantially higher than for conventional wetland resource mapping efforts. Costs to produce this type of detail, which may be required in court cases, have averaged approximately $1,000 per square mile (D.R. Williams, LESC, personal communication, 1990). In other instances, investigations are aimed at quantifying the loss of wetland resources over time by climatological effects. D.C. Williams and Lyon (undated) correlated changing Great Lakes water levels to wetland extent, and Grace (1985) investigated the effect of thermal discharges on seasonal wetland production.

A state agency in Georgia acquires aerial photography at predetermined intervals to monitor development along the Atlantic coastline (S. Stevins, Georgia Dept. of Natural Resources, personal communication, 1990). The U.S. Army Corps of Engineers performs similar monitoring in most of their districts, as they are responsible for making jurisdictional determinations of wetlands regulated under Section 404 of the Clean Water Act. The progressive destruction of wetlands in association with hazardous waste sites is also documented, and this information is used to plan clean-up activities at Comprehensive Environmental Response Compensation, and Liability Act of 1980 (CERCLA) sites (Norton and Prince, 1985).

At an agricultural conversion site in Florida seven sets of available historical photos, black-and-white and color photographs with scales ranging from 1:10,000 to 1:24,000, were acquired and interpreted. The purpose of the investigation was to assess and document the type and amount of wetland habitat destruction (Williams, 1981). Natural vegetation present before conversion, vegetation removal, and surface drainage network construction were documented over time. Ground information was collected prior to the photo analysis, and comparisons conducted to confirm map accuracy after completion. Frequently, photography is flown at periodic intervals to ascertain if a developer is in compliance with Federal and/or State issued permits. If such investigations become court cases, large courtroom displays of interpreted data are prepared with expert witness testimony usually given by the principal investigator or experienced photo-interpreter.

SATELLITE- OR AIRCRAFT-BASED SENSOR SYSTEMS

Digital Image Processing

Digital image processing of satellite or aircraft-acquired data into land cover categories involves the examination of the reflectance or spectral patterns of pixels contained within the image. Image data is organized into a matrix (i.e., raster format), with each pixel (or cell) covering a certain dimension on the ground. Each pixel contains one data value representing the spectral intensity of that location on the earth's surface for a particular wavelength (i.e., 7 wavelength bands, 7 intensity values). Fundamental energy-matter interactions control and influence the spectral characteristics of land cover types. The proportion of energy reflected, absorbed, and transmitted for each cover type varies depending on the material and ground conditions (Lillesand and Kiefer, 1987). Most of the earth's features have unique spectral characteristics, and thereby can be identified and mapped on the basis of their spectral signatures.

Pixels commonly correspond to more than one type of land cover and therefore the pixel value represents a weighted average. The result is that each raster pixel is a combined product of all the resources present at that ground location. Presence of "mixed pixels" causes problems when boundary or edge information is required. Boundary effects can take the form of mixed pixels occurring as transition areas between adjacent cover types. The spectral characteristics of those mixed pixels are unlike the land-cover types on either side, and depending on the classification technique utilized to process the digital data, may be incorrectly grouped as a separate class. Other possible sources of error or inconsistencies when dealing with satellite or aircraft data originate from atmospheric scattering, sun angle, topographic influences, and the process of geographical rectification of the imagery. The scattering, sun angle, and topography can be modeled or accounted for by assumptions and/or ancillary data, while the rectification process has a reportable positional error once the geometric transformation is completed.

In digital image processing, data can be analyzed in several ways, most commonly with either image enhancement or image classification algorithms. In order to summarize the remotely sensed data into land cover/land use, those processes which are addressed as classification techniques are employed. The general approach in the classification of digital imagery fall into two categories; supervised or unsupervised approaches. In a supervised classification, known land cover types such as forest, agriculture, wetland, and urban have been identified either through field work, aerial photography, maps, or personal experience (Heaslip, 1975). These sites are then located in the image and homogeneous examples are delineated. These delineated areas serve as training sites because their spectral characteristics are related to known resources, and can be used to "train" the classification algorithm (Jensen, 1986). Statistical information is generated for each identified training cover type and subsequently that information is used to classify all unknown pixels remaining in the image. Conversely, in unsupervised classifications, the location of particular land cover types is not known, and the statistical parameters required to define the training classes are then determined with clustering algorithms. It is then the responsibility of the analyst to label these clusters with the appropriate land cover category.

Evaluation of the land-cover classification accuracy requires comparison between the classification results and reference data for the area. This reference data can be taken from various sources such as existing specialized maps, photo-interpreted aerial photographs, and actual "field checks." However, the reference data which is utilized during classification should not be used in the accuracy assessment because it heavily biases the results. Congalton (1988) compared several sampling schemes for assessing the accuracy of classified remotely sensed data, and found that generally a 1% sample should be obtained. To assess the agreement between the classification and the reference data, site-specific comparison are made by calculating the frequency of coincident classes, point by point, on the reference data and the classification results. These values are reported in an error matrix (sometimes called a confusion matrix or contingency table). The total overall percent correct is the ratio of the sum of diagonal values to total number of cell counts in the matrix. Proportions of diagonal values to row sums are the category accuracy relative to errors of commission, and proportions of diagonal values to column sums as category accuracy relative to errors of omission (Story and Congalton, 1986). Detailed statements of accuracy are derived from the error matrix in the form of individual land-cover category accuracies. For each class, percent commission and percent omission are calculated from the error matrix as well as the overall accuracy.

Change Detection with Digital Imagery

Change detection projects which utilize digital imagery are generally more limited in the amount and type of historical data available. Digital change detection tasks are somewhat complex and require consistent multiple date imagery acquisition to ensure reliable mapping of baseline conditions and change trends. Several different image processing methods can be employed for change detection studies. These methods (i.e., image differencing, image ratioing, classification comparison, comparison of pre-processed imagery, and vector change analysis) are specific to the form in which the imagery is obtained, and to what precision the two images can be co-registered (Jensen, 1986). It should be noted that depending on whether the imagery used is "raw" or unclassified, or the result of classification or transformation algorithms, the different methods of digital change analysis will produce slightly different information. For example, image differencing results in an image (or map) that reveals only the locations that have undergone change, and is directly dependent on the spatial registration between the images. Conversely, comparisons between two classified images depict both the changed area locations and the nature of the change (Howarth and Wickware, 1981). The accuracy of this type of derived change detection map, and the statistics generated from it, are contingent on both the spatial registration between the maps and the individual classification accuracies of the maps (Hodgson et al., 1988). Furthermore, the composite map produced from digital change analysis will not have an accuracy greater than the least accurate map in the analysis (Newcomer and Szajgin, 1984).

Digital change detection offers an alternative to manual photo-interpretation, but can be more difficult to perform accurately due to the spatial, spectral, and temporal constraints placed on multidate imagery analysis. The results from photo interpretation of large-scale photographs may have higher accuracies, but are time consuming, difficult to replicate, prone to errors of omission, and costly in terms of data acquisition. Digital change detection studies must utilize analysts familiar with the environment to be inventoried, the quality/accuracy of the data set(s), and the characteristics of change detection algorithms in order to be successful (Jensen, 1986).

Satellite Sensors

There are two primary sources for obtaining satellite acquired digital imagery: Earth Observation Satellite Company (EOSAT) and Systeme Pour l'Observation de la Terre (SPOT) Corporation. EOSAT is a participant in a cooperative agreement between the National Oceanic and Atmospheric Administration (NOAA) and the USGS to facilitate the commercialization of satellite based technology to the private sector. At present there are two satellites being operated by EOSAT, Landsat 4 and 5. The Landsat systems carry two types of sensors: the multispectral scanner (MSS) and the thematic mapper (TM). Both the Landsat 4 and 5 satellites have exceeded their design lives by 3 years with Landsat 5 being the primary system utilized in North America. The MSS provides data in four spectral bands: green, red, and two reflected infrared bands at about 80-m ground resolution. The TM has six spectral bands with 30-m resolution and a thermal infrared band with 120-m resolution. The Landsat TM sensor possesses higher spectral resolution because of the seven individual bands which capture reflectance data in the visible through the thermal portions of the electromagnetic spectrum. These satellites are in polar orbits and are capable of collecting spectral information every 16 days over the same surface area.

France launched the first SPOT satellite in February 1986. This satellite has sensors on board which are capable of producing 10-m resolution black-and-white panchromatic images, and 20-m resolution three band multispectral images. SPOT sensors can be directed off-nadir to produce stereoscopic coverage and have a repeat collection cycle of 26 days. The spatial resolution of the SPOT satellite is finer than Landsat TM, but the lessor spectral resolution and smaller area per scene of SPOT imagery make comparisons difficult between the two sensors. The two systems could be considered complimentary in that the SPOT imagery has been merged with Landsat™ imagery, with resulting hybrid imagery that reflects the spatial resolution of SPOT with the spectral resolution of Landsat™ (Salvaggio and Szemkow, 1989; Welch and Ehlers, 1987).

Landsat MSS Previous Work

Landsat MSS has been investigated for its utility in wetland mapping and inventory updating of broad wetland communities (Jensen et al., 1980; Hardin, 1985; May, 1986). Some of the resource mapping problems encountered with the use of Landsat MSS data for wetlands originate from the low spatial resolution (80 m) of the MSS sensor which hampers detection of wetlands smaller than 1.6 hectares (4 acres), the inability to place boundaries with high reliability, and uncertain availability of multidate imagery (16-day repeat cycle and cloud free conditions required) for use in discrimination of different vegetation types based on phenology. A study by Jensen et al. (1980) compared Landsat MSS, color aerial photography, and radar images for the identification of giant kelp beds in southern California. It was found that kelp was separable from ocean in Landsat MSS, but that the acreage estimates generated from Landsat MSS were consistently underestimated in comparison to those derived from the color photography. Such underestimations were attributed to the photo interpreter's ability to identify less-dense areas of kelp on the photographs.

A study undertaken by Ernst-Dottavio et al. (1981) was aimed at determining whether Landsat MSS could identify inland wetlands in northern Indiana. In that study, spectral characteristics of inland wetlands were quantified utilizing a helicopter-mounted radiometer sensitive to the four Landsat MSS channels. The study found hardwood swamps, shallow marshes, and shrub swamps to be spectrally similar, therefore the classification resulted in low individual class accuracies. However, the deep marsh environment showed higher accuracies, and an overall accuracy of 71% was reported for the investigation.

In a study by Wood (1983), black-and-white photography, aerial 35mm color slides, CIR highlight photography, Landsat MSS, and Landsat TM simulator were used to document historical changes in wetlands in California. The Landsat MSS classification results showed considerable confusion when compared to land use maps complied from the CIR aerial photography. As a result, the wetland areas were greatly overestimated. Wood suggests that better classification accuracies could have been obtained with perhaps a different image date, or a multitemporal approach.

Hardin (1985) reported an overall accuracy of 72% for distinguishing wetland from nonwetland in Delaware utilizing multidate Landsat MSS images acquired in 1974. Classification accuracies ranged from 74 to 28% for individual marine species. Freshwater wetlands were less accurately classified, and most misclassifications occurred between the various wetland types. Hardin believes state statutory requirements make it unlikely for Delaware to ever discard the use of low altitude photography in favor of satellite data, even if accuracy were greatly improved. In another study using Landsat MSS data, May (1986) evaluated the feasibility of using automated classification techniques to update habitat maps based on the Cowardin system. In comparing the resulting Landsat MSS based habitat classifications with existing habitat maps prepared from aerial photographs, the study found a high degree of omission and commission errors, and low mapping accuracies for all habitat categories. It was noted that the fundamental problem encountered in the study was the inability to accurately co-register the Landsat MSS to the photo-derived habitat maps.

Landsat TM Previous Work

Landsat TM has improved wetland monitoring capabilities over the Landsat MSS system. In comparison to the Landsat MSS, the TM sensor provides seven narrower spectral bands, better spatial resolution, improved radiometric sensitivity, and a higher number of quantization levels (i.e., digital numbers for TM = 256 and for MSS = 128). TM is still subject to a repeat collection cycle of 16 days, and use of the data is contingent upon near cloud-free conditions. In a study by Wood (1983) Landsat TM Simulator data were acquired over California's

Central Valley to supplement existing aerial photography and Landsat MSS. The objective was to develop a series of historical maps to document the extent of wetland change between 1937 and 1982. Very little confusion was found between wetland and other land use classes based on unsupervised classification results. The comparison of four resulting classes (permanent wetland, seasonal wetland, native vegetation, and water) against the manual photo interpretation of CIR transparencies resulted in an overall agreement exceeding 90%. In a study by Jacobson et al. (1987) a waterfowl habitat inventory map was generated for an area northeast of Anchorage using Landsat TM acquired in August 1987. The result was a thematic map depicting six waterfowl habitat classes (deep freshwater, shallow freshwater, turbid water, aquatic bed, deep marsh, and shallow marsh). Direct comparisons between the TM classified waterfowl habitat information and the existing NWI for the area was not possible due to different classification strategies. The 51 unique wetland types identified by the NWI for this area were reduced to 24, and contained only those categories which were related to waterfowl habitat. Frequency analysis was performed between the two data sets. The TM results when compared to the reduced NWI data set as coincident, omitted, or committed categories, achieved an accuracy of over 90% for lacustrine systems and about 25% for seasonal and temporarily flooded wetlands. When point and linear wetlands were removed from the NWI data, the TM classification missed 78% of the wetlands under 2 acres. The NWI mapped the area as having 83% of the wetland basins under 2 acres in size. The authors conclude that Landsat TM can provide reliable wetland information for basins greater than 2 acres, and because of the computerized nature of the resulting data bases, updates can be incorporated when more detailed information becomes available. Generally, when current TM imagery and NWI information are combined, wetland preservation and planning issues can be addressed more readily.

A study to identify suitable wetland habitats for wood stork with Landsat TM data was undertaken by Hodgson et al. (1988) in east-central Georgia. In the analysis, a computer classified foraging habitat map was created using TM imagery collected for 2 years, with each year having imagery for both "wet" and "dry" seasons. Seven categories were generated: deep water, shallow water, macrophytes, cypress/mixed wetland, bottomland/hardwoods, pine/mixed uplands, and agriculture/clearings. Classification accuracy was evaluated by comparing the TM results to previously verified foraging sites. Overall accuracy results of 74 and 88% for the 2 years investigated were reported.

Pickus (1990) found Landsat TM to effective for delineating wetlands under the Environmental Protection Agency's (EPA) Section 404 Wetlands Advanced Identification program in southeastern Louisiana. Based on comparisons between field collected reference data and NWI data, correlation between the classified TM imagery and field verification resulted in an 85% accuracy in distinguishing wetlands. Accuracies of 86% and 71% were obtained in the identification of bottomland hardwoods and uplands, respectively. Results were poor for cypress/tupelo (48%) and water (56%) categories. When compared to digital NWI data, the classified TM wetlands identification had a very high correlation (90%).

Ancillary information was utilized in this project to create a GIS which contained in addition to the satellite-derived land cover, hydrography, soils, transportation, and elevation layers. Pickus concluded that for this project, digital analysis of Landsat TM, when incorporated into a GIS containing ancillary data, was sufficient for advanced wetlands identification.

In the state of Illinois, preliminary evaluations have begun to assess the potential of satellite data for updating NWI (Luman, 1990). TM data from May 1987 were classified into generic Anderson Level 1 land-cover categories (agriculture, grass, shrub, forest, wetland, and water) for a study site located in southern Illinois. Sample point comparisons were made against a variety of ancillary data, and the overall map accuracy was found to be 85%. When individual categories were evaluated, accuracies ranged from 52% for shrub, 76% for wetland, and 97% for nonturbid water. Luman concludes that TM data would be useful for detecting previously unmapped wetland habitats.

SPOT Previous Work

Very few references were located in which SPOT imagery was evaluated for wetland mapping applications. SPOT Corporation has published brochures describing applications in which SPOT imagery was used for wetland monitoring and vegetation inventorying projects. One project involved the detailed mapping of a national wildlife refuge into 18 classes based on density, species composition, and other factors. In this instance, SPOT panchromatic and multispectral imagery were merged for analysis with a resulting minimum mapping unit of 0.01 ha. No accuracy assessment was cited for this example. SPOT also lists projects where submerged aquatic vegetation was located and mapped successfully, and another where a wetland area dominated by a particular species was mapped and correlated to vegetation index factors for assessing biomass distribution. Again, no accuracy statements were made about either of these applications.

In a study by Mackey (1990), 11 dates of SPOT multispectral imagery collected over 3 years were evaluated to determine seasonal and annual changes in a 400-ha, southeastern freshwater marsh. Unsupervised classification techniques were used to generate land cover maps for each date. Satellite TM and aircraft multispectral imagery have been analyzed for this site in previous years with reported accuracies ranging from 70 to 85% for distinguishing open water, freshwater marsh, shrub-scrub, and cypress/tupelo cover types (Jensen et al., 1984, 1986, 1987). No accuracy assessment was performed, but the author contends that accuracies from this project would be consistent with previous work performed by Jensen. The resulting data were analyzed for trend analysis with a prediction for a drier, more persistent wetland community developing.

Luman (1990) analyzed two SPOT scenes (each a hybridization containing both panchromatic and multispectral information) for utility in updating the NWI in Illinois. The existing NWI information was regrouped to match a state-based

classification system for direct comparison against the results from digital classification. Good correspondence was found between the satellite-based classifications and the NWI maps, especially where the NWI had been simplified to the
state-based classification, and the NWI source information had been of good-to-
high quality. Luman concludes that the use of the two SPOT sensors merged
(MSS and panchromatic data) is acceptable for detailed analysis of wetland habitat structure, and for validating water regime modifiers within the Cowardin classification. The higher spatial resolution of SPOT appears to provide a tool in
which to monitor wetland change effectively.

Aircraft Multispectral Scanners

Aircraft multispectral scanners (aircraft MSS) are multiband sensors that can
have from 4 to 230 separate channels or bands in which the sensor collects information from various parts of the electromagnetic spectrum. These systems
are designed to measure and record the radiant energy reflected and emitted from
the ground, and have a wider range of spectral sensitivity than photographic systems. Their exact configurations vary by the spectral ranges which are sensed
and how narrow of a bandwidth the sensors are capable of recording. Very specific data acquisition parameters are required for planning aircraft MSS mis-
sions. The area to be covered must be specified in terms of location, size, and
boundaries, with the position of the flight lines clearly defined (usually lati-
tude/longitude coordinates). If mosaicking of the data is required, all flight lines
should be flown in the same direction to minimize illumination differences. Also
specified are the appropriate spectral channels, acquisition time (time/day/month/
year), and flying height which controls the resulting resolution and amount of
digital data obtained. Resolutions as fine as 1 m are possible from aircraft based
sensors. Supporting field data collection requirements should be clearly defined
prior to the mission. Following data acquisition, postprocessing is necessary to
calibrate the data for systematic scanner distortions. The resulting digital data
is in raster format. There are a limited number of these operational systems available for public use (EPA-EMSL, NASA-AMES, GEOSCAN). Landsat TM data
is more practical than airborne MSS because it is routinely collected and less
expensive. Conversely, aircraft MSS systems can offer flexibility in terms of
resolution, timing, and finer spectral sensitivity. Most of the aircraft-mounted
multispectral systems are also equipped with mapping cameras to allow simultaneous collection of aerial photography.

The EPA Environmental Monitoring Systems Laboratory—Las Vegas (EMSL-
LV) operates an aircraft-based MSS in which a highly precise navigation system has been incorporated to provide very accurate spatial registration. In the
past few years, the Air Force's Global Positioning System (GPS) has revolutionized positioning technology. The use of GPS adds a dimension of geometric control and correction that has not been available in the past. The major problem with aircraft scanners is poor geometry with data being adversely affected
by variations in aircraft attitude or deviations from the flight line. Relatively in-

expensive GPS receivers and computer software are available which allow positional accuracies ranging from 5 to 25 m to be obtained. This combined with recent developments in Digital Elevation Models (DEMs) offers multispectral data that is geometrically much superior to older previous data.

Applications for aircraft MSS systems include the following examples:

1. Vegetation classification
2. Land cover/Land use mapping
3. Water quality monitoring
4. Wetland mapping
5. Monitoring of heated water discharges
6. Underground fire monitoring
7. Map updating
8. Geothermal monitoring

Aircraft MSS Previous Work

Savastano et al. (1984) reported the results of a study utilizing aircraft MSS to classify and map diverse nearshore marine and estuarine habitats in St. Joseph Bay, FL. The objective was to develop methods for separating seagrass and macroalgal habitats from the ocean bottom using a four band (blue through near IR) multispectral scanner. The near IR provided the capability to discriminate between emergent vegetation and water, but contained no useful information on submerged habitats as near-IR energy is almost completely absorbed by water. The information content of the blue band, which can penetrate shallow water environments, was lacking due to a low instrument signal-to-noise ratio. When digitally classified data were compared to ground truth information, it was found that the classification subdivided the desired categories too finely, breaking out classes that could not be identified on the basis of existing information. The study concluded that although it was not possible to collect ground truth information as detailed as the aircraft MSS, the resulting habitat maps were found to be generally accurate.

In an EPA Environmental Monitoring Systems Laboratory (EMSL) study by Page (1982), an 11-channel aircraft MSS was used to map nearshore kelp beds off the southern California coast. The study site was selectively chosen to be an ideal example, as the kelp beds were large and well developed. Unsupervised classification was performed on the aircraft MSS imagery using four of the channels (green, red, reflected IR, and thermal IR), and results were compared to estimates derived from manual interpretation of concurrently collected CIR aerial photography. It was found that the aircraft MSS-derived kelp estimates were considerably lower than the photo-derived estimates. This was attributed to the inherent generalization of mapping boundaries in the photo interpretation process, where the kelp beds were identified as large, visually distinct units. It was concluded that the photo interpretation process resulted in the inclusion of nonkelp features (i.e., ocean) within the kelp mapping units. Page summarized that air-craft MSS data could accurately survey, map, and compile acreage estimates for nearshore kelp resources. It appeared that the digital approach yielded results which had excellent positional agreement with maps produced from photo in-terpretation, while providing better estimates of kelp bed areal extent.

Extensive work has been done utilizing aircraft MSS data for inland wet-lands mapping in South Carolina (Jensen et al., 1984, 1986, 1987; Grace, 1985). In one investigation, 11-channel aircraft MSS data were collected in March 1981, with 3-m resolution over the Savannah River floodplain (Jensen et al., 1986). The data were analyzed to determine the optimum combination of channels for separability of wetland types with green, red, and two near IR bands being ul-timately selected. Supervised classification techniques were employed to map the data into the Cowardin classification system. Categories identified were: per-sistent emergent marsh, nonpersistent emergent marsh, shrub/scrub, algal mat, mixed deciduous upland forest, and mixed deciduous swamp forest. An overall accuracy of 83% was reported when comparisons were made on a pixel-to-pixel basis to ground transects. Another aspect conducted under this research was an effort to relate the types of vegetation present with water temperature. Daytime thermal-IR data were collected during the same mission, and compared against the classified maps constructed from the four channel subset. A statistical analy-sis was performed in which it was determined that associations did exist between vegetation types and their apparent temperature class intervals. However, Jensen concedes that other factors besides temperature, such as sedimentation and oxy-gen stress, may affect the establishment of different plant communities.

In a study undertaken at EPA-EMSL by Mynar (1990), an 11-channel air-craft MSS was utilized to map nearshore habitats in Puget Sound, WA. The pro-ject objective was to develop aircraft MSS data processing protocols with pre-liminary work focusing on seven test sites. A regional Cowardin classification was developed for the Puget Sound marine estuarine system by the Washington State Natural Heritage Program. MSS data collection corresponded with low tide conditions in July 1988, with CIR aerial photography collected simultaneously at a scale of approximately 1:13,000. The resolution of the MSS data was ap-proximately 5 m. The final MSS-derived classifications for two sites were ana-lytically compared with classifications derived from manual interpretation of the simultaneous CIR aerial photography. The overall comparable at the two sites were quite low, 55 and 53%. Mynar attributes the low accuracies to poor field verification data, and lack of accurate co-registration between the two databases, i.e., MSS classification and photo-interpreted CIR photography. It should be noted that lack of agreement does not necessarily indicate the MSS classifica-tions did not correctly depict the habitats. A larger "minimum mapping unit" re-sults from photo interpretation methods in which resources are interpreted and outlined on photographs. Low accuracy statistics have previously resulted when manually interpreted results are compared to high resolution MSS classified data where each individual pixel is evaluated and categorized (Page, 1982). The rec-tification process consisted of three separate, yet related components: MSS data

registration to an earth coordinate system (UTM); aerial photo-interpreted results registered to UTM coordinates; and the co-registration of the photo-interpreted results and the MSS data. Due to the nature of the estuarine environment,
reliable ground control points were difficult to locate (i.e., shifting sand, mud,
and water). Mynar concludes that MSS acquisition must be timed accordingly
to capture the resource(s) under investigation, that ground verification data must
be compatible with remotely sensed data, and that image rectification needs to
be more precise in order to access whether MSS imagery is appropriate for nearshore
habitat inventories.

Radar Applications

Radar can be useful for identifying broad wetland classes over large areas,
particularly if the area is perpetually cloudy. The longer wavelength radar systems have the ability to penetrate clouds and herbaceous vegetation canopies.
Radar is known to be sensitive to differences in the dielectric constant of surface materials, and produces intense backscatter or return signals from some wet
forested areas. Disadvantages of using radar are the high costs, limited data availability, and the complexity of the imagery. In a study conducted by Place (1985),
Seasat radar images were evaluated for their ability to improve wetland mapping when combined with conventional sources, i.e., aerial photography. The
Seasat satellite was launched in June 1978 and failed later that year in October.
It was an L-band (23.5 cm) Synthetic Aperture Radar (SAR), having H-H po-
larization and 25-m resolution. If vegetation or moisture conditions are of interest, Seasat is useful only for areas which are generally flat. Place compared
Seasat interpreted imagery and NWI data at four test sites on the coastal plain
between Maryland and Florida. Place concluded, based on statistical comparisons, that photo interpreters who used Seasat radar images to compliment their
conventional sources were able to map forested wetland more accurately (greater
than 85%), than those who did not.

Videography

Airborne videography is not a replacement for aerial photography, but rather
a substitute or complimentary data source when the quality and/or cost of an aerial survey can not be supported. Scientists at the USDA Agricultural Research
Service Unit at Welasco, TX, have been doing basic research on applications of
both normal color and infrared video for rangeland and other vegetation, and the
U.S. Fish and Wildlife Service in North Dakota have been using video to assist
in assessments of wetland and riparian ecosystems (Driscoll, 1990). Advantages
of video systems are: immediate availability of imagery; in-flight error-proofing
capabilities: ability to use narrow band filters for finer spectral resolution; ability to function in a wider range of atmospheric conditions than conventional pho-

tography; and ability to utilize satellite-type data processing software for analysis. Disadvantages include a lower resolving power than aerial photography and difficulties with calibration. Several video camera systems have been developed ranging from black-and-white with visible/NIR capabilities to multispectral false color systems. A comprehensive description of the various video systems available can be found in Everitt and Escobar (1989). Imagery from these systems has been used successfully to detect and assess ecological conditions such as plant communities and species, insect pests, soil moisture, soil drainage and salinity, grass phytomas levels, and burned areas (Everitt and Escobar, 1989). High-resolution airborne video data have also been utilized to update cartographic databases (Ehlers et al., 1989). Video image data lends itself to the utilization of automated image processing techniques, which can then be linked to GIS and used to update information layers. Wu (1989) utilized a commercially available VHS-format camcorder as an alternative to conventional color photography for a multisensor mapping effort of coastal wetlands in Louisiana. The video images were used as ground-truth in conjunction with SPOT, airborne MSS, and radar imagery. The airborne video images captured the dynamics, volume, and spatial distribution of the wetland vegetation very well. Wu concludes however, that further study of the spectral response characteristics of vegetation and surface features is needed to determine which video systems have the spectral bands most responsive to the target/scene.

GEOGRAPHIC INFORMATION SYSTEMS

Spatial relationships are apparent when viewed as cartographic products, but to analyze and model spatial processes quantitatively, the information needs to be in digital form. Geographic information systems (GIS) are distinguished from other database management systems by their ability to perform spatial analyses with multiple levels of data in a selected geographic area. Computerized GIS are widely utilized to store, query, retrieve, display, and manage large amounts of digital data assembled from many sources. These data are typically geographic, environmental, cultural, statistical or political. The technological migration from costly minicomputers to inexpensive workstations has greatly expanded the user base of GIS. Use of remotely sensed data has become more common in GIS applications as many commercial GIS packages have image display capabilities and some rudimentary image processing functions. Geographic data can be represented using either vector/polygon or raster/grid data formats. GIS does not generate primary data, but rather captures and processes information in a spatial context and serves as a platform on which decisions can be made. The ability to geometrically transform and integrate multiple data types is very important when accounting for differences in scale, projections, resolutions, and coordinate systems. The analysis of the data can be as simple as measuring line lengths and areas, or as complex as using modeling techniques to create "what if" scenarios. Analyses which are commonly used include measurements (distance, area, volume), interpretation or processing of a basic layer of data to create additional layers, boolean analytical processing, overlay analysis, distance searches, statistical calculations, and report generation. It should be noted there are functions which are difficult to perform in an image processing system (raster) that are relatively easy in a GIS (vector) and vice versa. For example, geometric or overlay operations are easier to perform in the raster domain whereas network analysis or topologic operations are more suited to the vector domain (Ehlers et al., 1989).

Both vector and raster methods of representing the spatial extent of geographical information can be translated or interchanged through the use of data exchange formats. The conversion of raster or other forms of spatial data into vector form can be approached in several ways. The most common is manual digitization from existing paper sources which is labor intensive, expensive, and requires in-place quality control procedures to reduce error propagation. Data exchange formats are a common approach to solving the solution of integrating various digital source data. Some GIS software packages offer conversion packages which transform raster or vector data into a Single Variable File (SVF) format. The SVF format is one type of bridge between vector and raster file structures.

The simultaneous display of raster images with vector cartographic data offers the capability to perform change detection analyses. In many applications where GIS databases exist, this tool can be used to update map resource information on a regular basis, or for rudimentary location and query functions. Often just a visual interpretation of geocoded data allows an index to be placed on the surface of the image, which then becomes available for interactive queries in the GIS. For example, a transportation network vector layer might be overlaid on a raster satellite image. The vector file then could be edited and new attributes connected to features for updating statistical information if construction altered the transportation network. Over time newly acquired images could be utilized in the GIS with minimal cost and labor. Burrough (1986) lists the following recommendations for the use of vector and raster data structures in a GIS environment:

> *Vector:* vector for archiving phenomenologically structured data (e.g., soils, land-use units, etc.), for network analysis (e.g., transportation, etc.), for high quality line drawing, and digital terrain modeling.
> *Raster:* raster for map overlays, map combinations, spatial analysis, altitude matrices, and for simulations and modeling when surfaces are encountered.
> *Combination Vector/Raster:* for plotting high quality lines with vectors, in combination with efficient area filling in color with raster structures such as run length codes or quadtrees.

Vector GIS Systems

The most dominant type of GIS are vector based and have a data structure based on topology to store the relationships among various spatial objects. Cartographic information is characterized by point, line, and area features which are further defined by information about what is on either side, and how it is connected to the other lines. This coordinate information is then cross-referenced to attribute files which contain spatial location in relation to other descriptive attributes. Vector/polygon data structures describe the unique lines or forms of specific geographic features (streets, lakes, rivers, etc). The coordinate space is presumed to be continuous, not quantized as with raster data, which allows all positions, lengths, and dimensions to be defined with vector data structures (Burrough, 1986). This vector representation of data is a way to replicate the feature as exactly as possible in a digital form.

Vector GIS have the capability to organize diverse types of data into a single database which contains the coordinate location of the feature, and its geographical, cultural, or scientific attributes. These attributes are the key identifiers in organization and description of the data layers contained within the GIS. The relational database capabilities allow for access to non-GIS attribute data. Raster data are processed into a single layer or dimension with one pixel value before becoming a layer in a vector-based GIS. Maintenance of these spatial descriptors as a part of the database allows GIS to perform normal database management functions as well as spatial manipulations.

Raster GIS Systems

In a raster environment, each pixel (or cell) represents one data type spatially, so if a house, tree, and road intersect at a pixel those elements are put into separate layers. Raster-based systems are compatible with the data produced by satellite or aircraft based sensors, and raster-based scanned maps and photographs. These systems produce imagery in raster form comprising a rectangular array of pixels, which is then analyzed and stored to create a grid map. The processed data may be in a digital form such as a thematic map created by computer classification, or in graphic form such as a paper map derived from visual image interpretation. Raster systems, because of data storage configuration can take longer to process and display, and usually require more computer disk space. Moreover, raster data may not explicitly represent feature boundaries, and instead have a stair-stepped appearance.

Digital Data Sources

Many digital spatial data sources have come into existence because the use of such sources compared with generating new data is almost always less expensive and time consuming. Remote sensing can provide a means for acquiring very recent land cover/land use data, and when merged with other data (i.e., soils, hydrology, roads, etc.) allows the greatest use to be made not only of the remotely sensed data, but all data available to the investigator from other sources. For some applications, maps, field data, and/or aerial photography would need to be scanned or manually digitized into a machine-readable form for integration with existing digital products. All data layers, whether procured, generated through digital image processing, or manual digitizing, should have reported spatial and thematic accuracies associated with the real world. When acquiring digital data from reliable sources, minimum standards for accuracy are usually ensured. Data of questionable integrity should be eliminated from consideration unless the intent is to use the information as a data layer on which updates, refinements, or corrections are to made later when more accurate information becomes available. The overall accuracy of any GIS database depends on the data layer with the lowest accuracy or resolution. The integration of various types of spatial and geographic information enables remote sensing to provide useful input for a broad range of applications in a cost-effective manner. For many environmental applications, remotely sensed data represents only one source of input for studying complex environmental problems. Some other sources of spatial data include maps, aerial photographs, census information, field measurements, and meteorological records. The challenge then, is not only where to get information but how to integrate information flow between different geoprocessing technologies. Once a GIS is functioning, information can be drawn from a single, consistent, uniform database which avoids the possibility of having separate collections of conflicting data. A comprehensive GIS will support various computer mapping and graphic products.

GIS provides a convenient and organized method for analysis of wetland ecosystems. GIS provides flexibility in which to build a cohesive database on which additional information can be incorporated, and refinements made if project objectives change. Relationships can be modeled and different scenarios can be put forth that may need to be addressed for impact assessments or trend analyses on the wetland ecosystem. The results of such capabilities provide the resource/project manager with the best information available for better protection and management of wetland resources.

EXAMPLE PROJECT COSTS

Each wetland resource mapping project is highly individualized depending upon legal requirements and/or the needs of the agency conducting the investigation. The following estimated project costs should be regarded with caution as differing study objectives result in varying costs. Communication must begin in the planning phase of any inventory project, and focus on data acquisition requirements and subsequent processing/interpretation to ensure that the final product will meet the user's needs. Decisions made regarding the scope of such endeavors must include the RS/GIS specialists so any preconceived expectations can be realistically evaluated and modified if necessary. Table 2 contains generic information regarding remote sensing system specifications and acquisition costs. Table 3 contains summary information pertaining to described projects/programs found described later in this section.

Table 2　Remote Sensing Sensor Specifications and Acquisition Costs

Sensor costs (sq. mile)	Spatial resolution	Temporal	Coverage (sq. mile)
AVHRR $0.00017	1.1 × 1.1 km	2/d	1,500,000
Landsat TM $0.32	30 × 30 m	16 d	10,000
SPOT $1.83	10 × 10 m	26 d	1200
CIR Photo $10–$15	1–3 m (1:40,000)	variable	32
Aircraft MSS and CIR Photo $15–$20	15 × 15 m 1–3 m	variable	variable

Table 3　Project and Program Summary Table

Project	Data source	Analysis	Map transfer	Encoded	Minimum map unit	Cost per acre
EPA-EMSL Projects						
Clark Fork	AP/MSS	WL,NPS	No	No	0.5 acre	$0.06
LPO, ID	TM/AP	LC	No	No	1–2 acre	$0.03
BLH, IL	TM/AP	WL, LC	Yes	Yes	1–2 acre	$0.19
Pearl River	TM	WL	No	No	1–2 acre	$0.21
Other Federal Programs						
NWI	AP	WL	Yes	Yes	1–2 acre	$0.06
USACE	AP	LC, LU	Yes	Yes	1–2 acre	$0.07
State Programs						
Florida	TM/AP	WL	No	No	1 acre	$0.02
Georgia	AP	WL	No	No	0.5–1 acre	N/A
Michigan	AP	LC, LU	Yes	Yes	1 acre	$0.06
Wisconsin	AP	WL	Yes	Yes	2 acre	$0.07

Note: AP, Aerial photography; MSS, Aircraft multispectral scanner; TM, Landsat Thematic Mapper; LC, Land Cover; LU, Land Use; WL, Wetland; NPS, Non-point source.

EPA-EMSL Projects

The following resource mapping projects were undertaken at EPA's Environmental Monitoring Systems Laboratory in Las Vegas.

Aquatic Macrophyte Mapping—Clark Fork and Tributaries, MT

Color infrared photography, 1:18,000 scale, was collected July 1988 concurrently with aircraft multispectral scanner data. The purpose of the project was shoreline interpretation of wetlands and associated vegetation (algal blooms, rooted macrophytes, and riparian tracts), and nonpoint pollution source features (cropland, pasture, confined feeding, landfills, waste water treatment facilities, outfall locations, timber harvests, golf courses/urban recreation, nursery/orchard, industrial, commercial, and residential). The areal coverage and interpretation efforts centered on several river courses located in Montana, Idaho, and Washington. Only the area contained within the photographs were analyzed and mapped for the above class groupings. Project costs reflect EMSL project management, photo acquisition and film processing, manual photo-interpretation of 740 flight line miles, graphic personnel support, and a deliverable of bound reports (3 volumes/set) containing the photo-interpreted overlays attached to the photographs. Not included are costs associated with aircraft usage, flight crew, and aircraft maintenance. No map transfer or digitization of the photo-interpreted data was performed. The cost for this project is estimated at $37.00 per square mile ($0.06/acre).

Lake Pend Oreille Watershed Characterization Using Landsat TM Data

A Landsat Thematic Mapper subscene (100 km × 100 km) dated July 1989 was used in this analysis. An Anderson Level 1 land cover characterization (forest, agriculture, range land, barren, water, wetland, and urban) was derived using unsupervised image processing techniques. The resulting overall accuracy reported for only the forest, agriculture, range land, and barren categories was 78% (Lee, 1990). Statistical data were generated regarding areal extent of land cover resources for approximately 1000 square miles (640,000 acres). This information was used by the State of Idaho for nonpoint pollution modeling efforts for the Lake Pend Oreille Watershed. Project costs reflect EMSL project management, TM imagery acquisition, supporting NHAP photographs, image processing, accuracy assessment, final report, and graphic products (slides and map acetate overlays at 1:100,000 scale). The cost is estimated at $20.62 per square mile ($0.03/acre).

Bottomland Hardwoods Identification Using Landsat TM Data

Subsets of two Landsat Thematic Mapper scenes dated April 1988 and August 1988 were used in a pilot project to assist in advanced identification of wetlands. The analysis focused on four 7.5-min quadrangles (approximately 240 sq miles) located in southern Illinois in which several different image processing methodologies were tested for optimum bottomland hardwood wetland identification. Once completed, the next phase will apply the derived results to a watershed level analysis of wetland resources (not included in cost estimation). An Anderson Level 1 land cover was derived, except for the wetland class which was further subdivided into a Level 2 classification (forested vs. nonforested wetlands). The overall agreement between the satellite-based classification and photo-interpreted CIR photography collected in August 1988 was approximately 81%. Project costs reflect EMSL project management, TM imagery acquisition, supporting photo-interpretation of existing CIR imagery (1:24,000), image processing, accuracy comparisons between the TM classification and photo-interpretation, final report, and graphic products (slides and plots). The cost is estimated at $122.00 per square mile ($0.19/acre).

Pearl River Wetlands Advanced Identification

A GIS database was constructed for the Pearl River Watershed (approximately 200 sq miles) located in southern Louisiana in which layers were generated for hydrography, soils, land cover, transportation, elevation, and field reference data. The data layers were integrated or compiled from digital and map sources obtained from Landsat TM imagery (land cover), NWI, SCS, and USGS. Two dates of Landsat TM imagery were obtained from precopyrighted archived data. Wetlands were identified as wetland vs. nonwetland, and further subdivided into local species such as bottomland hardwoods and cypress/tupelo groupings. Overall accuracy between the GIS based wetland identification and the field verification information was 85% (Pickus, 1990). Project costs reflect EMSL project management, TM image processing, GIS data layer digitization or integration, development of a user friendly GIS interface (ARC/INFO), and a formal project report. The cost is estimated at $133.00 per square mile ($0.21/acre).

USFWS NWI Program

The production of NWI information is performed by blocks of USGS quadrangles. USGS color-infrared NHAP and NAPP photography are photo-interpreted according to the Cowardin classification (Cowardin et al., 1979) with interpreted results transferred to the appropriate quadrangles. These maps are available in paper form by quadrangle, and subsequently become digitized in selected areas where cost-sharing between the USFWS and the customer/user is provided. To date, 60% of the conterminous United States, and 20% of Alaska have been completed. The average cost per acre ranges from $0.03 to $0.06, depending on the number and complexity of wetlands found within the quadrangle (D. Woodward, USFWS-NWI, personal communication, 1990).

USACE Project

The following project was performed by the U.S. Army Corps of Engineers, Detroit District, Engineering Division, Great Lakes Hydraulics and Hydrology Branch.

Resource Analysis and Land Cover/Current Use Inventory Data Base for the U.S. Side of the Great Lakes, 1989–1992

A GIS incorporating land cover/current use inventory information was created with source information derived from photo-interpreted CIR aerial photography (1:24,000 dated Aug/Sept 1988) for approximately 1500 linear miles of the U.S. Great Lakes shoreline excluding the State of Michigan. The State of Michigan has been constructing a digital land cover data base over the past 10 years utilizing photo-interpreted 1:24,000 scale CIR photography. The land cover/current use classification used in this Corps project is based on the classification system used by the State of Michigan, Department of Natural Resources called MIRIS (Michigan Resource Inventory System). The MIRIS classification is a refined Anderson Level 3 which subdivides the wetland category into wooded wetlands (wooded and shrub/scrub classes), nonwooded (aquatic beds, emergents, flats), and Great Lakes coastal submergent classes. The final photo-interpreted map overlays were digitally scan-encoded into vector format. Project costs are based on contracts which required photo acquisition and film processing, manual interpretation, map transfer, and digital scan encoding/quality control to vector form (Intergraph). The costs do not include Corps district project management or additional coordination provided by the State of Michigan. The cost is estimated at $42.00 per square mile ($0.07/acre) (R.L. Gauthier, USACE, personal communication, 1990).

State Mapping Programs

Florida

The Florida Department of Transportation (DOT), State Topographic Office mapped the state's wetland resources under contract to the Florida Freshwater Gaming Commission. Landsat TM data from 1986 to 1988 for the entire state was classified to a user-specified Anderson Level 3. The TM data analysis was supplemented with soils data and existing photography (CIR 1:40,000; B&W 1:24,000) to allow refinement of classes during production. The classification categories included: coastal strand, saltmarsh, wet and dry prairies, swamps (mangrove, cypress, hardwood, bay), pine forest, mixed hardwoods, upland forest, exotic species, disturbed communities, bare ground, and open water. The classified data is available only in digital form with inherent 30-m pixel resolution. Project costs reflect TM data acquisition, DOT image analysis, and limited field reviews. The costs do not include Freshwater Gaming Commission project coordination, or their independent field reviews of the final classification. The cost is estimated at $9.50 per square mile or $0.02/acre (G. Maudin/A. Shopmyer, Florida DOT, personal communication, 1990).

Georgia

In 1972, Georgia passed the Marshland Enforcement Act which mandated an aerial surveillance program be established to monitor and regulate the state's coastal estuaries. Black-and-white aerial photography is periodically acquired at a scale at 1:40,000. In addition, routine light aircraft and helicopter flights collect 35 mm photos to monitor shoreline development, to assure compliance to state laws, and to locate unauthorized developments. Reports are generated on a case-by-case basis for unauthorized activities. No standard minimum mapping unit was given, but is estimated to be between 0.5 and 1 acre (S. Stevins, GDNR, personal communication, 1990). No cost data were available.

Michigan

In Michigan, the entire state has been encoded into a digital GIS (Intergraph) by the Michigan Department of Natural Resources (MDNR), Michigan Resource Inventory System (MIRIS). Color-infrared aerial photography (dated 1979 and 1985, 1:24,000) were photo-interpreted for land cover/land use with a minimum mapping unit of 1 acre. Wetlands were mapped to a revised Anderson Level 3 (lowland hardwood, shrub wetland, emergent wetland, and aquatic beds). Data is available by county in a variety of formats (digital, map, and statistical). The cost is estimated at $39.00 per square mile or $0.06/acre (M. Scieszka, MDNR, personal communication, 1990).

Wisconsin

In Wisconsin, the entire state's wetlands were mapped in approximately 5 years and was completed in 1984 by the Wisconsin Department of Natural Resources (WDNR). Black and white infrared photography at a scale of 1:20,000 were photo-interpreted with acquisition costs averaging $30,000 per county. Interpretative data was transferred to 1:24,000 scale township-centered rectified base maps with a minimum mapping unit of 2 acres. Data are compiled on a county-by-county basis with three to four counties revised annually. All current work is performed by contractors. The state is in the process of creating a GIS which incorporates both past and ongoing work. The costs reflect Wisconsin DNR project management, photo acquisition and film processing, photo-interpretation, map transfer, and manual digitization of the map overlays. Maps are available in paper and digital formats where digitization is completed. The cost is estimated at $43.00 per square mile or $0.07/acre (L. Stoerzer, WDNR, personal communication, 1990).

ACKNOWLEDGMENT

The author wishes to acknowledge the help and technical assistance provided by Mr. David R. Williams of Lockheed Environmental Systems and Technologies Company. Mr. Williams has many years of experience in photo-interpretation of wetland systems. He contributed to the Aerial Photography Application Section and supplied several of the personal references cited under the Example Project Costs Section.

The information in this document has been funded wholly or in part by the United States Environmental Protection Agency under contract number 68-C0-0050 to Lockheed Environmental Systems and Technologies Company. It has been subjected to the Agency's peer and administrative review, and has been approved for publication as an EPA document. Mention of trade names or commercial products does not constitute endorsement or recommendation for use.

REFERENCES

Anderson, J. R., Hardy, E. E., Roach, J. T., and Witmer, R. E., A land use cover classification system for use with remote sensor data, U.S. Geological Survey Professional Paper 964, 1976, 27pp.

Brown, W. W., Wetland mapping in New Jersy and New York, *Photogramm. Eng. Remote Sensing*, 44(3) 303–314, 1978.

Burrough, P. A., *Principles of Geographic Information Systems for Land Resources Assessment*, Oxford University Press, New York, 1986.

Carter, V., Application of remote sensing to wetlands, in *Remote Sensing for Resource Management*, Johannsen, C. J., and Sanders, J. L., Eds., Soil Conservation Society of America, Ankeny, IA, 1982.

Carter, V., Malone, D. L., and Burbank, J. H., Wetland classification and mapping in western Tennessee, *Photogramm. Eng. Remote Sensing*, 45(3) 273–284, 1979.

Cowardin, L. M., Carter, V., Golet, F. C., and LaRoe, E. T., Classification of wetlands and deepwater habitats of the United States, U.S. Fish and Wildlife Service, Report No. FWS/OBS-70/31, 1979, 103pp.

Duggan, J. S., Puget Sound wetlands inventory: photography volumes 1–6, EPA-Environmental Monitoring Systems Laboratory, Las Vegas, NV, Report No. TS-AMD-82072, 1983.

Driscoll, D., Remote sensing: USFS pest management group *GIS World Magazine*, Oct/Nov. 1990.

Ehlers, M., Edwards, G., and Bedard, Y., Integration of remote sensing with geographic information systems: a necessary evolution, *Photogramm. Eng. Remote Sensing*, 55(11) 1619–1627, 1989.

Ehlers, M., Hintz, R. J., and Greene, R. H., High resolution airborne video system for mapping and GIS applications, Proceedings of 12th Biennial Workshop on Color Aerial Photography and Videography, American Society of Photogrammetry and Remote Sensing, Falls Church, VA, 1989, 171–177.

Ernst-Dottavio, C. L., Hoffer, R. M., and Mroczynski, R. P., Spectral characteristics of wetland habitats, *Photogramm. Eng. Remote Sensing*, 47(2) 223–227, 1981.

Everitt, J. H. and Escobar, D. E., The staus of video systems for remote sensing applications, Proceedings of 12th Biennial Workshop on Color Aerial Photography and Videography, American Society of Photogrammetry and Remote Sensing, Falls Church, VA, 1989, 6–29.

Federal Interagency Committee for Wetland Delineation, Federal Manual for Identifying and Delineating Jurisdictional Wetlands. U.S. Army Corps of Engineers, U.S. Environmental Protection Agency, U.S. Fish and Wildlife Service, and USDA Soil Conservation Service. Cooperative technical publication, 1989, 76 pp. plus appendices.

Gammon, P. T. and Carter, V., Vegetation mapping with seasonal color infrared photographs, *Photogramm. Eng. Remote Sensing*, 45(1) 87–97, 1979.

Grace, J. B., Historic macrophyte development in par pond, submitted to the U.S. Dept. of Energy, Environmental Sciences Division, Savannah River Laboratory, Report No. DPST-85-841, Aieken, SC, 1985.

Hardin, D. L., Remote sensing of wetlands for fish and wildlife habitat management in Delaware—a comparison of data sources, Integration of Remotely Sensed Data in GIS for Processing of Global Resource Information, CERMA Proceedings, Washington, D.C., 1985.

Hardisky, M. A. and Klemas, V., Tidal wetlands natural and human-made changes from 1973 to 1979 in Delaware: mapping techniques and results, *Environ. Manage.*, 7(4) 339–344, 1983.

Heaslip, G. G., *Environmental Data Handling*, John Wiley and Sons, New York, 1975.

Hodgson, M. E., Jensen, J. R., Mackey, H. E., and Coulter, M. C., Monitoring wood stork foraging habitat using remote sensing and GIS, *Photogramm. Eng. Remote Sensing*, 54(11) 1601–1607, 1988.

Howarth, P. J. and Wickware, G. M., Procedures for change detection using landsat digital data, *J. Remote Sensing*, 2, 277–291, 1981.

Howland, W. G., Multispectral aerial photography for wetland vegetation mapping, *Photogramm. Eng. Remote Sensing*, 46(1) 87–99, 1980.

Jacobson, J. E., Ritter, R. A., and Koeln, G. T., Accuracy of thematic mapper derived wetlands as based on national wetland inventory data, ASPRS/ACSM/WFPLS Fall Convention, American Society of Photogrammetry and Remote Sensing, Falls Church, VA, 1987.

Jensen, J. R., Ramsey, E. W., Mackey, H. E., Christensen, E. J., and Sharitz, R. R., Inland wetland change detection using aircraft MSS data, *Photogramm. Eng. Remote Sensing*, 53(5) 521–529, 1987.

Jensen, J. R., Hodgson, M. E., Christensen, E. J., Mackey, H. E., Tinney, L. R., and Sharitz, R. R., Remote sensing inland wetlands: a multispectral approach, *Photogramm. Eng. Remote Sensing*, 52(1) 87–100, 1986.

Jensen, J. R., Introductory Digital Image Processing, Prentice Hall, New Jersey, 1986.

Jensen, J. R., Hodgson, M., Christensen, E. J., Mackey, H. E., and Sharitz, S. S., Multispectral remote sensing of inland wetlands in South Carolina: selecting the appropriate sensor, submitted to the U.S. Dept. of Energy, Savannah River Laboratory, Aieken, SC., 1984.

Jensen, J. R., Christensen, E. J., and Sharitz, R., Non-tidal wetland mapping in South Carolina using airborne multispectral scanner data, *Remote Sensing Environ.*, 16, 1–12, 1984.

Jensen, J. R., Estes, J. E., and Tinney, L., Remote sensing techniques for kelp surveys, *Photogramm. Eng. Remote Sensing*, 46(6) 743–755, 1980.

Lee, K. H., Internal report-watershed characterization using Landsat Thematic Mapper (TM) satellite imagery: Lake Pend Oreille, Idaho, EPA-Environmental Monitoring Systems Laboratory, Las Vegas, NV, Report No. TS-AMD-90C10, 1990.

Lillesand, T. M. and Kiefer, R. W., Remote Sensing and Image Interpretation, John Wiley and Sons, New York, 1987.

Luman, D. E., The potential for satellite-based remote sensing update of the Illinois portion of the national wetlands inventory, Quarterly Report, submitted to the Illinois Department of Conservation, IL., 1990.

Mack, W. M., Aerial survey of Utah wetlands, EPA-Environmental Monitoring Systems Laboratory, Las Vegas, NV, Report No. TS-AMD-08047, 1980.

Mackey, H. E., Jr., Monitoring seasonal and annual wetland changes in a freshwater marsh with SPOT HRV Data, Technical Papers 1990 ASCM-ASPRS Annual Convention, American Society of Photogrammetry and Remote Sensing, Falls Church, VA., 1990, 283–292.

Martin, A. C., Hotchkiss, N., Uhler, F. M., and Brown, W. S., Classification of wetlands of the United States, Special Science Report Number 20. U.S. Fish and Wildlife Service, Washington, D.C., 1953, 14 pp.

May, L. N., Jr., An evaluation of Landsat MSS digital data for updating habitat maps of the Louisiana coastal zone, *Photogramm. Eng. Remote Sensing*, 52(8) 1147–1158, 1986.

Mynar, F., II, Classification of Puget Sound nearshore habitats using aircraft multispectral scanner imagery, EPA, Environmental Monitoring Systems Laboratory, Las Vegas, NV, Report TS-AMD-90C11, 1990.

Newcomer, J. A. and Szajgin, J., Accumulation of thematic map error in digital overlay analysis, *The American Cartographer*, 11(1) 58–62, 1984.

Niedzwiedz, W. R. and Batie, S. S., An assessment of urban development into coastal wetlands using historical aerial photography: a case study, *Environ. Manage.*, 8(3) 205–214, 1984.

Norton, D. J., Suitability study of Chincoteague Wetlands, Virginia, EPA-Environmental Photographic Interpretation Center, Vint Hill, VA, Report No. TS-AMD-85037, 1986a.

Norton, D. J., Initial feasibility study: wetlands identification Sussex County, Delaware, EPA-Environmental Photographic Interpretation Center, Vint Hill, VA, Report No. TS-AMD-86071, 1986b.

Norton, D. J., Engle, S. W., and Simmons, J. D., Water resources management applications of remote sensing at EPA's environmental photographic interpretation center, Proceedings of the 51st Meeting American Society of Photogrammetry and Remote Sensing, Washington, D.C., 1985.

Norton, D. J. and Prince, J., Using remote sensing for wetlands assessment in superfund hazardous waste sites, Proceeding of the 19th International Symposium on Remote Sensing of Environment, ERIM, Ann Arbor, MI, 1985.

Page, S. H., Remote sensing of kelp beds demonstration report: Channel Islands National Park, California, EPA-Environmental Monitoring Systems Laboratory, Las Vegas, NV, Report No. TS-AMD-82003, 1982.

Place, J. L., Mapping of forested wetland: use of Seasat RADAR images to complement conventional sources, *Prof. Geographer*, 37(4) 463–469, 1985.

Pickus, J., Pearl River wetlands advanced identification: a geographical information systems demonstration project, EPA-Environmental Monitoring Systems Laboratory, Las Vegas, NV, Report No. 215-90C05, 1990.

Roller, N. E. G., Remote sensing of wetlands, Environmental Research Institute of Michigan, (NASA-CR-153282), Ann Arbor, MI, 1977, 165 pp.

Salvaggio, C. and Szemkow, P., Generation of high-resolution hybrid multiband imagery from existing high-resolution panchromatic and low-resolution multispectral images, Technical Papers, 1989 ACSM-ASPRS Annual Convention, American Society of Photogrammetry and Remote Sensing, Falls Church, VA, 1989.

Savastano, K. J., Faller, K. H., and Iverson, R. L., Estimating vegetation coverage in St. Joseph Bay, Florida with an airborne multispectral scanner, *Photogramm. Eng. Remote Sensing*, 50(8) 1159–1170, 1984.

Shaw, S. P. and Fredine, C. G., Wetlands of the United States, U.S. Fish and Wildlife Service, Circular 39, 1956, 67 pp.

SPOT Image Corporation, 1897 Preston White Drive, Reston, VA, 22091, (703)620-2200.

Story, M. and Congalton, R. G., Accuracy assessment: a user's perspective, *Photogramm. Eng. Remote Sensing*, 52(3) 397–399, 1986.

Swartwout, D. J., MacConnell, W. P., and Finn, J. T., An evaluation of the national wetlands inventory in Massachusetts, In-Place Resource Inventories Workshop, University of Maine, Orono, ME, 1981, 685–691.

Welch, R. and Ehlers, M., Merging multiresolution SPOT HRV and Landsat TM data, *Photogramm. Eng. Remote Sensing* 53(3) 301–303, 1987.

Williams, D. C. and Lyon, J. G., Use of a geographic information system data base to measure and evaluate wetland changes in the St. Marys River, Michigan, Detroit District, U. S. Army Corps of Engineers, submitted.

Williams, D. R., Aerial photographic analysis of Utah sand and gravel company settling ponds, Salt Lake County, Utah, EPA-Environmental Monitoring Systems Laboratory, Las Vegas, NV, Report No. TS-PIC 88218, 1989.

Williams, D. R., Current and historical photographic inventory of agricultural development, Great Cedar Swamp, Middlesborough County, Massachusetts, EPA-Environmental Monitoring Systems Laboratory, Las Vegas, NV, Report No. TS-AMD-85532, 1985.

Williams, D. R., Wetland delineation and photography: Piceance Creek, Colorado and James River, South Dakota, EPA-Environmental Monitoring Systems Laboratory, Las Vegas, NV, Report No. TS-AMD-82002, 1983.

Williams, D. R., Historical inventory of development activity M & K Ranches, Gulf County, Florida, EPA-Environmental Monitoring Systems Laboratory, Las Vegas, NV, Report No. TS-AMD-8125, 1981.

Wood, B. L., Wetland mapping in Colusa County, California, NASA/AMES International Renewable Resource Inventions for Monitoring Conference, Corvallis, OR, 1983.

Wu, S. T., Utility of a digital video and image analysis system for forest and coastal wetland mapping, Proceedings 12th Biennial Workshop on Color Aerial Photography and Videography, American Society of Photogrammetry and Remote Sensing, 1989, 164–170.

BIBLIOGRAPHY

Alam, M. S., Shamsuddin, S. D., and Sikder, S., Application of remote sensing for monitoring shrimp culture development in a coastal mangrove ecosystem in Bangladesh, Technical Papers, 1990 ACSM/ASPRS Annual Convention, American Society of Photogrammetry and Remote Sensing, Falls Church, VA, 23–32, 1990.

Balogh, M. E. and Becker, D. L., Riparian vegetation inventory: Parker II division of the lower Colorado River, Blythe, California, EPA-Environmental Monitoring Systems Laboratory, Las Vegas, NV, Report No. TS-AMD-85561, 1986.

Bartlett, D. S., Klemas, V., Crichton, O. W., and Davis, G. R., Low-cost aerial photographic inventory of tidal wetlands, University of Delaware, College of Marine Studies, Newark, DE, Submitted to Dept. of Natural Resources and Environmental Control, State of Delaware, 1976.

Benson, A. S. and DeGloria, S. D., Interpretation of Landsat-4 thematic mapper and multispectral scanner data for forest surveys, *Photogramm. Eng. Remote Sensing*, 51(9) 1281–1289, 1985.

Best, R. G., Wehde, M. E., and Linder, R. L., Spectral reflectance of hydrophytes, *Remote Sensing Environ.*, 11, 27–35, 1981.

Bogucki, D. J., Remote sensing to identify, assess, and predict ecological impact on Lake Champlain Wetlands, State University of New York, College at Plattsburgh, NY, 1978.

Boule, M. E. and Shea, G. B., Snohomish estuary wetlands study, Northwest Environmental Consultants Inc., Seattle, WA, Submitted to Seattle District, U.S. Army Corps of Engineers, WA, 1978.

Byrne, G. F., Crapper, P. F., and Mayo, K. K., Monitoring land-cover change by principal components analysis of multi-temporal Landsat data, *Remote Sensing Environ.*, 10, 175–184, 1980.

Carter, V. and Smith, D. G., Utilization of remotely sensed data in the management of inland wetlands, U.S. Geological Survey, Contracts IN-385 and I-414, 1973, 15 pp.

Chavez, P. S., Guptill, C., and Bowell, J. A., Image processing techniques for thematic mapper data, Technical Papers, 50th Annual Meeting of The American Society of Photogrammetry, 50(2) 728–742, 1984.

Civco, D. L., Topographic normalization of Landsat thematic mapper digital imagery, *Photogramm. Eng. Remote Sensing*, 55(9) 1303–1309, 1989.

Civco, D. L., Kennard, W. C., and LeFor, M. W., A technique for evaluating inland wetland photointerpretation: The Cell Analytical Method (CAM), *Photogramm. Eng. Remote Sensing*, 44(8) 1045–1052, 1978.

Congalton, R. G., A comparison of sampling schemes used in generating error matrices for assessing the accuracy of maps generated from remotely sensed data, *Photogramm. Eng. Remote Sensing*, 54(5) 593–600, 1988.

Cowardin, L. M. and Myers, V. I., Remote sensing for identification and classification of wetland vegetation, *J. Wildlife Manage.*, 38(2) 308–314, 1974.

Dottavio, C. L. and Dottavio, F. D., Potential benefits of new satellite sensors to wetland mapping, *Photogramm. Eng. Remote Sensing*, 50(5) 599–606, 1984.

Ernst, C. L. and Hoffer, R. M., Using Landsat data with soils information to identify wetland habitats, in *Satellite Hydrology*. Proceedings, Pecora V Symposium. American Water Resources Association, Minneapolis, MN, 1981.

Farmer, A. M. and Adams, M. S., A consideration of the problems of scale in the study of the ecology of aquatic macrophytes, *Aquatic Botany*, 33, 177–189, 1989.

Federal Interagency Committee for Wetland Delineation, Federal Manual for Identifying and Delineating Jurisdictional Wetlands. U.S. Army Corps of Engineers, U.S. Environmental Protection Agency, U.S. Fish and Wildlife Service, and U.S.D.A. Soil Conservation Service, Washington, D.C. Cooperative technical publication, 1989, 76 pp. plus appendices.

Frayer, W. E., Monahann, T. J., Bowden, D. C., and Graybill, F. A., Status and trends of wetlands and deepwater habitats in the conterminous U.S. 1950's to 1970's, U.S. Fish & Wildlife Service, National Wetlands Inventory, St. Petersburg, FL, 1983.

Gilmer, D. S., Work, E. A., Colwell, J. E., and Rebel, D. L., Enumeration of prairie wetlands with landsat and aircraft data, *Photogramm. Eng. Remote Sensing*, 46(5) 631–634, 1980.

Hall, L. B., Clar, R. C., Von Loh, J. D., Halls, J. N., Pucherelli, M. J., and McCabe, R., The use of remote sensing and GIS techniques for wetland identification and classification in the Garrison Diversion Unit-North Dakota, Technical Papers 1988 ACSM-ASPRS Annual Convention, American Society of Photogrammetry and Remote Sensing, Falls Church, VA, 1988.

Jaynes, R. A., Clark, L. D., Jr., and Landgraf, K. F., Inventory of wetlands and agricultural land cover in the Upper Sevier River Basin, Utah, University of Utah, Center for Remote Sensing and Cartography, Salt Lake City, UT.

Jensen, J. R., and Davis, B. A., Remote sensing of aquatic macrophyte distribution in selected South Carolina reservoirs, Technical Papers 1987 ASPRS-ACSM Annual Convention, Vol. 1, American Society of Photogrammetry and Remote Sensing, Falls Church, VA, 1987, 57–65.

Jensen, J. R., Estes, J. E., and Tinney, L., Remote sensing techniques for kelp surveys, *Photogramm. Eng. Remote Sensing*, 46(6) 743–755, 1980.

Johnson, M. O. and Goran, W. D., Sources of digital spatial data for geographic information systems, U.S. Army Corps of Enginners, Construction Engineering Research Laboratory, Technical Report No. N-88-01, 1987, 33 pp.

Karaska, M. A., Walsh, S. J., and Butler, D. R., Impact of environmental variables on spectral signatures acquired by the Landsat thematic mapper, Technical Papers, 53rd Annual Meeting of the American Society of Photogrammetry and Remote Sensing, Vol. 1, 1987, 371–384.

Klemas, V., Evaluation of spatial radiometric and spectral thematic mapper performance for coastal studies, Quarterly Status Report, Delaware University, DE, 1983, 4pp.

Klemas, V. and Hardisky, M. A., Remote sensing of estuaries: an overview, University of Delaware, College of Marine Studies, Newark, DE.

Laboratory for Application of Remote Sensing (LARS) Purdue University, Application of remote sensing technology to the solution of problems in the management of resources in Indiana, W. Lafayette, IN, 1979.

Long, K. S., Remote sensing of aquatic plants, USACE Waterways Experiment Station Technical Report No. A-79-2, Vicksburg, MS, 1979.

Lyon, J. G., Remote sensing analysis of coastal wetland characteristics: St. Clair Flats, Michigan, Proceedings of the 13th International Symposium on Remote Sensing of Environment, ERIM, Ann Arbor, MI, 1979, 1117–1129.

Mackey, H. E., Jr. and Jensen, J.R., Macrophyte mapping with video technology in a fresh water lake, Proceedings of the First Workshop on Videography, American Society of Photogrammetry and Remote Sensing, Vol. 265, 1988, 86–71.

Mackey, H. E., Jr., and Jensen, J. R., Remote sensing of wetlands applications overview, Proceedings of the First Workshop on Videography, American Society of Photogrammetry and Remote Sensing, 1988, 32–33.

Mackey, H. E., Jr., Jensen, J. R., Hodgson, M. E., and O'Cuillin, K. W., Color infrared video mapping of upland and wetland communities, Proceedings of the 11th Biennial Workshop Color Aerial Photography and Videography in the Plant Sciences. American Society of Photogrammetry and Remote Sensing, 1987, 252–260.

Masry, S. E. and MacRitchie, S., Different considerations in coastal mapping, *Photogramm. Eng. Remote Sensing*, 46, 521–528, 1980.

McEwen, R. B., Kosco, W. J., and Carter, V., Coastal wetland mapping, *Photogramm. Eng. Remote Sensing*, 42(2) 221–232, 1976.

Mead, R. A. and Gammon, P. T., Mapping wetlands using orthophotoquads and 35mm aerial photographs, *Photogramm. Eng. Remote Sensing*, 47(5) 649–652, 1981.

Meisner, D. M., Fundamentals of airborne video remote sensing, *Remote Sensing Environ.*, 19, 63–79, 1986.

Meisner, D. M. and Lindstrom, D. E., Design and operation of color infrared aerial video, *Photogramm. Eng. Remote Sensing*, 51(5) 555–560, 1985.

Mouchot, M. C., Sharp, G., and Lambert, E., Thematic cartography of submerged marine plants using the fluorescence line imager, Canada Centre for Remote Sensing, Ottawa, ON.

Nelson, R. W., Logan, W. J., and Weller, E. C., Playa wetlands and wildlife on the southern Great Plains: a characterization of habitat, U.S. Fish and Wildlife Service, Fort Collins, CO, Report No. FWS/OBS-83/28, 1983, 163 pp.

Newbury, G. E., Changes in the wetlands of Hunting Creek, Fairfax County, Virginia, U.S. Army Engineer Topographic Labs, Fort Belvoir, VA, 1981.

Nixon, P. R., Escobar, D. E., and Menges, R. M., Use of multi-band video system for quick assessment of vegetal condition and discrimination of plant species, *Remote Sensing Environ.*, 17, 203–208, 1985.

Pelletier, R. E., James, R. T., and Smoot, J. C., Evaluation hydrologic modeling components of the Florida Everglades with AVHRR and Ancillary data, Technical Papers, 1990 ASCM-ASPRS Annual Convention, Vol. 4, American Society of Photogrammetry and Remote Sensing, Falls Church, VA, 1990, 321–330.

Sabins, F. F., Jr., Remote Sensing Principals and Interpretation, W.H. Freeman and Company, New York, 1987.

Scarpace, F. L., Quirk, B. K., Kiefer, R. W., and Wynn, S. L., Wetland mapping from digitized aerial photography, *Photogramm. Eng. Remote Sensing*, 17(6) 829–838, 1981.

Scarpace, F. L., Kiefer, R. W., Wynn, S. L., Quirk, B. K., and Frederichs, G. A., Quantitative photo-Interpretation for wetland mapping, Proceedings of the 41st Meeting of the American Society of Photogrammetry, American Society of Photogrammetry and Remote Sensing, Falls Church, VA, 1975, 750–771.

Schowengerdt, R. A., Techniques for Image Processing and Classification, Academic Press, Orlando, FL, 1983, 249pp.

Sharp, G., Carter, J., Roddick, D. L., and Carmichael, G., The utilization of color aerial photography and ground truthing to assess subtidal kelp (Laminaria) resources in Nova Scotia, Canada, Technical Papers 1981 ASPRS-ACSM Annual Convention, American Society of Photogrammetry and Remote Sensing, Falls Church, VA, 1981.

Shima, L. J., Anderson, R. R., and Cater, V. P., The use of aerial photography in mapping the vegetation of a freshwater marsh, *Chesapeake Sci.*, 17(2) 74–85, 1976.

SPOT Image Corporation, 1897 Preston White Drive, Reston, VA, 22091, (703)620-2200.

Steffenson, D. A., and McGregor, E. E., The application of aerial photography to estuarine ecology, *Aquat. Botany*, 2, 3–11, 1976.

Steward, W. R., Carter, V., and Brooks, P. D., Inland (non-tidal) wetland mapping, *Photogramm. Eng. Remote Sensing*, 46(5) 617–628, 1980.

Stohr, C. J. and West, T. R., Terrain and look angle effects upon multispectral scanner response, *Photogramm. Eng. Remote Sensing*, 51(2) 229–235, 1985.

U.S. Army Corps of Engineers, Environmental impact statement for operations, maintenance and minor improvements of the federal facilities at Sault Ste. Marie, Michigan, Appendix F: sediment aquatic plant, and bathymetry mapping from airborne scanner data, Detroit District Corps of Engineers, Detroit, MI, 1988.

Wallsten, M. and Forsgren, P. O., The effects of increased water level on aquatic macrophytes, *J. Aquat. Plant Manage.*, 27, 32–37, 1989.

Weismiller, R. A., The application of remote sensing technology to the solution of problems in the management of resources in indiana, Semiannual Status Report, Purdue University, IN, 1979.

Weismiller, R. A., Kristof, S. J., Scholz, D. K., and Anuta, P. E., Change detection in coastal zone environments, *Photogramm. Eng. Remote Sensing*, 43(12) 1533–1539, 1977.

Wicker, K. M. and Meyer-Arendt, K. J., Utilization of remote sensing in wetland management, Proceedings, Pecora VII Symposium, Falls Church, VA, American Society of Photogrammetry, 1981, 217–229.

Williamson, F. S. L., Investigations on classification categories for wetlands of chesapeake Bay using RS data, Smithsonian Institution Annual Report, Oct. 1972–Oct. 1973, 1974, 98 pp.

Wobber, F. J., Remote sensing trends in state resources management, *Photogramm. Eng. Remote Sensing*, 40(9), 1974.

Work, E. A. and Gilmer, D. S., Utilization of satellite data for inventorying prairie ponds and lakes, *Photogramm. Eng. Remote Sensing*, 42(5) 685–694, 1976.

CHAPTER 19

Analyzing the Cumulative Effects of Forest Practices: Where Do We Start?

Kass Green, Steve Bernath, Lisa Lackey, Matthew Brunengo, and Stuart Smith

ABSTRACT

Forest practices related to timber harvesting—clear-cutting, building roads, spraying herbicides, and so on—have both short- and long-term cumulative effects on the environment. Forestry practices affect wildlife habitat, riparian areas, snow and rainfall runoff, fisheries—that is, entire ecosystems. In Washington State, a Forest Practices Board has enacted regulations directed at assessing the cumulative effects of forest management. The state's Department of Natural Resources has charged Pacific Meridian Resources with creating a land-cover Geographic Information System (GIS) layer for 20 million acres of private and state lands. The GIS will be used to monitor and predict changes resulting from forest practices. Initially, GIS and remote sensing are being used to prioritize hydrologic basins for assessment based on their susceptibility to cumulative effects.

INTRODUCTION

One of the most compelling challenges facing land managers today is how to identify, measure, and monitor the cumulative effects of land management activities across both time and space. In no field are those challenges more intense than in forestry, where management activities—primarily forest harvesting and road building—can affect large expanses of land over long periods of time and where competing claims on the multiple resources of forests create conflicting demands for forest use.

Cumulative effects are changes to the environment caused by the spatial and temporal interaction of natural ecosystem processes resulting from two or more forest practices. Concerns about the potential cumulative effects of forest practices on wildlife, hydrology, fish, and erosion and sedimentation have become critical throughout the western United States. Although timber harvesting has

From *Geo Info Systems*, Feb. 1993. With permission.

been and still is a significant economic activity in the state of Washington, timber management has transformed the forest environment of the Pacific Northwest during the past century from a nearly continuous stand to a patchwork of urban and agricultural areas, harvest units, and mature trees. The political and regulatory environment of forestry has also changed because of increased attention to protecting woodland resources, potential off-site effects of forest practices, and an expanding regional population increasingly interested in the recreational and aesthetic resource provided by the state's forests. Cumulative effects analysis has focused primarily on the potential for fragmenting wildlife habitat, accelerating erosion, increasing flooding, deteriorating water quality, and changing stream morphology.

In the past, analyses of effects of forest practices relied primarily on evaluating *local* conditions—a forest harvest unit and areas directly adjacent to it. Considerations of ecosystems and watersheds typically were absent, partially because the information and methods needed to assess cumulative problems generally were inadequate or nonexistent. In addition, no standardized measure of cumulative effects could be communicated to interested parties.

For the past decade Washington State's Forest Practices Board has recognized concerns about cumulative effects and tried to comprehensively address them (Geppert et al., 1984). In 1989 several interest groups, including Indian tribes, state agencies, forest land owners, environmental organizations, and local governments, participated in the Sustainable Forestry Round Table. The round table explicitly recognized cumulative effects as an important forest issue and asked, "Given current knowledge, what kind of technical approach could be used to begin dealing with cumulative effects?"

Those discussions and increasing public interest led the state legislature to provide incentives to Washington's Forest Practices Board to pass a regulation addressing cumulative effects. Subsequently, the Forest Practices Board directed the Department of Natural Resources (DNR) to consult with timber, fish, and wildlife interests to provide a scientifically based tool for analyzing cumulative effects resulting from forest practices (Bernath et al., 1992). This effort culminated in a proposed technical framework to address cumulative effects on the 20 million acres of private and state forest lands in Washington on a watershed basis.

A basic premise of cumulative effects analysis is that landscapes differ in their sensitivities to forestry activities. For example, the effects of forest harvesting on water quality will vary spatially because of differences in climate, geologic materials, vegetative cover, and terrain. Because of the diversity of operations and conditions, all areas are not equally sensitive to any particular forest practice and probably no area is sensitive to all possible negative practices. Consequently, the risks to water quality, fish, and wildlife habitat associated with different forest management activities vary from place to place.

Cumulative effects assessment will eventually be implemented within all watershed basins containing state and private forest land. But staff and funds are not available to perform all work on all basins in a short time. Because all watersheds cannot be examined simultaneously, watershed screening—the identification and ranking of basins according to their susceptibility to cumulative effects—will determine the order in which basins will be visited for cumulative effects assessment.

Four types of screens have been identified for key categories of hazard and risk: slope instability, fisheries, wildlife, and hydrology. Fisheries and slope instability screens, as well as the final prioritizing of basins, were done by DNR staff. In spring 1991, DNR retained Pacific Meridian Resources (Emeryville, CA) to create a land-cover GIS layer for Washington's forested areas and to use the land-cover layer and other GIS data to create the DNR-designed wildlife and hydrologic screens. This article focuses on how GIS and remote sensing are being used to prioritize basins across the state.

METHODS

The project was completed in two phases. Phase 1 involved building the GIS coverages required to create the screens. In the second phase, the screens were actually created. All data and results were delivered to DNR as GIS files.

Phase 1: Data Capture

Eight GIS coverages are required to prioritize the watersheds for further examination:

- watershed basins
- forest type
- current land cover (for both the wildlife and hydrology screens and to be used as baseline vegetation data for future projects, including long-term monitoring of cumulative effects)
- precipitation (for the hydrology screen)
- rain on snow (also for the hydrology screen)
- soil hazard rating
- slope
- fish presence or absence and hatchery locations.

Because time and funds are limited, existing data had to be used wherever possible, and new layers had to be developed quickly and at minimum cost. A major challenge of the project was integrating data from various sources into a reliable and working analytical GIS.

Watershed Basins

The state of Washington has established hydrologic administrative units, called Water Resource Inventory Areas (WRIAs), across the state. Basins for the screening analysis are subdivisions of WRIAs and are fourth- to fifth-order in size, averaging 100,000 to 150,000 acres. A total of 204 basins have been designated. Each WRIA has 1 to 7 basins. Approximately 30 of the basins were excluded from this analysis because they do not contain significant tracts of commercial forest. DNR provided the basin and WRIA GIS coverages.

Forest Type

The project focused on the effects of cumulative practices on forest-type land. *Forest lands* are nonresidential or agricultural acres that are capable of growing deciduous or conifer trees, including acres in harvest units that do not presently support a canopy of trees. Therefore, it was necessary to build a GIS layer that characterized forest type (a land-use designation) vs. current forest cover (a vegetation-type classification). U.S. Geological Survey (USGS) land use–land cover digital maps were used to determine forest type. These 1:250,000-scale coverages have a minimum mapping unit of 10 acres for urban areas, water, and mines and quarries, and 40 acres for all other categories. They are derived from aerial photography and other remotely sensed data; dates of classification vary from 1973 to 1983.

Current Forest Cover

The first step in developing the forest-cover layer was to specify cover classes. A simple classification system was agreed on that would identify vegetation in critical stages of maturity for wildlife and hydrology and make it possible to use existing U.S. Forest Service (USFS) GIS forest-cover data for national forest and national parklands. Table 1 describes the classes.

GIS forest-cover data for national forest and national parklands were derived from size and crown closure classifications previously developed for USFS's Region 6 by Pacific Meridian Resources (Teply and Green, 1991). Those existing data covered 15.2 million acres of Washington State forested lands (including some private lands), or 48% of the total study area. Forest-cover data for the remaining 16.6 million acres of private, tribal, state, and other government lands were captured by classifying portions of 15 terrain-corrected and georeferenced LandSat Thematic Mapper (TM) scenes (EOSAT, Lanham, MD). All processing was performed using image-processing software (ERDAS, Atlanta, GA) in conjunction with ARC/INFO GIS software (Environmental Systems Research Institute, Redlands, CA).

To use remotely sensed data to map vegetation, correlations must be established between the variation found in the remotely sensed data (aerial photography or satellite imagery) and the variation that occurs in vegetation on the ground. To establish such links, we used a multistage approach that relied on integrating image processing; interpreted high-altitude, 1:67,600-scale, black-and-white aerial photography; field reconnaissance; and review of draft maps by DNR regional foresters.

Table 1   Description of Land Cover Classes

Class	Definition	Wildlife equivalency	Hydrologic equivalency
1	>10% tree crown closure in trees ≥21" DBH with >70% total crown closure and <75% of the crown in hardwoods or shrubs	Late seral stage	Hydrologically mature (100% mature)
2	<10% tree crown closure in trees ≥ 21" DBH with >70% total crown closure and <75% of the crown in hardwoods or shrubs	Mid-seral stage	Hydrologically mature (100% mature)
3	10–70% total crown closure and <75% of the crown in hardwoods or shrubs	Early seral stage	Hydrologically immature (50% mature)
4	<10% crown closure and/or >75% of the crown in hardwoods or shrubs	Cleared forest	Cleared forest (0% mature)
5	Water	Open water	Open water
6	Nonforested	Nonforest land	Nonforest (100% mature)

First, the seven bands of the TM data were reduced to Bands 3, 4, 5, and a ratio of Band 3 to Band 4. The first three bands were used because they are most indicative of vegetation. The ratio compensated for confusion caused by shadows on north slopes. Next, each scene was cropped to show only the area being classified. The relevant area of each scene was determined by:

- relationship with adjacent overlapping scenes
- major ecological regions (so that areas in different ecological regions of the state were classified separately)
- USFS and National Park Service lands already classified
- extent of the study area

An ISODATA unsupervised clustering algorithm was then run on the imagery for each area. Unsupervised classes (or spectral clusters) were identified using photo interpretation and minimal field reconnaissance. Spectral clusters that obviously corresponded to one of the five classes in the classification system were colored in a dull color that symbolized the class represented. Unknown or confused spectral clusters were given bright, garish colors.

The colored spectral clusters were then plotted as hard-copy maps at the same scale as the high altitude photography to allow direct comparison. The maps were reviewed for misidentification and to identify the brightly colored unknown spectral classes. Following review by Pacific Meridian staff, the plots were reviewed by experienced DNR regional foresters. DNR foresters were asked to delineate misidentified areas and answer questions about confused or mislabeled areas, which were then corrected.

For analysis, the raster coverages were then transformed into polygons for each WRIA. Polygons simplify the vegetation coverage, imitate the photo interpreter's delineation of vegetation types, and eliminate the noise associated with the image pixel classification. The minimum size for polygons was 10 acres for noncontrasting land-cover types and 5 acres for contrasting types (e.g., nonforested areas next to closed canopy stands).

Accuracy assessment of the new forest-cover GIS was performed by DNR's Forest Practices Division staff (Harper, 1991). DNR divided the state into 15 ecoregions, and each scene was divided into the ecoregions within it. Polygons of at least five acres were delineated on every twentieth 1:12,000 aerial photo for four selected townships within each scene–ecoregion combination. Photo interpretation of the polygons was compared with the correlating classification. An error matrix with user's and producer's accuracy (Story and Congalton, 1985) was created for each ecoregion–scene combination. Table 2 lists the results of overall accuracy for each scene–region.

The priorities for cumulative-effects analysis may need to be changed over the long term as forest cover changes in each basin. Another step in cumulative-effects analysis will be to maintain the forest-cover GIS. Because the forest-cover GIS coverage is derived from digital satellite imagery, changes can be monitored by directly comparing past images with newly acquired, up-to-date images. Multidate Landsat TM imagery provides an efficient means to identify and measure change in land-cover over time, particularly change from loss in vegetation cover (Maus et al., 1992). As part of a National Aeronautics and Space Administration (NASA)–funded contract, Pacific Meridian Resources has developed production methods for using satellite imagery to detect land-cover and land-use change. DNR staff were trained to use these methods during a two-day, on-site training session.

Table 2   Results of Accuracy Assessment

Scene, path-row	Ecoregion	Number of samples	Overall accuracy (%)
43-27	Palouse	36	83.3
43-28	Blue Mountains	12	91.7
44-26 and 44-27	Pend Orielle	30	93.3
45-27	Southeast Cascades	39	97.4
45-28	Southeast Cascades (a)	30	93.3
45-28	Southeast Cascades (b)	30	96.7
46-27	Puget lowland	30	86.7
46-27	Central Cascades	33	87.9
46-28	Southwest Washington	33	93.9
47-26	North Cascades	33	97.0
47-26	Puget lowland	24	95.8
47-27	Southwest Washington	36	88.9
47-28	Southern coast	30	92.2
48-26 and 48-27	Northwest coast	36	88.9
Overall	Washington private and tribal lands	432	91.7

Precipitation and Rain on Snow

Two of the GIS coverages used in the analyses were precipitation and rain on snow. The precipitation coverage was digitized by DNR using a National Oceanic and Atmospheric Administration (NOAA) 10-year, 24-h precipitation, isohyetal map (Miller et al., 1973). The rain-on-snow coverage was created by DNR as a map of five precipitation zones likely to have various amounts of snow on the ground in early January. The zones are as follows:

- lowland—no snow, only rain
- rain dominated—some snow, but mostly rain
- rain on snow—approximate amount of snow potentially melted by the worst possible 24-h storm in a 10-year period
- snow dominated—some rain, but mostly snow
- highlands—no rain, only snow

National Weather Service, cooperative snow survey, and elevation data were all used to delineate the rain-on-snow zones. Both the precipitation and the rain-on-snow coverages were reprojected into UTM zones and clipped to each WRIA for analysis.

Soil Hazard and Slope

Soil-hazard ratings are available for state and private commercial forest lands in a soils layer stored on DNR's existing GIS. Because digital soil-hazard information is not available for federal lands, an average soil-hazard rating was assumed on USFS and National Park Service lands. Slope classes were derived from statewide, USGS, 3-arc-second, digital elevation data. Five classes were used: 0 to 5% from 5 to 27%, from 27 to 49%, from 49 to 70%, and more than 70%.

Fish Presence and Absence and Hatcheries

The Washington Rivers Information System, a GIS database maintained by the Department of Wildlife, contains site-specific information about state, tribal, and federal hatchery facilities and the presence and absence of fish species. An updated and verified version of those data is maintained by DNR and was used as input data in the fisheries screen.

Phase 2: Analysis

The wildlife and hydrology screens were created using ERDAS's raster GIS module on individual WRIAs. The slope instability screens were created in ARC/GRID on a statewide basis. The fisheries screen was computed statewide using ARC/INFO.

Wildlife Screen

The wildlife screen scores each WRIA for six factors:

* number of acres of forest land
* proportion of the forested type in late-seral-stage habitat
* proportion of the forested type in mid-successional forest cover
* proportion of the forested type in functional, late-seral-stage habitat (defined as areas having more than 60 continuous acres in the late-seral cover type)
* proportion of the forested type in large patches of late-seral-stage habitat (defined as areas having more than 640 continuous acres in the late-seral cover type)
* distribution of late-seral-stage habitat by basin.

Focusing on late-seral forest in the wildlife screen emphasizes providing diversity of habitat for species, such as the spotted owl, not found in earlier successional stages. The size and distribution of such patches are important because these factors influence the dispersal of wildlife. Consideration of the midsuccessional stage reflects the need to preserve natural diversity in a basin because it identifies the habitat that will soon grow into the late-seral stage.

Hydrology Screen

The primary hydrologic problem related to forest management in Washington is how timber harvesting affects runoff during rain-on-snow events. In the Pacific Northwest, the heaviest rains generally occur when cyclic-frontal winter storms bring warm, moist air from the southwest. Because some snow likely will be on the ground at the middle and higher elevations when such storms occur, snowmelt can combine with rainfall to aggravate floods and trigger landslides. The distribution of forest stands having varying degrees of crown closure and their similarity to stands having natural runoff characteristics can be used to rate the basins for their vulnerability to altered peak flows and channel morphology.

The basic purpose of the hydrology screen is to calculate the change in the amount of available water (WA) during a hypothetical 10-year, 24-h, rain-on-snow storm (Brunengo, 1992) from an assumed fully forested condition to the current forested condition. The formula emphasizes the significance of rain-on-snow events and is based on snowpack, precipitation, temperature, and wind speed.

The GIS model calculates the area-weighted averages of storm precipitation plus snowmelt for each basin using two scenarios: (1) all forest types support hydologically mature vegetation, and (2) the pattern of vegetation is as it exists on the current land-cover maps. The difference between these calculated values for a given basin represents the change as a result of timber harvest in water available for runoff that could occur in a storm. The greater the difference, the greater the potential for increased peak flows and damaging effects downstream. The data produced from this screen were normalized to values between 0 and 100.

Slope Instability Screen

Slope instability is directly related to soil stability, slope, and hydrology. The slope instability screen is a combination of the soil hazard, slope, and rain-on-snow GIS layers. To facilitate a statewide assessment, all analyses were performed in ARC/GRID. First, the GIS layers for soil hazard and slope were combined to produce one layer that indicates where steep and unstable areas are located. That layer was then combined with the rain-on-snow layer to show areas in which storms will increase the probability of slides. The combined data were area-weighted for each basin and normalized to result in basin values between 0 and 100.

Fisheries Screen

Fish production requires continuous supplies of reliable, high-quality water. The fisheries screen recognizes the importance of streams that supply water to fish hatcheries or support the state's fish populations. The screen uses data from the Washington Rivers Information System to calculate the proportion of fish to the total mileage of the river systems in each basin. Additional weight is given to basins having one or more fish hatcheries present. Most values for this screen were fractions of one. A few basins that had a high proportion of active fish habitat and hatcheries had values greater than one.

RESULTS

Wildlife Screen

Figure 1 displays the distribution of land-cover types across the state of Washington. Table 3 lists the acres of each land-cover class for the study area. Figures 2 and 3 present wildlife screens for the Hoh River WRIA on the Olympic Peninsula.

The percentage of late-seral habitat to total forest type in a WRIA ranges from 0 to 60%. Fifteen of the 55 WRIAs in the study area have more than 25% of their forest-type acres in late-seral habitat. Of those 15, 11 have more than 25% of forest-type acreage in continuous blocks of 640 acres or more.

In general, a very strong correlation seems to exist between the proportion of late-seral-stage habitat and the amount of national forest or National Park Service lands in a WRIA. WRIAs having the highest amount of seral habitat included areas of the Olympic and North Cascades national parks and the wilderness areas of the Mount Baker–Snoqualmie and Gifford Pinchot national forests. WRIAs having less than 5% late-seral-stage forest comprise primarily private, state, and tribal lands that have been heavily harvested.

The wildlife screen results are being forwarded to the Wildlife Committee of the Forest Practices Board for consideration in their exploration of methods to develop landscape-based regulation of forest practices for the protection of wildlife.

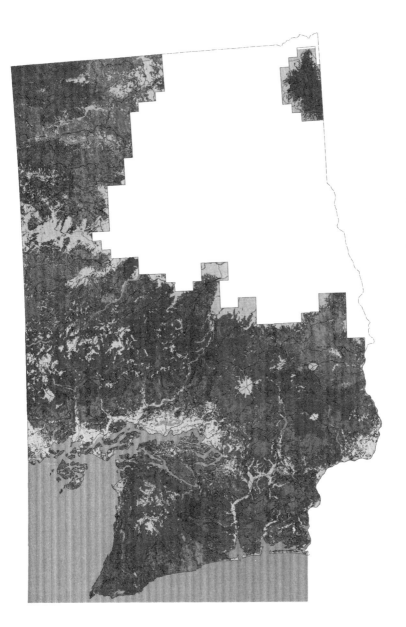

Figure 1   Land-cover types for Washington State study area.

Table 3   Acres of Each Land-Cover Class for the Study Area

Class	Description	Acres
1	Late seral stage	5,074,485
2	Mid-seral stage	6,759,677
3	Early seral stage	4,881,117
4	Other	8,808,365
Total forest type		25,523,644
5	Water	604,195
6	Nonforest	5,612,231
Total for study area		31,740,070

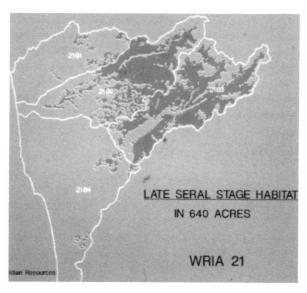

Figure 2   Wildlife screen for Hoh River WRIA on the Olympic Peninsula in 640-acre units.

Watershed Prioritization

Figures 4 through 6 present an example of the hydrologic screens for the Hoh River WRIA on the Olympic Peninsula. The hydrologic, slope instability, and fisheries screens were combined to prioritize the forest basins of the state for watershed analysis. The following equation was used:

$$C = (H + S)/2 * F$$

where, C = the combined ranking, H = the value from the hydrologic screen, S = the value from the slope instability screen, and F = the value from the fisheries screen. Figure 7 illustrates the results of the combined ranking of watershed prioritization across the state of Washington.

Between the start of the project, when screening was first envisioned, and the present, a cumulative-effects regulation was adopted and cumulative-effects

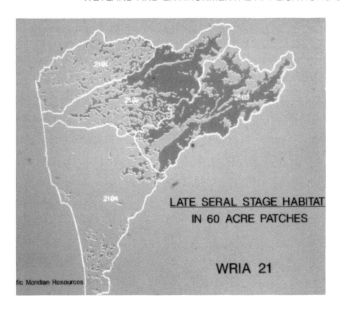

Figure 3   Wildlife screen for Hoh River WRIA in 60-acre units.

Figure 4   Water available for runoff in fully forested conditions for the Hoh River WRIA.

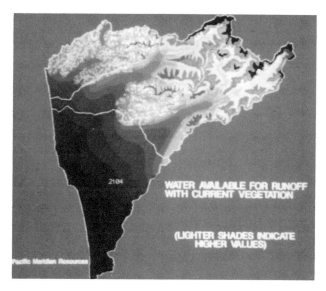

Figure 5 Water available for runoff with current vegetation conditions in the Hoh River WRIA.

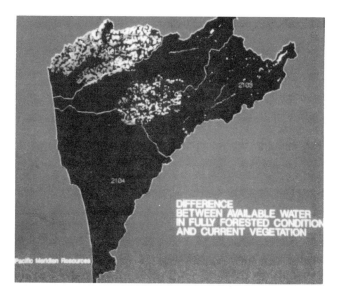

Figure 6 The difference between water available for runoff in fully forested conditions and current conditions in the Hoh River WRIA.

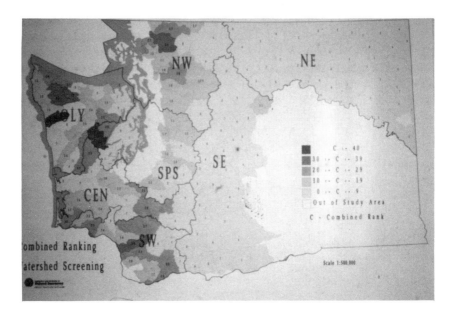

Figure 7   Combined rankings of watershed prioritization throughout Washington State.

analyses are being implemented in Washington's forest basins. In the process of negotiating the final rules and investigating available technologies for cumulative-effects assessment, the size of a basin in which the assessment would be performed was reduced from a fourth- to fifth-order stream averaging 100,000 to 150,000 acres in size, to basins averaging 10,000 to 50,000 acres. As a result, the watershed screening is being recalculated for each of the smaller basins, a process that is made easier by the existence of all data layers and analysis models in a GIS. The watershed screen is where the DNR will first perform cumulative-effects assessment on regional and statewide bases.

CONCLUSIONS

Concern about the cumulative effects of forest practices will continue to affect forest management decisions and regulations. Public attention is focused in this arena and will continue to be vigilant. Because of the high values at stake, methods for identifying, measuring, and monitoring cumulative effects must be scientifically objective and produce management and regulatory strategies that are effective, efficient, and fair.

Responding to directives from policy-making authorities, Washington's DNR staff and cooperators in the state's timber, fish, and wildlife agreement have begun to analyze and regulate forest practices based on their potential to cause significant cumulative effects. GIS and remote sensing are critical tools in this process. Performing the analysis in a GIS provides methods and results that are easily tested for their sensitivity to assumptions, repeatable, and accessible to interested parties. Using satellite imagery to create the forest-cover GIS saved both time and money and produced a coverage that can be easily updated and assessed for change.

The assessment of basinwide hazards and forestry activities likely to affect them can help resource managers minimize both local and cumulative problems by aiding in the design of prescriptions for watershed-specific forest practices. The resulting watershed prioritization provides a physical basis for identifying where DNR will direct its limited resources. Subsequent forest practices in areas of identified sensitivity will incorporate forest-management prescriptions specified to ensure that fisheries, water quality, and public facilities will be protected and restored.

Evaluating the risk to public resources helps establish criteria for decision making based on standards for water and habitat quality. Appropriate changes in forest practices to minimize the potential for significant adverse effects is the result. In addition, the assessment can be used to identify opportunities for monitoring the effectiveness of management and regulatory actions in meeting public resource objectives.

ACKNOWLEDGMENTS

The authors thank the staffs of both the Department of Natural Resources and Pacific Meridian Resources for their hard work, which made this project possible. We are particularly indebted to Deanna Harper, Paul Hardwick, and the DNR regional foresters, who so patiently reviewed the forest-cover classifications.

REFERENCES

Bernath, S., et al., Using GIS and image processing to prioritize cumulative effects assessments, GIS '92 Symposium Proceedings, Vancouver, BC, Canada, 1992.

Brunengo, M., et al., Screening for watershed analysis: a GIS method of modeling the water input from rain on snow storms for management and regulation of clearcut forest harvest, 1992 Western Snow Conference Proceedings. Jackson Hole, WY, 1992.

Geppert, R., et al., *Cumulative Effects of Forest Practices on the Environment—A State of Knowledge*, Ecosystems, Olympia, WA, 1984.

Harper, D., *Accuracy Assessment Procedures*, Washington State Department of Natural Resources, Olympia, WA, 1991.

Maus, P., et al., Utilizing Satellite Data and GIS to Map Land Cover Change, GIS '92 Proceedings, Vancouver, BC, Canada, 1992.

Miller, J., et al., *Precipitation Frequency Atlas of the Western United States. Volume IX— Washington State.* Atlas 2, National Oceanic and Atmospheric Administration, Washington, D.C., 1973.

Story, M. and Congalton, R., Accuracy assessment: a user's perspective, *Photogramm. Eng. Remote Sensing* 52(3), 397–399, 1986.

Teply, J. and Green, K., Old growth forests: how much remains? *Geo Info Systems* 1(4), 22–31, 1991.

CHAPTER 20

Errors in GIS: Assessing Spatial Data Accuracy

Paul V. Bolstad and James L. Smith

INTRODUCTION

Effective use of Geographic Information Systems (GIS) requires adjustments in how resource professionals collect and document their spatially referenced data. Prior to GIS, maps were used mainly for tactical planning; high spatial accuracy was not required, so errors in spatial data were ignored or considered acceptable.

With GIS, many heretofore costly or time-consuming spatial analyses are now practical. However, this increases the need to measure and document spatial data accuracy (Smith et al., 1991). As models become more sophisticated and rely on site-specific data from many sources, the combined effects of spatial or attribute errors may limit the value of model predictions. For example, errors in stand volume from a GIS-based growth simulator might be due to errors in the growth model, errors in GIS stand area, or erroneous inventory data. We need to know the accuracy of our spatial data before we can effectively perform such spatially based analyses.

This article discusses spatial data accuracy in a resource management setting, beginning with a brief description of spatial data models and an overview of spatial data accuracy (including definition, identification, quantification, and documentation). It concludes with a review of accuracies associated with various common data sources and a prescription for improvements in the application of GIS technologies.

DATA MODEL

The data model defines how real-world spatial entities (stands, roads, lakes) are represented in the GIS. Both spatial (location, size, geometry) and tabular (timber type, forage value, site index) characteristics must be represented. Digital

Reprinted from the *Journal of Forestry* 90(11) published by the Society of American Foresters, 5400 Grosvenor Lane, Bethesda, MD 20814-2198. Not for further reproduction.

"objects" are usually represented by a set of coordinates defining the spatial characteristics and by attribute or tabular data describing the tabular characteristics. Most data models involve several data layers, each corresponding to a different type of data (Figure 1).

Once the data model has been specified, the area is "digitized"—that is, a digital representation is recorded for the area coordinates used to represent each spatial entity, as well as the tabular data associated with the spatial data. For example, the boundaries and tabular data describing all stands, lakes, roads, and property boundaries might be entered. Data entry and update are likely to be the largest share of GIS investments (Congalton and Green, 1992). This fact underscores the need to control and document the accuracy of the spatial data.

As described earlier, the real world is represented by both a spatial component (coordinates) and a tabular component (attribute data). Because errors may occur in both, we must discuss two types of accuracy: positional and attribute. Positional accuracies are used to gauge how well the location, size, and shape of real-world features are represented in the database. Attribute accuracies reflect how well the tabular characteristics represent reality.

POSITIONAL ACCURACY

Positional accuracy can be viewed as a measure of how well the coordinates in the data layer correspond to the "true" coordinates of an entity on the ground. Accuracies, often defined by some measure of positional error, are usually described separately for the horizontal and vertical dimensions.

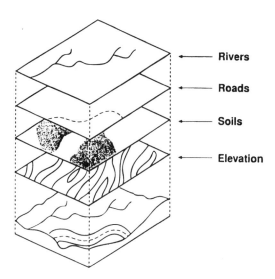

Rivers

Roads

Soils

Elevation

Figure 1 A geographic information system can combine many layers of spatial information into one database.

Positional inaccuracies can be derived from several sources. For example, manual digitizing from paper maps is a common method for entering spatial data (Burrough, 1986). Errors may originate in the field surveys, in drafting or map production, in nonuniform distortion of the paper map, in the digitizing equipment itself, or in the activities of the operator. Positional error may also be added in data processing steps. It is possible to measure error associated with each step; but errors may be nonadditive, so it is best to measure the accuracy of the resultant digital data layers.

Ideally, positional errors should be documented for each separate data layer. "True" coordinates would be determined for a number of real-world entities and compared to the coordinates of their corresponding digital objects. "True" coordinates should be accurately determined—at least to the level of accuracy required by all intended geographic analyses. For example, a random sample of public land survey system (PLSS) section corners, paired with the corresponding database coordinates for the digital representations, would allow error calculation for each point, which could then characterize the positional error in the data layer.

While rigorous empirical tests are desirable for all data layers, they are often not performed due to insufficient knowledge, time, funds, or numbers of suitable checkpoints. Many GIS users are not familiar with data accuracy concepts and measurements. In addition, obtaining precise locations for check features requires field surveys.

In the PLSS example cited above, the section corners must be selected and identified in the field, and the coordinates determined. Specialized equipment and skilled personnel are often required to obtain desired accuracies. Multiply this effort by the number of data layers and the costs often become quite large. Finally, many features represented in a digital database are difficult to identify in the field. For example, stand boundaries interpreted from aerial photographs may be difficult to identify on the ground. Despite these difficulties, there are few viable substitutes for empirical tests of database accuracy.

ERROR MODELS

Positional accuracy may be modeled in various ways. Individual points defining geographic features may be considered as a population with both vertical and horizontal error distributions. If the vertical error (the difference between true and database elevation) is assumed to follow a random normal distribution, the mean and variance of this distribution then characterizes the population of vertical point errors. A commonly used metric is the vertical root mean-square error, $RMSE_v$:

$$RMSE_v = [e_r^2/n]^{1/2} \qquad (1)$$

where e_r is the residual error for each measured point (the difference between the true and database elevations), and n is the number of test points. The validity of the distribution model, and the parameters that define the probability density function, can be determined through sampling. Horizontal error may also be modeled statistically, although more distributions and approaches have been proposed because horizontal error may be considered either univariate or bivariate, with interactions between x and y errors. One common metric is the horizontal root-mean square error:

$$RMSE_h = [(e_x^2 + e_y^2)/n]^{1/2} \qquad (2)$$

where e_x is the error measured in the x direction and e_y is the error in the y direction. Another common measure is the horizontal circular standard error, which can be approximated by $RMSE_h/1.4142$ (Thompson and Rosenfield, 1971).

Describing the positional error for linear or areal features is a bit more complicated. The difference between the "true" and digital data line may vary along the length of the line (Figure 2a). Errors will range from zero at line crossings to high values at the largest separations. One widely applied notion, the epsilon model, defines a distance measured at right angles to the line direction (Perkal, 1956; Chrisman, 1982). A band around each true line is defined by an epsilon distance (Figure 2b). Various probability distributions have been postulated (Dunn et al., 1990). The proportion or probability of lines within a specified band width can be used to assess line accuracy—for example, 95% of the digital line coordinates may be required to be within an epsilon band of 15 ft for a given data layer.

ERROR SOURCE AND MAGNITUDE

Digital spatial data are derived from a number of sources including maps, aerial photographs, satellite imagery, traditional traverse and leveling surveys, and global positioning system surveys. Each source involves a number of steps and transformations from the original field measurements to the final digital coordinates. The origin and magnitude of common error sources are summarized in Table 1.

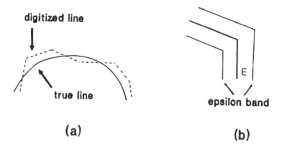

Figure 2   Errors in linear features can be described by (a) the difference between true and digital line location and (b) the epsilon band concept.

Field Measurements

All positional information ultimately relies on field measurements. These may be very precise and painstakingly obtained, such as those that define legal property boundaries, or they may be quite rough, such as the "eyeball" location of an inventory plot on topographic maps. While accuracies for eyeball estimates are in practice unknowable, those determined from the reduction of closed-loop traverse and leveling survey data can be accurately determined (Wolf and Brinker, 1989). Traditional surveying practices (theodolite or transit with stadia) commonly achieve accuracies to 1 in 10,000. Since survey points are generally within a few miles of high-accuracy geodetic control monuments, locations surveyed using traditional methods should provide better than 1-ft accuracy. Unfortunately, traditional surveying methods are too expensive and time-consuming for most data layers.

Global Positioning System (GPS) technology, another "direct" method, provides lower costs and higher throughput than traditional surveying in many situations. GPS consists of a control segment, a constellation of satellites, and GPS receivers. Positions are determined based on signals broadcast by each satellite. Code-based GPS methods (most commonly applied in a resource setting) provide measurement accuracies of 50 to 150 ft with one receiver and 5 to 15 ft with two receivers. Carrier-phase receivers, while more expensive and time-consuming, are capable of providing one-inch RMSEs (Wolf and Brinker, 1989).

Even though GPS technologies have reduced field survey costs dramatically, they are only appropriate for a limited number of data layers. GPS field measurements are best applied where high accuracies are required (e.g., the control data layer) or for sparse linear networks that may be traversed at a high rate of speed such as roads or trails. They are also appropriate when accuracy is re-

Table 1   Reported *RMSE* Ranges for Common Spatial Data Sources Based on Reports of Best "Commercial" Practices

Accuracy component	*RMSE* range (ft)
Traditional transit surveys	0.1–2
GPS	
Carrier phase	0.1–1
Code-based, stand-alone	50–150
Code-based, differential	5–15
Maps	
1:24,000 before digitization	10–35
Manually digitized from 1:24,000 maps	20–55
Digital line graph data	20
Aerial photographs	
Uncorrected 9 in. flat terrain	10–50
Uncorrected 9 in. steep terrain	125–250
Corrected 9 in. steep or flat terrain	0.5–5
Uncorrected 35 or 70 mm	25–300
Corrected 35 or 70 mm	3–15
Satellite data	
Landsat TM	15–45
SPOT HRV, multispectral	10–30

quired but alternative methods are unavailable—for example, sites far from control points, or remote locations under forest cover. However, GPS is currently not practical where large areas need to be covered in detail.

Maps

Manual or automated map digitization is currently the most common form of spatial data entry, and as such has the greatest impact on spatial data accuracy. Manual digitizing usually involves putting a paper or mylar map on a digitizing surface and tracing the features to be entered.

Original map accuracy is an important determinant of spatial data accuracy. In the United States, national map accuracy standards, or NMAS (Wolf and Brinker, 1989), require that no more than 10% of all "well-identified" points exhibit more than 1/50th of an inch positional error for map scales less than 1:20,000, or more than 1/30th of an inch for scales greater than 1:20,000. "Well-defined" points include road intersections, road/stream crossings, and wells. For 1:24,000-scale maps, this implies a ground-measured *RMSE* of less than 24 ft (univariate normal error model) or 26 ft (bivariate model). How well do available maps meet these standards? In one evaluation (Thompson and Rosenfield, 1971), approximately 83% of the USGS 1:24,000-scale maps tested met NMAS criteria.

Three cautions are in order. First, these test results apply for well-defined points, and many features digitized from maps (rivers, forest boundaries, contours) are not well defined. Second, these results were reported for high-quality maps produced by the U.S. Geological Survey; these accuracies may not hold for other sources or other map series. Finally, no limits are implied for the 10% of points that may be in error.

Digitization

Positional accuracies during digitization are affected by the equipment and by operator skill or state of mind. Currently digitizers report precision and/or accuracies better than 0.001 in. One rigorous evaluation (Warner and Carson, 1991) identified equipment errors of 0.0035 in., which at a 1:24,000 scale corresponds to 7 ft. Operator errors vary widely but are generally larger than digitizing table errors. One small study documented manual digitizing precision of approximately 0.0025 in. for well-defined points, approximately twice the observed equipment error (Bolstad et al., 1990). This translates to 4.8 ft on a 1:24,000-scale map and 50 ft on a 1:250,000 map.

Different errors may be observed for digitized arcs and polygons. Linear features are often depicted with line widths of 0.01 to 0.04 in. In addition, generalization of smooth curves introduces additional errors. One study found *RMSEs* of 22 ft for points from USGS digital line graph data (Vonderhoe and Chrisman, 1985). Another study measured approximate average epsilon distances below 0.005 in., equivalent to 10 ft on 1:24,000 maps (Dunn et al., 1990). Furthermore, when epsilon ranged from 0.005 to 0.04 in., polygon area error ranged from 1.6

to 16% and was inversely related to polygon size. In another study, polygon area error varied from 5 to 15% of polygon area under realistic assumptions about spatial errors (Prisley et al., 1989).

Coordinate Registration

Registration involves converting from digitizer coordinates to the coordinate system defined by the map projection used for printing the source map. Positional accuracies may suffer at any of the steps involved: identifying control points in both the geographic and digitizer space; obtaining coordinates for the control points in both coordinate systems; choosing a mathematical transformation and estimating coefficients; and applying the transformation to the digitized data, thus producing the output layer.

While large blunders are easily detected, small blunders or random errors are not. Control point coordinates must be obtained from ground surveys or from controls drafted on the source map. When field measurements are lacking, control is commonly digitized from geographic coordinate points drafted on the map, e.g., Universal Transverse Mercator graticule intersections drafted on 1:24,000-scale base maps. These controls will contain the positional errors described above. Empirical tests have documented map-derived control errors ranging from 7 to 25 ft (Norberto Fernandez et al., 1991) and 5 to 279 ft (Bolstad et al., 1990). GPS technology should facilitate the collection of accurate control data, thus reducing the impact of control uncertainty.

Imagery

Imagery is a common source for natural resource spatial data, both for initial database development and for updates. Most current imagery comes from aerial cameras and satellite scanners, although systems based on video cameras and airborne scanners may soon become popular.

Aerial photographs for resource mapping have been routine for the last 50 years. During this time, correction methods and accuracies of photo-based mapping have been thoroughly documented. Unfortunately, these error correction methods are often ignored. In the past it was not considered a problem because area errors were less than those associated with the technology used to measure map area (e.g., dot grids or planimeters) and the maps were not used in spatial analyses such as multilayer overlays. However, these errors are becoming more obvious and objectionable in a GIS framework.

Tilt and terrain distortion are the two major causes of positional inaccuracies in large-format (9-in.) aerial photographs. Elevation variation causes radial displacement away from the photo perspective center, while camera tilt causes perspective distortion (Wolf, 1983). Tilt distortion may be present even in "vertical" aerial photographs, because vertical photos are commonly defined as those with tilts less than 3°. If not removed, these distortions may result in large positional errors in data layers derived from the photographs (Figure 3). For ex-

ample, one simulation study of "vertical" photographs reported average positional errors in GIS data layers of 13 to 52 ft over flat terrain and 125 to 240 ft
over steep terrain (Bolstad, 1992). The same study documented area errors of 0
to 9% on 9-in. vertical aerial photos.

Terrain effects may be removed by stereoscopic viewing. Stereo interpretation recreates a three-dimensional model, which may then be projected onto a
planar surface (Wolf, 1983). Stereomodels may be projected on a base map, as
with a stereo zoom transfer scope, in which case the quality of the data can only
be as good as the map.

While manual interpretation of stereopairs will remove terrain distortion, it
will not remove tilt effects (Figure 3, relief = 0). The stereomodel will have at
least the minimum of the tilts in the two stereopair photos. Tilt can be removed
through a three-dimensional projective transformation (which requires a minimum of four control points), but will not be removed with the two-dimensional
transformation commonly used to register digitized spatial data. Commercial
software has been marketed with "tilt compensation." However, it does not remove tilt errors; rather, it averages them over the image using a two-dimensional affine transformation. Simultaneous removal of tilt and terrain distortion
requires a more rigorous analytical photogrammetric model based on projection geometry.

Terrain- and tilt-corrected digital data may be compiled directly from the
stereomodel using microcomputer-based analytical stereoplotters (Warner, 1990).
Although these devices are more expensive than stereo zoom transfer scopes,

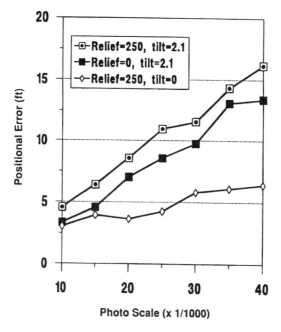

Figure 3 The effects of tilt, terrain, and photo scale on horizontal positional error.

they are more accurate and flexible. Instrument errors may be less than two feet for PC-based analytical stereoplotters, so most error depends on control-point accuracy and the operator. These systems, when used with GPS for control, are capable of providing highly accurate data over broad areas.

Other factors may affect the accuracy of photo-derived data layers. For example, lens or camera distortion, while usually quite small in mapping cameras, may be large in many small-format (35 or 70 mm) systems. Decentering, radial displacement, or local irregularities may be quite large in lenses or camera systems that are not produced expressly for mapping.

Satellite-based scanners are another source for spatial data, especially when large areas need be mapped. Automated classification of satellite imagery is an established method of land cover mapping (Lillesand and Kiefer, 1987). Classification converts multiband reflectance data to a single-layer land cover map. This land cover map is registered to a geographic coordinate system. Thus the accuracy of class boundary location is a function of classification accuracy, geometry of the image, and quality of the registration.

Empirical studies indicate most systematic geometric errors can be removed from current high-resolution satellite data. Linear registration of Landsat Thematic Mapper data resulted in *RSMEs* of approximately 23 to 46 ft over flat terrain and 45 to 90 ft in areas of approximately 3000 feet of relief (Welch et al., 1984). Positional accuracies for data from the SPOT satellite system are also in the 16- to 83-ft range. Much of this error is due to control-point identification error, since it is difficult to locate subpixel points consistently in a digital image. Positional accuracies in high-relief areas can be improved through digital stereopair analysis or through geocoding based on digital elevation models (Labovitz and Marvin, 1986; Labovitz and Wolf, 1988).

ATTRIBUTE ACCURACY

Where does tabular (attribute) information come from? Generally from either remote sensing (cameras, scanners, video) or field inventory. Forest stand boundary lines are drawn on aerial photographs, and the cover type is identified by the photointerpreter. Soil type characteristics are based on field information, but soil type boundaries are interpreted on aerial photographs. Satellite image classifications result in identified categories, such as homogeneous regions of land use or land cover. Road and stream attributes are typically collected in the field or from local experience. For decades, forest inventory was the accepted method for gathering mensurational information. None of these methods produces information that exactly represents actual feature attributes.

Photointerpretation

Photointerpretation quality results from complex interaction among the skill and background of the person performing the task, the methods used, the classes or categories identified, and the characteristics of the aerial photography (Avery and Berlin, 1985). With all these factors, it is impossible to present general conclusions about the quality of photointerpretation.

Few studies have assessed the accuracy of photointerpretation in operational forestry settings (Biging et al., 1991), although methods have been developed (Congalton and Mead, 1983) wherein a number of photointerpreted sites are checked on the ground and the results are cross-tabulated into an error or confusion matrix. In the example presented in Table 2, 11 sites photointerpreted as being conifer actually were conifer, but two photointerpreted conifer sites were actually hardwood and four were mixed. The overall photointerpretation accuracy (67.2%), accuracy for individual categories from both the map reader's (user) and map maker's (producer) perspective, and the types of photointerpretation errors committed can be determined. The user must then evaluate the quality of photointerpretation under the specific conditions encountered, and remember that this quality is translated into the tabular component of the GIS database.

Digital Imagery

Since the use of digital images in natural resource management is rising, the accuracy of attributes derived from these sources must be considered. Factors such as sensor type, classification method, season of acquisition, and categories to be identified affect the quality of information developed from digital images. The practicality of satellite image data has been debated for nearly 20 years, even as methods are developed to ease and improve their use; thus confusion matrices are routinely applied to satellite-derived digital maps. Many accuracy studies have used photointerpretation as "ground-truth." This misnomer has clearly had a negative impact on the results of satellite-image classification accuracy. Methods are available for incorporating these accuracy results into particular applications, including GIS (Prisley and Smith, 1987; Czaplewski, 1992). The need to evaluate the quality of nonphotographic, remote sensing information has been recognized by researchers; a GIS practitioner who uses digital imagery as a source of attribute information must do the same.

Field-Collected Information

The quality of field-collected attribute information used in a GIS must also be considered. Resource managers generally realize that since most inventory information is collected using a sampling scheme, there is inherent variation in the resulting data. This variation is reported in the form of a sample variance. For instance, the results of a forest cruise might be reported as 100,000 cubic feet plus or minus 8000 cubic feet. This is entirely appropriate, but the standard deviation figure almost never becomes part of an attribute database. In addition, field information is often averaged over a large region and then applied to smaller areas.

Road attributes are rarely inventoried—they are usually determined either from existing maps or from local experience. This kind of information is often

Table 2   An Error or Confusion Matrix of Photointerpreted Points

Photo-interpreted information	Ground-checked information			User accuracy
	Hardwood	Conifer	Mixed	
Hardwood	10	5	3	10/18 = 55.6%
Conifer	2	11	4	11/17 = 64.7%
Mixed	4	1	18	18/23 = 78.3%
Producer accuracy	10/16 = 62.5%	11/17 = 64.7%	18/25 = 72.0%	39/58 = 67.2%

out of date, which introduces the attribute quality of timeliness. Clearly, even the quality of field-collected attributes must be evaluated when using a GIS database.

CONCLUSIONS

The foundation of a geographic information system is spatial and tabular data. Most of the time and money in a GIS are spent on basic data. Data quality affects virtually every use of the GIS to either a small or large extent. The various components or aspects of GIS database quality discussed in this article need to be recognized so that system developers, managers, and users can apply the database appropriately and effectively to natural resource management problems.

REFERENCES

Avery, T. and Berlin, G., *Interpretation of Aerial Photographs*, Burgess Publ. Co., Minneapolis, MN, 1985, 554 pp.

Biging, G. S., Congalton, R. G., and Murphy, E. C., A comparison of photointerpretation and ground measurements of forest structure. *In* Technical papers, ACSM/ASPRS annual convention, pp. 6–15, Am. Congr. Surv. & Map./Am. Soc. Photogramm. & Remote Sens., Bethesda, MD, 1991.

Bolstad, P. V., Geometric errors in natural resource GIS data: tilt and terrain effects in aerial photographs. *For. Sci.* 38(2):367–80, 1992.

Bolstad, P. V., Gessler, P., and Lillesand, T. M., Positional uncertainty in manually digitized map data, *Int. J. Geogr. Inf. Syst.* 4:399–412, 1990.

Burrough, P. A., *Principles of Geographical Information Systems for Land Resource Assessment*. Clarendon Press, Oxford, UK, 1986, 194 pp.

Chrisman, N. R., A theory of cartographic error and its measure in digital databases. *In* Proceedings, Auto-Carto 5, pp. 159–64, Am. Congr. Surv. & Map./Am. Soc. Photogramm. & Remote Sens., Bethesda, MD, 1982.

Congalton, R. and Green, K., The ABCs of GIS: an introduction to geographic information systems, *J. For.* 90(11):13–20, 1992.

Congalton, R. and Mead, R., A quantitative method to test for consistency and accuracy in photointerpretation, *Photogramm. Eng. & Remote Sens.* 49:69–74, 1983.

Czaplewski, R. L., Misclassification bias in areal estimates, *Photogramm. Eng. & Remote Sens.* 58:189–92, 1992.

Dunn, R., Harrison, A. R., and White, J. C., Positional accuracy and measurement error in digital databases and land use: an empirical study, *Int. J. Geogr. Inf. Syst.* 4:385–98, 1990.

Labovitz, M. L. and Marvin, J. W., Precision in geodetic correction of TM data as a function of the number, spatial distribution, and success in matching of control points: a simulation. *Remote Sens. of Environ.* 20:237–52, 1986.

Labovitz, M. L. and Wolf, R. E., A methodology to account for terrain relief distortion in earth viewing satellite imagery. *In* Technical papers, 1988 annual convention, ACSM/ASPRS, pp. 137–49, Am. Congr. Surv. & Map./Am. Soc. Photogramm. & Remote Sens., Bethesda, MD, 1988.

Lillesand, T. M. and Kiefer, R. W., *Remote Sensing and Image Interpretation*, John Wiley & Sons, New York, 1987.

Norberto Fernandez, R., Lozano-Garcia, D. F., Deeds, G., and Johannsen, C. J., Accuracy assessment of map coordinate retrieval, *Photogramm. Eng. & Remote Sens.* 57:1447–52, 1991.

Perkal, J., On epsilon length. *Bull. Acad. Polonaise Sci.* 4:399–403, 1956.

Prisley, S., Gregoire, T., and Smith, J., The mean and variance of area estimates computed in an arc-node geographic information system, *Photogramm. Eng. & Remote Sens.* 55:1601–12, 1989.

Prisley, S. and Smith, J., Using classification error matrices to improve the accuracy of weighted land cover models. *Photogramm. Eng. & Remote Sens.* 53:1259–63, 1987.

Smith, J., Prisley, S., and Weih, R., Considering the effects of spatial data uncertainty on forest management decisions, *In* Proceedings, GIS/LIS '91. pp. 286–92. Am. Congr. Surv. & Map./Am. Soc. Photogramm. & Remote Sens., Bethesda, MD, 1991.

Thompson, M. M. and Rosenfield, G. H., How accurate is that map? *Surv. & Map.* 31:57–64, 1971.

Vonderhoe, A. H. and Chrisman, N. R., Tests to establish the quality of digital cartographic data: some examples from the Dane County Land Records Project, *In* Proceedings of Auto-Carto 7, pp. 552–59, Am. Congr. Surv. & Map./Am. Soc. Photogramm. & Remote Sens., Bethesda, MD, 1985.

Warner, W. S. A PC-based analytical stereoplotter for wetland inventories: an efficient and economical photogrammetric instrument for field offices. *For. Ecol. & Manage.* 33/34:571–81, 1990.

Warner, W. S. and Carson, W. W., Errors associated with a standard digitizing tablet., *ITC J.* 91(2):82–85, 1991.

Welch, R., Jordan, T. R., and Ehlers, M., Comparative evaluation of geodetic accuracy and cartographic potential of Landsat-4 and Landsat-5 Thematic Mapper image data, *Photogramm. Eng. & Remote Sens.* 51:1249–62, 1984.

Wolf, P. R., *Elements of Photogrammetry*, Ed. 2, McGraw-Hill. New York, 1983, 628 pp.

Wolf, P. R. and Brinker, R. C., *Elementary Surveying*, Ed. 8, Harper & Row, New York, 1989, 696 pp.

To Save a River: Building a Resource Decision Support System for the Blackfoot River Drainage

David E. James and Mason J. Hewitt III

ABSTRACT

To support the Section 525 Amendment to the Clean Water Act, a geographic information system (GIS) was initiated to help U.S. Environmental Protection Agency (EPA) personnel in regions 8 and 10 and water quality authorities in Montana, Idaho, and Washington understand the water quality issues associated with Idaho's Lake Pend Oreille hydrologic system. Scientists from the EPA Environmental Monitoring Systems Laboratory–Las Vegas spatial analysis team demonstrated the utility of a watershed-scale information management system for the Blackfoot River component of the lake's hydrologic system. Concurrently, they developed a nonpoint source pollution modeling approach to help local resource managers identify silvicultural resource management alternatives for the area and reduce the bias associated with selecting those alternatives. The project has yielded several recommendations for further development of nonpoint source pollution models using GIS technology, and basinwide expansion of the role of GIS and decision-support tools to address water quality issues within the Columbia River Basin and other parts of the United States.

INTRODUCTION

The Blackfoot River in western Montana, a major tributary of the Clark Fork of the Columbia River, drains approximately 1.5 million acres of mostly forested lands before emptying into the Clark Fork River. The drainage is surrounded by steep, highly dissected mountainous areas that have intermittent areas of bottomland agriculture and pasture. Most commercial activities within the Blackfoot River drainage focus on the timber industry and a limited amount of mining, agriculture, and recreation.

From *Geo Info Systems*, December 1992. With permission.

According to surveys conducted by the Montana Department of Fish, Wildlife, and Parks, as well as reports from members of the Big Blackfoot River Chapter of Trout Unlimited, the quality of the Blackfoot River fishery has been in decline since the 1970s. Don Peters of the Montana Department of Fish, Wildlife, and Parks (1991) has reported that almost all sections of the river have fewer trout than could be expected. A recent sampling showed native fish—bull and cutthroat trout—in such low numbers that viable wild populations are in serious jeopardy. That decline has prompted the Montana State Water Quality Bureau and the U.S. Environmental Protection Agency's (EPA's) Region 8 Montana Operations Office to use geographic information systems (GIS) technology to investigate nonpoint source pollution within the drainage.

A LEGISLATIVE MANDATE

The Federal Water Pollution Control Act Amendments of 1972 established definite goals for restoring and maintaining the physical, chemical, and biological integrity of the nation's waters. Section 208 of the act requires nonpoint sources of water pollution to be identified and controlled to the extent feasible. Nonpoint sources associated with *silviculture*—that is, forestry—are specifically mentioned in the act as areas to be addressed by water quality management. *Nonpoint sources* of pollution result from natural causes, human actions, and the interactions between natural events and conditions resulting from human use of natural resources. To control these sources of pollution, EPA has used federal regulation to develop the concept of Best Management Practices (BMP). As defined by EPA, BMP are practices that are determined—after assessing the problem and initiating appropriate public participation—to be the most effective, practicable means of preventing or reducing the amount of pollution generated by nonpoint sources (EPA, 1980).

The 1987 Amendments to the Clean Water Act added Section 525, which specifically directs the EPA Administrator to study sources of pollution for Idaho's Lake Pend Oreille (Figure 1), and to "consider existing studies, surveys and test results" in the process of "identifying the sources of pollution." GIS technology seems to fit this need because of its capability to organize, analyze, and display the collected data.

Because Lake Pend Oreille is in EPA Region 10 and the contributing river is in EPA Region 8, a joint investigation committee was formed that included representatives from two EPA Regions and three states. Scientists at EPA's Environmental Monitoring Systems Laboratory in Las Vegas, Nevada (EMSL-LV), then developed a GIS to help EPA personnel in Regions 8 and 10 and water quality authorities in Montana, Idaho, and Washington understand water quality issues associated with the Lake Pend Oreille hydrologic system.

This pilot project focused on a proof-of-concept study to address if and how a GIS could be used in a water quality decision-support framework. The constraints of time and money required that we address only a portion of the Lake

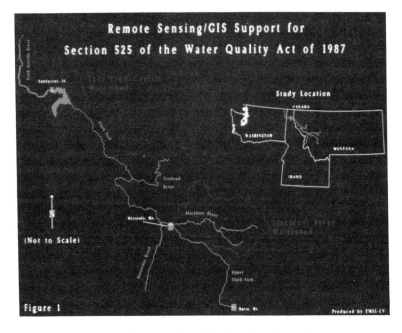

Figure 1   The Lake Pend Oreille hydrologic system.

Pend Oreille watershed. However, the pilot was the beginning of a long-term effort to use GIS as a management tool for understanding and managing the water quality of the Lake Pend Oreille watershed. The reasons for choosing the Blackfoot River drainage as the test-bed for the study stem from its status as a historic trout fishery and the heavy recreational use it receives because of its proximity to Missoula, Montana. Additionally, the ongoing GIS efforts of the EPA Region 8 Montana Operations office, the Montana Department of Health and Environmental Sciences, and the Montana State Library regarding the Upper Clark Fork River Superfund sites offered the opportunity for beginning a Clark Fork River Basin information management system.

THE BLACKFOOT RIVER GIS

The Blackfoot River GIS project demonstrates the utility of a watershed-scale information management system. Begun as a GIS data base for the Blackfoot River drainage area, the system has developed into a nonpoint source pollution potential model geared toward silvicultural resource management issues and a decision-support mechanism capable of assessing the suitability and feasibility of different resource management alternatives. A partial list of management issues pertaining to the Blackfoot River includes:

- comparison of silvicultural prescriptions for reducing sediment yield
- investigations of water quality for managing fisheries
- environmental assessment of potential effects of proposed mining activities

The information management system has been made available to resource managers interested in the Blackfoot River drainage. A user interface has been developed to provide access to baseline thematic data such as soils, hygrography, elevation, and land cover; the results of the nonpoint modeling effort; and opportunities for additional analytical processing. By offering a method for analyzing different resource management approaches before they are applied, this information management system can help decision makers reduce the risk associated with making resource management choices.

The Blackfoot River GIS will become an important component of a Clark Fork River Basin information management system. As the spatial data base evolves and is integrated with other resource data bases, the analytical modeling effort that this project initiated will help scientists and others address many resource-management needs. In time, the ability to model many of the diverse relationships found within the watershed will be placed in the hands of the resource managers through the Blackfoot River GIS user interface.

PROJECT COMPONENTS

The GIS for the Blackfoot River drainage was developed using EPA's standard GIS software (ARC/INFO, Environmental Systems Research Institute, Redlands, California). The project's goal was to provide an information management system that could be applied to a variety of resource management needs. Three project development tracks were identified to characterize and generate the products necessary to achieve the project goals:

- a GIS database
- analytical modeling
- the user interface

Those tracks allowed concurrent development of separate products that together created the desired decision-support mechanism.

The GIS spatial data base was the focus of project data acquisition activities and contains the information necessary to model alternatives and provide ancillary data of interest to resource managers. The database, which resides in the Montana State Library's Natural Resource Information System (NRIS), is available to the Montana Water Quality Bureau, resource managers, and other users in the Blackfoot River drainage. During the project, the database evolved into a watershed-based modeling structure that allows for timely and efficient movement of data through alternative management scenarios. In particular, the data can be used to evaluate proposed silvicultural alternatives and the potential effect of those alternatives on the Blackfoot River's water quality. As the system's user base expands, the type and complexity of the resource management issues will broaden beyond silvicultural applications. As the spatial database evolves and becomes integrated with other resource databases, it will become a tool capable of addressing many resource management needs.

The user interface is the access point between the integrated spatial database–modeling alternatives and project users. The interface consists of a series of ARC Macro Language (AML) programs that provide interactive graphic tools used to browse through the database, to inspect modeling parameters, and to perform alternative management analyses. The ability to analyze alternative management scenarios quickly and consistently may reduce the risks involved in making resource management decisions. The user interface has also been transferred to NRIS and will be available to the public through the state library's "GIS in Libraries" program on a SUN workstation (Sun Microsystems, Mountain View, CA).

What is a Decision Support System?

Depending on what you read, a decision support system (DSS) is defined in many ways. The trend toward DSS seems to lead back to a management information systems approach from the 1970s. The most popular literature today generally defines a DSS by a set of characteristics, rather than by a strict description. The DSS literature is full of phrases such as "interactive, user approach, problem finder, solution investigation," and "synergistic." Within the scope of the Blackfoot River project, we have used the term DSS, but have never completely defined what we were building. Before, during, and after the project we offered several definitions to add some direction to our efforts. The descriptions are as follows:

- a system to promote arguments among resource managers who rarely talk to each other;
- a structured system for evaluating environmental data in a common framework; and
- a forward-looking, what-if simulator.

Although we did not completely agree on the definition, we did define the components of a DSS. A DSS has a data base component that is closely linked to a model component and has a user interface. The resulting system offers a data analysis capability in a geographic framework having a what-if simulator designed to promote arguments among resource managers.

MODEL SELECTION

The database created for the Blackfoot River GIS focused on the elements necessary to develop a nonpoint source pollution model. Preliminary database elements were identified through a literature search of current research and discussions with natural resource management personnel. Identified data base elements routinely required for nonpoint source pollution modeling include hydrography, soil composition, climatic data, topography, and land cover.

Next, a procedure developed by the U.S. Department of Agriculture Forest Service and EPA was adopted for modeling in the Blackfoot River Basin (EPA, 1980). Commonly referred to as the Water Resources Evaluation of Nonpoint Silvicultural Sources (WRENSS) model, this procedure provides an analysis method that can be used to describe and evaluate watershed changes resulting from nonpoint silvicultural activities, including changes in streamflow, surface erosion, movement of soil mass, and total potential sediment discharge.

For this study, only model parameters pertaining to surface erosion and sediment delivery were investigated. Because this effort was geared toward the development of a basinwide information management system, the modeling effort was not expected to be as robust as necessary to form a definitive opinion about nonpoint source modeling using GIS.

The user interface and the implementation of the WRENSS model have been presented at several conferences whose participants were interested in watershed modeling, including the EPA-U.S. Forest Service "Technical Workshop on Sediments," the NCASI "Cumulative Effects Workshop," and the "On Common Ground" conference sponsored by *Geo Info Systems, GPS World,* and *CADalyst* magazines. Those presentations have generated professional interaction among a large group of scientists and interested parties.

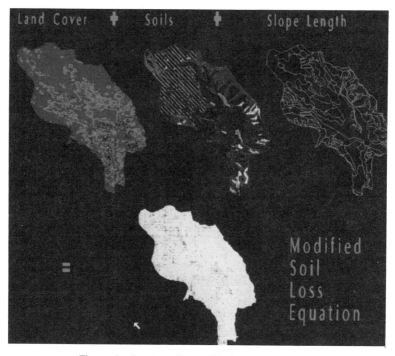

Figure 2   Inputs to the modified soil loss equation.

MODELING DATABASE DEVELOPMENT

The WRENSS model adopted for this project incorporates a modified universal soil loss equation (USLE) (Wischmeier and Smith, 1968) to estimate potential soil loss. This modified USLE is tied to four key elements: rainfall, soil erodability, slope length, and vegetation or land cover (Figure 2). Together, those elements provide the information necessary to solve the USLE and predict baseline soil erosion for a given tract of land. During silvicultural practices, a wooded tract will be harvested (generally clear-cut), and the effects will include an increase in the amount of soil lost through surface erosion. The USLE also can be solved for the harvested area. The difference between the two calculations would estimate the increase in soil lost through surface erosion. This value is further modified by a sediment delivery factor based on the size of the subwatershed (Roehl, 1962), resulting in an estimate of the increase in sediment delivered to the critical reach of the third-order watershed.

A *third-order watershed* (Figure 3) is the land surface area contributing to a third-order stream. Strahler defines stream orders in reference to the number of tributaries intersecting to form a downstream segment. In this method, all segments having no tributaries are referred to as first-order. When two first-order segments intersect, the downstream segment is considered to be a second-order stream. A third-order stream would be formed by the intersection of two second-order streams (Strahler, 1952).

The Blackfoot River GIS project has used WRENSS soil-loss-estimation parameters as a paradigm for developing the GIS database and the associated mod-

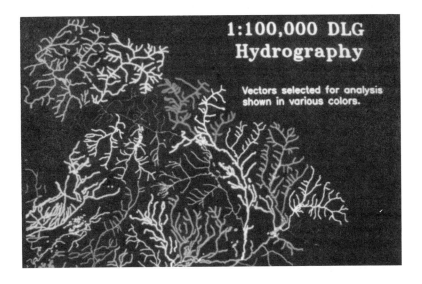

Figure 3   Third-order watershed.

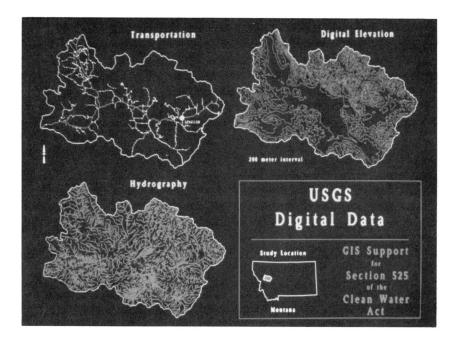

Figure 4   Data layers acquired from the U.S. Geological Survey for the Blackfoot project.

eling software. Rainfall, soils, slope length, and land-cover data were assembled basinwide, spatially divided into subsets, and joined together to form a single modeling data set or coverage for each subwatershed. This coverage, named modified soil loss equation (MSLE), contains all of the individual polygons that compose each key modeling element, and it provides the necessary attributes to calculate WRENSS soil loss.

We undertook the following data acquisition activities to establish the GIS database and satisfy the modeling requirements (Figure 4).

Rainfall

Precipitation data are necessary for understanding the potential for soil loss from nonpoint sources. Two sources were investigated for precipitation data. The isoerodent map (USDA, 1977) provides coarse information for the rainfall factor of the WRENSS model on a nationwide basis and was used to characterize the Blackfoot River Basin as a whole. The isoerodent estimations were refined using historic snow survey data available from the Soil Conservation Services's Central Forecast System located at the West National Technical Center in Portland, OR. These data allowed for corrections based on thaw and snowmelt (EPA, 1980).

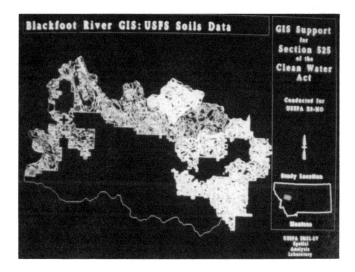

Figure 5   Soils data from the U.S. Forest Service for Lola and Helena National forests.

Soils

Various soils data were collected for the Blackfoot River GIS database. As many sources of soils data as possible were included to compare their relative usefulness for modeling. Additionally, making the range of soils data available to resource managers through the GIS database and user interface will enhance interaction between agencies that have similar resource interests.

Soil Conservation Service (SCS) data were available for the study area at two scales: 1:24,000-scale county soil survey data and 1:250,000-scale State Soil Geographic (STATSGO) series soils data. The county-level 1:24,000-scale data contained the appropriate modeling attributes for soil erosion but were available only for that portion of the study area found in Missoula County—a total of 18 U.S. Geological Survey (USGS) 7.5-min quadrangles. The 1:250,000-scale STATSGO database provides full basin coverage but aggregates the soil erosion attributes, making it less applicable in a modeling scenario.

Limited soils data also were obtained from the U.S. Forest Service. Two data sets were obtained in digital format, both representing areas of the Helena National Forest (Figure 5). Additional soils data were obtained from the Lolo National Forest in the form of land-type designations drawn on 1:24,000-scale topographic sheets and digitized. These data combine specific vegetative, geologic, and land-form information and include a range of more general attribute information.

In future modeling efforts, these data sets will be compared for information quality, ease of use for modeling, and cost-benefit in terms of acquisition costs versus resource management needs.

Slope Length

Calculating slope data was an integral part of creating the model. The only elevation data available for the study area were the 1:250,000-scale digital elevation models (DEM) from USGS. These data are available as one-degree blocks covering the entire nation.

The topographic factor derived for the WRENSS model attempts to estimate the distance water will travel down a slope without interruption. Scientists from the EMSL-LV spatial analysis team (SAT) developed custom software to generate the slope-length value and the topographic factor used in WRENSS modeling. The topographic-factor data sets were generated using the GIS software's GRID module (Version 6.0) because the input data were acquired in a raster format. These data had to be converted to a vector format to be compatible with other modeling data sets. To do so, the topographic-factor output had to be resampled to a larger cell size and stratified into classes of slope length prior to vectorization. Future efforts will be developed using the GRID module because of its enhanced modeling capabilities.

Land Cover

The type of land cover and the way in which it changes over time form the basic premise for calculating increases in nonpoint source pollution from silvicultural activities. Three activities are accounted for in the model selected for the Blackfoot River GIS project: silviculture, road building, and fires. The road-building activities will require input on a large scale from land managers. Land-cover changes resulting from silviculture and fires are detectable from satellite imagery, which has been chosen as the source for the Blackfoot River drainage land cover. A Landsat Thematic Mapper (TM) (EOSAT, Lanham, Maryland) satellite scene for July 1989 was selected to produce the land-cover data set.

Model input for land cover pertains to information that can only be inferred from classical TM data. For the WRENSS model, crown closure and crown height were the pertinent inputs to developing the vegetation management factor. Identifying the relationship between a standard land-cover classification scheme and these extremely detailed modeling requirements would require extensive on-site investigations. For example, the modeling requirements for land cover include estimations for percentage of crown closure, crown height, and percentage of ground cover. To effectively estimate these parameters would require a significantly funded research effort beyond the scope of this project. Therefore, a generalized land-cover data set was generated to establish basinwide data at a less than desirable level of resolution for performing detailed modeling.

Although this attempt at silvicultural decision support is not expected to be used for final resource management decisions or "prescriptions," the level of land-cover detail may be acceptable. If not, detailed information about timber stands from responsible agencies and private landowners may be incorporated into the data base for modeling purposes.

Hydrography

To calculate changes in nonpoint source pollution, the drainage basin's hydrographic structure must be understood. Digital line graph (DLG) data at 1:100,000 scale were obtained from USGS to provide hydrography data for the Blackfoot River drainage. Using processing techniques developed by EMSL-LV SAT, the DLGs were edgematched and assembled into a GIS coverage. The DLG hydrography data combined with elevation data were used to generate third-order watersheds using methods described below. These subwatersheds were used as the base units for the modeling effort and as the tiling structure or partitions for the GIS data base library (Figure 6).

DATABASE STRUCTURE

During initial project scoping, the size of the completed database for the Blackfoot River GIS was projected to be about 200 MB: in fact, the completed database is more than 300 MB. The elevation and land-cover data sets contain such massive amounts of information that these data could not be processed on a watershed basis, even using workstation technology. The prospect of assembling and processing the entire data set seemed unreasonable given the available

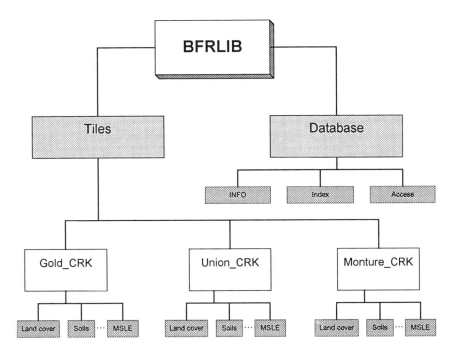

Figure 6 The Blackfoot River GIS library structure.

computer resources at EMSL-LV or the Montana State Library. This situation was mitigated by the fact that the WRENSS model is intended to be applied to third-order or smaller watersheds. Using the third-order watersheds as a tiling structure allowed an appropriate data base design to be developed to support the Blackfoot River GIS database and the associated modeling effort using the LIBRARIAN module of the ARC/INFO software. This tiling structure allowed the massive raster-based slope length and land-cover data sets to be divided into subsets and resampled before conversion to a vector format and use in the WRENSS modeling using the GIS software.

To identify third-order watersheds and the area of contribution to the critical stream reach, innovative techniques were developed using the DEMs and the DLG hydrography coverage. The DEM processing programs available at EMSL-LV include a watershed delineation technique based on programs developed by S. Jensen at the EROS Data Center (Jensen and Dominique, 1988), which were modified for integration with the GIS software by SAT scientists (Pickus and Fisher, 1990). Using the three-arc-second elevation model combined with a raster version of the DLG hydrography coverage, areas of contribution to each identified third-order reach were established. The hydrologic modeling tools developed by Jensen have since been integrated into the GIS software GRID module in Version 6.1 to support watershed delineation and other hydrologic modeling efforts.

USER INTERFACE

The user interface (Figure 7) is the access point to the GIS database and modeling software for the Blackfoot River project. The interface was developed on a SUN SPARCstation (Sun Microsystems) using ARC/INFO Version 5.0 GIS software and has recently been converted to run under the latest version of that software. Identified third-order watersheds form the structure of the GIS data base using tiling. The AML-based interface consists of a series of menus and graphic displays that provide access to the database and allow the user to perform modeling runs. On-line help is available as a menu selection. The interface software was written to incorporate the LIBRARIAN module capabilities, which include the use of the LIBRARIAN nomenclature and data handling techniques that provide efficient access to the database.

Map libraries organize spatial data into a logical, manageable framework. LIBRARIAN uses a hierarchical data model to keep track of available map libraries and the coverages that comprise each individual tile (ESRI, 1991). As shown in Figure 6, the Blackfoot River GIS library structure comprises standard ARC/INFO data structures—coverages and INFO files—which allow the user to take advantage of the GIS software's toolbox for updating, managing, and accessing map library data.

The GIS database is accessed at either the Blackfoot River Basin level or the individual subwatershed level. The major difference between the two levels is

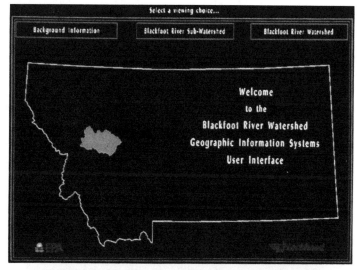

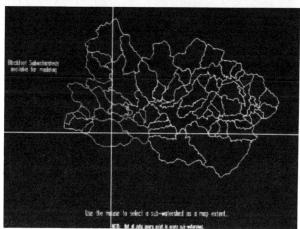

Figure 7   The above screen (A) shows the beginning of a user interface. Users select the watershed to be modeled (B) and select elements that contribute to developing a modified soil loss equation (C).

the availability of the modeling software. At the basin level, the interface is geared toward browsing, inspecting, and querying the entire database. Modeling is performed at the subwatershed level. Both levels use the menu-driven access system. Menu selections allow the user to move in and out of windows, display various data sets, document data, and query the data set's attribute database.

Modeling Alternative Management Prescriptions

For the user interface, the modeling was designed to take place on a subwatershed basis (Figure 8). Potential management alternatives or silvicultural

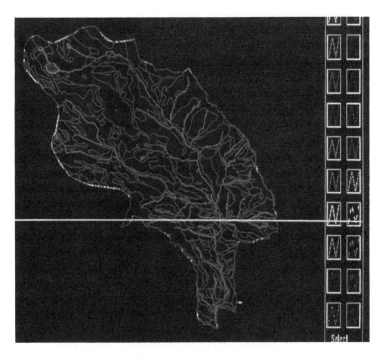

Figure 7   *(Continued)*

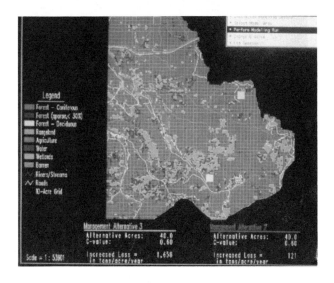

Figure 8   A modeling run produces two management alternatives.

prescriptions are tracked for each modeling session; each time a series of management alternatives is identified within a particular watershed, the software maintains individual records for each identified alternative. Each alternative can be thought of as a silvicultural prescription projecting a clear-cut harvest on the identified area. During each modeling session, the MSLE coverage and accompanying software generate an estimate of the increase in sediment delivered to the critical reach of the third-order watershed for each alternative management prescription identified.

CONCLUSIONS

The Blackfoot River project database and user-interface were delivered to Montana State Library NRIS personnel in Helena, MT, for installation on their GIS. NRIS is the logical home for the system because it provides a mature GIS environment within which the Blackfoot River GIS may expand. As the Montana State Water Quality Bureau and other resource managers interested in the Blackfoot River drainage become familiar with the database and user interface, use can be expected to increase. Continued use of the system will help identify what other data should be added to the database as well as how to enhance the project's user interface and modeling capabilities. We hope this project will introduce local resource managers to GIS and automated spatial analysis and foster an environment conducive to refining the model.

Although the project's ultimate goal was to develop a Decision Support System (DSS) to explore management alternatives, the authors elected to design the database–information system first to provide a test-bed for decision support. A true DSS would be better created by designing the system to meet identified needs of the decision makers and then building the system; however, given that resource management decisions and end users were not defined, the project to date has been designed toward the needs of a specific model. We hope the results of the Blackfoot River GIS project will support individual user needs and provide a platform for examining the large variety of data important to the diverse resource managers and decision makers interested in the Blackfoot River.

Nonpoint Source Modeling

The data assembled as input into the GIS modeling database were, in all likelihood, too coarse to produce accurate modeling results. In a project such as this, the goal is as much to demonstrate the capability of GIS technology as it is to develop the model. Nonpoint source pollution modeling is an inexact science that has much room for discussion and additional research; even the most detailed nonpoint source pollution models are used only as a relative measure of disturbance between alternative management activities. In that sense, this project has tried to join GIS technology and modeling processes in a consistent manner as a mechanism for removing some of the potential for bias in decision making.

RECOMMENDATIONS

Recommendations for future activities involving modeling for nonpoint source pollution using GIS technology include

1. Developing the nonpoint source pollution modeling in an ARC/INFO GRID format, which may provide a data model that is more conducive to flow modeling. Additionally, certain file-size limitations involved in converting raster-format data into vector-format coverages would be alleviated. This limitation was a major obstacle in developing the Blackfoot River GIS database.
2. Identifying more resolute databases for use with the modeling software. The classified landcover database (satellite derived) was too generalized to support the modeling effort to its maximum capability. Also, the elevation data used to generate the slope-length calculations were too coarse to provide accurate input to the modeling runs. It would be appropriate to try to assemble a database that has similar scales of input data. The differences between the 1:24,000-scale soils data and the 1:250,000-scale digital elevation model are substantial enough to cast doubt on the suitability of using them together;
3. Taking advantage of advancements in user-interface tools. Enhanced software capabilities can provide a user interface that is easier for people who do not have GIS training to use. New graphical user interface development tools are much more consistent with workstation-based "windows" management systems; and
4. Continuing to develop decision support system tools for basinwide planning that support water quality initiatives and establishing appropriate guidance for environmental assessment, management, and enforcement. The time and technology are available for EPA to promote a multistate, multiagency dialogue about water quality issues involving the Columbia River Basin. The lessons learned in this type of cooperative venture could provide valuable input for developing basinwide water quality planning efforts throughout the United States.

Notice

The information in this document has been funded (wholly or in part) by the U.S. Environmental Protection Agency under contract 68-C0-0050 to Lockheed Engineering and Sciences Company, Inc. It has been subjected to Agency review. Mention of trade names or commercial products does not constitute endorsement or recommendation for use.

REFERENCES

ESRI, Using Map Libraries. *ARC/INFO Users Guide*. Redlands, California, 1991.
Jensen, S. and Dominique, J., Extracting Topographic Structures from Digital Elevation Data for Geographic Information System Analysis, *Photogrammetric Engineering and Remote Sensing* 54(11):1593–1950, 1988.
Peters, D., The Blackfoot Challenge. *Montana Outdoors* September/October, 1991.

Pickus, J. and Fisher, L. T., EMSL-LV DEM Processing Standards for Topographics Analysis. Unpublished.

Roehl, J. W., *Sediment Source Areas, Delivery Rations, and Influencing Morphological Factors*, Publication 59:202–213. United Kingdom: International Association of Scientific Hydrology, Commission on Land Erosion, 1962.

Strahler, A. N., Quantitative Analysis of Watershed Geomorphology. *Trans A, Geophysical Union* (38):913–920, 1957.

U.S. Environmental Protection Agency An Approach to Water Resources Evaluation of Nonpoint Silvicultural Sources. Procedural handbook. EPA-600/880-12. Athens, Georgia.

U.S. Department of Agriculture, Soil Conservation Service, *Procedure for Computing Sheet and Rill Erosion on Project Areas*. Technical release No. 41, Revision 2. Washington, D.C.: U.S. Government Printing Office, 1977.

Wischmeier, W. H. and Smith, D. D., A Universal Soil-Loss Equation to Guide Conservation Farm Planning, *Transactions of International Congress on Soil Science* (1):418–425, 1968.

CHAPTER 22

Glossary of GIS Terms

Bruce L. Kessler

This glossary presents many of the words and phrases that a Geographic Information System (GIS) user may encounter. Some terms have been simplified to limit the overwhelming feeling that many people get when first presented with a GIS. Also, an effort was made to eliminate bias toward any GIS software package. If there is some residual slant, the author does not intend to imply that one GIS package is better or worse than the other. Words set in boldface are defined elsewhere in the glossary. Numbers refer to the reference list at the end of the glossary.

absolute map accuracy
The accuracy of a map in relationship to the earth's geoid. The accuracy of locations on a map that are defined relative to the earth's geoid are considered absolute because their positions are global in nature and accurately fix a location that can be referenced to all other locations on the earth. Contrast *absolute map accuracy* with *relative map accuracy*.[3]

acceptance test
A set of particular activities performed to evaluate a hardware of software system's performance and conformity to specifications.

accuracy
If applied to paper maps or map databases, degree of conformity with a standard or accepted value. *Accuracy* relates to the quality of a result and is distinguished from **precision**.[5] If applied to data collection devices such as digitizers, degree of obtaining the correct value.

address matching
A mechanism for relating two files using address as the key item. Geographic coordinates and attributes subsequently can be transferred from one address to the other.[2]

Reprinted from the *Journal of Forestry* (1992), vol. 90, no. 11, pp. 37–45; published by the Society of American Foresters, 5400 Grosvenor Lane, Bethesda, MD 20814–2198. Not for further reproduction.

algorithm

A step-by-step procedure for solving a mathematical problem. For instance, the conversion of data in one map projection to another map projection requires that the data be processed through an algorithm of precisely defined rules or mathematical equations.

aliasing

The occurrence of jagged lines on a raster-scan display image when the detail exceeds the resolution on the screen.[1]

American National Standards Institute (ANSI)

An institute that specifies computer system standards. The abbreviation is often used as an adjective to computer systems that conform to these standards.

American Standard Code for Information Interchange (ASCII)

A set of codes for representing alphanumeric information (e.g., a byte with a value of 77 represents a capital M). Text files, such as those created with a computer system's text editor, are often referred to as ASCII files.[2]

AM/FM

See **automated mapping/facilities management**.

analog map

Any directly viewable map on which graphic symbols portray features and values; contrast with **digital map**.[4]

annotation

Text on a drawing or map associated with identifying or explaining graphics entities shown.[5]

ANSI

See **American National Standards Institute**.

application

A program or specially defined procedure, generally in addition to the standard set of basic software functions supplied by a GIS. Historically, an application was developed by the vendor or by a third party and purchased separately. Developed to perform a series of steps, these applications may create specialized reports, complex map products, or lead an operator through a decision process. Some of the more common applications are now becoming part of the basic software functions.

arc

See **line**.

arc-node structure

The coordinate and topological data structure used by some GISs. Arcs represent lines that can define linear features, the boundary of areas, or polygons. In arc-node structures, there is an implied direction to the line so that it may have a left and right side. In this way, the area bounded by the arc can also be described, and it is not necessary to double-store coordinates for arcs that define a boundary between two areas.

architecture

In computers, the architecture determines how the computer is seen by someone who understands its internal commands and instructions and the design of its interface hardware.[5]

area

A closed figure (polygon) bounded by one or more lines enclosing a homogeneous area and usually represented only in two dimensions. Examples are states, lakes, census tracts, aquifers, and smoke plumes.

ASCII

See **American Standard Code for Information Interchange**.

aspect

A position facing a particular direction. Usually referred to in compass directions such as degrees or as cardinal directions.

attribute

(1) A numeric, text, or image data field in a relational database table that describes a spatial feature such as a point, line, node, area, or cell. (2) A characteristic of a geographic feature described by numbers or characters, typically stored in tabular format and linked to the feature by an identifier. For example, attributes of a well (represented by a point) might include depth, pump type, location, and gallons per minute.

automated mapping/facilities management (AM/FM)

A GIS technology focused on the specific segment of the market concerned with specialized infrastructure and geographic facility information applications and management, such as roads, pipes, and wires.

axis

A reference line in a coordinate system.[5]

band

One layer of a multispectral image representing data values for a specific range of the electromagnetic spectrum of reflected light or heat. Also, other user-specified values derived by manipulation of original image bands. A standard color display of multispectral image displays three bands, one each for red, green, and blue. Satellite imagery such as Landsat™ and SPOT provide multispectral images of the earth, some containing seven or more bands.[2]

base map

A map showing planimetric, topographic, geological, political, and/or cadastral information that may appear in many different types of maps. The base map information is drawn with other types of changing thematic information. Base map information may be as simple as major political boundaries, major hydrographic data, and major roads. The changing thematic information may be bus routes, population distribution, or caribou migration routes.

benchmark tests

Various standard tests, easily duplicated, for assisting in measurement of product performance under typical conditions of use.[5]

binary large object (BLOB)

The data type of a column in a relational database management system (RDBMS) table that can store large images or text files as attributes.[2]

bit

The smallest unit of information that can be stored and processed in a computer. A bit has two possible values, 0 or 1, which can be interpreted as YES/NO, TRUE/FALSE, or ON/OFF.[2]

BLOB

See **binary large object**.

Boolean expression

(1) A type of expression based upon, or reducible to, a true or false condition. A Boolean *operator* is a key word that specifies how to combine simple logical expressions into complex expressions. Boolean operators negate a predicate (NOT), specify a combination of predicates (AND), or specify a list of alternative predicates (OR). For example, the use of AND in "DEPTH > 100 and GPM > 500." (2) Loosely, but erroneously, used to refer to logical expressions such as "DEPTH greater than 100."[2]

breakline

A line that defines and controls the surface behavior of a triangulated irregular network (TIN) in terms of smoothness and continuity. Physical examples of breaklines are ridge lines, streams, and lake shorelines.

buffer

A zone of a given distance around a physical entity, such as a point, line, or polygon.

bundled

Refers to the way software is sold. In the early days of computers, software products were sold integrated with hardware, that is, they were "bundled." *See also* **unbundled**.[5]

byte

A group of contiguous bits, usually eight, that is a memory and data storage unit. For example, file sizes are measured in bytes or megabytes (1 million bytes). Bytes contain values of 0 to 255 and are most often used to represent integer numbers or ASCII characters (e.g., a byte with an ASCII value of 77 represents a capital M). A collection of bytes (often 4 or 8 bytes) is used to represent real numbers and integers larger than 255.[2]

CAD

See **computer-aided design**.

cadastre

A record of interests in land, encompassing both the nature and extent of interests. Generally, this means maps and other descriptions of land parcels as well as the identification of who owns certain legal rights to the land (such as ownership, liens, easements, mortgages, and other legal interests). Cadastral information often includes other descriptive information about land parcels.[3]

CAE

See **computer-aided engineering**.

CAM

See **computer-aided mapping**.

cardinal

Refers to one of the four cardinal directions—north, south, east, or west.

Cartesian coordinate system

A concept from French philosopher and mathematician Rene Descartes (1596–1650). A system of two or three mutually perpendicular axes along which any point can be precisely located with reference to any other point; often referred to as x, y, and z coordinates.[5] Relative measures of distance, area, and direction are constant throughout the system.

cell

The basic element of spatial information in a grid data set. Cells are always square. A group of cells forms a grid.

centroid

The "center of gravity" or mathematically exact center of a regularly or irregularly shaped polygon; often given as an x,y coordinate of a parcel of land.[5]

chain

See **line**.

character

(1) A letter, number, or special graphic symbol (*, @, –) treated as a single unit of data. (2) A data type referring to text columns in an attribute table (such as NAME).[2]

clip

The spatial extraction of physical entities from a GIS file that reside within the boundary of a polygon. The bounding polygon then works much like a cookie cutter.

cluster

A spatial grouping of geographic entities on a map. When these are clustered on a map, there is usually some phenomenon causing a relationship among them (such as incidents of disease, crime, pollution, etc.).

COGO

See **coordinate geometry**.

column

A vertical field in a relational database management system (RDBMS) data file. It may store from one to several bytes of descriptive information.

command

An instruction, usually one word or concatenated words or letters, that performs an action using the software. A command may also have extra options or parameters that define more specific application of the action.

computer-aided design (CAD)

A group of computer software packages for creating graphic documents.

computer-aided engineering (CAE)

The integration of computer graphics with engineering techniques to facilitate and optimize the analysis, design, construction, nondestructive testing, operation, and maintenance of physical systems.[5]

computer-aided mapping (CAM)

The application of computer technology to automate the map compilation and drafting process. Not to be confused with the older usage, computer-aided manufacturing; usually associated with CAD, as in CAD/CAM.[5]

configuration

The physical arrangement and connections of a computer and its related peripheral devices. This can also pertain to many computers and peripherals.

conflation

A set of functions and procedures that aligns the arcs of one GIS file with those of another and then transfers the attributes of one to the other. Alignment precedes the transfer of attributes and is most commonly performed by rubber-sheeting operations.[2]

conformality

Small areas on a map are represented in their true shape and angles are preserved—a characteristic of a map projection.

connectivity

(1) The ability to find a path or "trace" through a network from a source to a given point. For example, connectivity is necessary to find the path along a network of streets to find the shortest or best route from a fire station to a fire. (2) A topological construct.

contiguity

The topological identification of adjacent polygons by recording the left and right polygons of each arc.

continuous data

Usually referenced to grid or raster data representing surface data such as elevation. In this instance, the data can be any value, positive or negative. Sometimes referred to as real data. *See also* **discrete data**.

contour

A line connecting points of equal value. Often in reference to a horizontal datum such as mean sea level.

conversion

(1) The translation of data from one format to another (.e.g., TIGER to DXF; a map to digital files); (2) Data conversion occurs when transferring data from one system to another (e.g., SUN to IBM); (3) *See also* **data automation**.

coordinate

The position of a point in space with respect to a Cartesian coordinate system (x, y, and/or z values). In GIS, a coordinate often represents locations on the earth's surface relative to other locations.

coordinate geometry (COGO)

A computerized surveying-plotting calculation methodology created at MIT in the 1950s.[5]

coordinate system

The system used to measure horizontal and vertical distances on a planimetric map. In a GIS, it is the system whose units and characteristics are defined by a map projection. A common coordinate system is used to spatially register geographic data for the same area. *See* also **map projection**.

coterminous

Having the same or coincident boundaries. Two adjacent polygons are coterminous when they share the same boundary (such as a street centerline dividing two blocks).[3]

curve fitting

An automated mapping function that converts a series of short, connected straight lines into smooth curves to represent entities that do not have precise mathematical definitions (such as rivers, shoreline, and contour lines).[3]

dangling arc

An arc having the same polygon on both its left and right sides and having at least one node that does not connect to any other arc.[2]

data

A general term used to denote any or all facts, numbers, letters, and symbols that refer to or describe an object, idea, condition, situation, or other factors. It may also refer to line graphics, imagery, and/or alphanumerics. It connotes basic elements of information which can be processed, stored, or produced by a computer.[5]

data automation

Mostly the same as digitizing, but can also mean using electronic scanning for data collection.

data dictionary

A coded catalog of all data types or a list of items giving data names and structures. May be on-line (referred to as an automated data dictionary), in which case the codes for the data types are carried in the database. Also referred to as DD/D for data dictionary/directory.[2]

data integration

The combination of databases or data files from different functional units of an organization or from different organizations that collect information about the same entities (such as properties, census tracts, and street segments). In combining the data, added intelligence is derived.

data model

(1) A generalized, user-defined view of the data related to applications: (2) A formal method for arranging data to mimic the behavior of the real-world entities they represent. Fully developed data models describe data types, integrity rules for the data types, and operations on the data types. Some data models are triangulated irregular networks (TINs), images, and georelational or relational models for tabular data.[2]

database

Usually a computerized file or series of files of information, maps, diagrams, listings, location records, abstracts, or references on a particular subject or subjects organized by data sets and governed by a scheme of organization. **Hierarchical** and **relational** define two popular structural schemes used in a GIS.[5] For example, a GIS database includes data about the spatial location and shape of geographic entities as well as their attributes.

database management system (DBMS)

(1) The software for managing and manipulating the whole GIS including the graphic and tabular data. (2) Often used to describe the software for managing (e.g., input, verify, store, retrieve, query, and manipulate) the tabular information. Many GISs use a DBMS made by another software vendor, and the GIS interfaces with that software.

datum

A set of parameters and control points used to accurately define the three-dimensional shape of the earth (e.g., as a spheroid). The corresponding datum is the basis for a planar coordinate system. For example, the North American datum for 1983 (NAD83) is the datum for map projections and coordinates within the United States and throughout North America.

DBMS

See **database management system.**

DEM

See **digital elevation mode.**

densify

A process of adding vertices to arcs at a given distance without altering the arc's shape. *See also* **spline** for a different method for adding vertices.

digital

Usually refers to data that is in computer-readable format.

digital elevation model (DEM)

(1) A raster storage method developed by the U.S. Geological Survey (USGS) for elevation data. (2) The format of the USGS elevation data sets.

digital exchange format (DXF)

(1) ASCII text files defined by Autodesk, Inc. (Sausalito, CA) at first for CAD, now showing up in third-party GIS software.[5] (2) An intermediate file format for exchanging data from one software package to another, neither of which has a direct translation for the other but where both can read and convert DXF data files into their format. This often saves time and preserves accuracy of the data by not reautomating the original.

digital line graph (DLG)

(1) In reference to data, the geographic and tabular data files obtained from the U.S. Geological Survey (USGS) that may include base categories such as transportation, hydrography, contours and public land survey boundaries. (2) In reference to data format, the formal standards developed and published by the USGS for exchange of cartographic and associated tabular data files. Many non-DLG data may be formatted in DLG format.

digital map

A machine-readable representation of a geographic phenomenon stored for display or analysis by a digital computer; contrast with **analog map**.[4]

digital terrain model (DTM)

A computer graphics software technique for converting point elevation data into a terrain model displayed as a contour map, sometimes as a three-dimensional "hill and valley" grid view of the ground surface.[5]

digitize

A means of converting or encoding map data that are represented in analog form into digital information of x and y coordinates.

digitizer

(1) A device used to capture planar coordinate data, usually as x and y coordinates from existing analog maps for digital use within a computerized program such as a GIS. Also called a *digitizing table*. (2) A person who digitizes.

DIME

See **geographic base file/dual independent map encoding**.

Dirichlet tesselation

See **Thiessen polygons**.

discrete data

Categorical data such as types of vegetation or class data such as speed zones. In geographic terms, discrete data can be represented by polygons. Sometimes referred to as integer data. *See also* **continuous data**.

distributed processing

Where computer resources are dispersed or distributed in one or more locations. The individual computers in a distributed processing environment can be linked by a communications network to each other and/or to a host or supervisory computer.

DLG

See **digital line graph**.

dots per inch (dpi)
Often referred to in printing/plotting processes and relates to how sharply an image may be represented. More dots per inch implies that edges of images are more precisely represented.

double-precision
Refers to a level of coordinate accuracy based on the possible number of significant digits that can be stored for each coordinate. Whereas single-precision coverages can store up to 7 significant digits for each coordinate and thus retain a precision of 1 m in an extent of one million m, double-precision coverages can store up to 15 significant digits per coordinate (typically 13 to 14 significant digits) and therefore retain the accuracy of much less than 1 m at a global extent.[2]

dpi
See **dots per inch**.

DTM
See **digital terrain model**.

DXF
See **digital exchange format**.

eastings
The *x*-coordinates in a plane-coordinate system. *See* also **northings**.

edge match
An editing procedure to ensure that all features crossing adjacent map sheets have the same edge locations, attribute descriptions, and feature classes.

feature
A representation of a geographic entity, such as a point, line, or polygon.

file
A single set of related information in a computer that can be accessed by a unique name (e.g., a text file created with a text editor, a data file, or a DLG file). Files are the logical units managed on disk by the computer's operating system. Files may be stored on tapes or disks.[2]

flat file
A structure for storing data in a computer system in which each record in the file has the same data items or fields. Usually, one field is designated as a "key" that is used by computer programs for locating a particular record or set of records or for sorting the entire file in a particular order.[3]

font
A logical set of related patterns representing text characters or point symbology (e.g., A,B,C). A font pattern is the basic building block for markers and text symbols.[2]

foreign key

In relational database management system (RDBMS) terms, the item or column of data that is used to relate one file to another.

format

(1) The pattern in which data are systematically arranged for use on a computer. (2) A file format is the specific design of how information is organized in the file. For example, DLG, DEM, and TIGER are geographic data sets in particular formats available for many parts of the United States.

Fourier analysis

A method of dissociating time series or spatial data into sets of sine and cosine waves.[1]

fractal

An object having a fractional dimension; one that has variation which is self-similar at all scales, in which the final level of detail is never reached and never can be reached by increasing the scale at which observations are made.[1]

gap

The distance between two objects that should be connected. Often occurs during the digitizing process or in the edge-matching process.

GBF/DIME

See **geographic base file/dual independent map encoding**.

generalize

(1) Reduce the number of points, or vertices, used to represent a line. (2) increase the cell size and resample data in a raster format GIS.

geocode

The process of identifying a location as one or more x,y coordinates from another location description such as an address. For example, an address for a student can be matched against a TIGER street network to locate the student's home.[2]

geodata set

A collective term for all geographically located data structured in raster, arc-node, polygon, triangulated irregular network, and other formats.

geographic base file/dual independent map encoding (GBF/DIME)

A data exchange format developed by the U.S. Census Bureau to convey information about blockface/street address ranges related to 1980 census tracts. These files provide a schematic map of a city's streets, address ranges, and geostatistical codes relating to the Census Bureau's tabular statistical data. *See also*

topologically integrated geographic encoding and referencing (TIGER) data created for the 1990 census.

geographic data
The composite locations and descriptions of geographic entities.

geographic database
Efficiently stored and organized spatial data and possibly related descriptive data.

geographic information retrieval and analysis (GIRAS)
Data files from the U.S. Geological Survey (USGS). GIRAS files contain information for areas in the continental United States, including attributes for land use, land cover, political units, hydrologic units, census and county subdivisions, and federal and state landownerships. These data sets are available to the public in both analog and digital form.

geographic information system (GIS)
An organized collection of computer hardware, software, geographic data, and personnel designed to efficiently capture, store, update, manipulate, analyze, and display all forms of geographically referenced information. Certain complex spatial operations are possible with a GIS that would be very difficult, time-consuming, or impractical otherwise.[2]

geographic object
A user-defined geographic phenomenon that can be modeled or represented using geographic data sets. Examples of geographic objects include streets, sewer lines, manhole covers, accidents, lot lines, and parcels.[2]

geographical resource analysis support system (GRASS)
(1) A public-domain raster GIS modeling product of the U.S. Army Corp of Engineer's Construction Engineering Research Laboratory (CERL). (2) A raster data format that can be used as an exchange format between two GISs.

georeference
To establish the relationship between page coordinates on a paper map or manuscript and known real-world coordinates.[2]

GIRAS
See **geographic information retrieval and analysis**.

GIS
See **geographic information system**.

graduated circle
A circular symbol whose area, or some other dimension, represents a quantity.

graphical user interface (GUI)

A graphical method used to control how a user interacts with a computer to perform various tasks. Instead of issuing commands at a prompt, the user is presented with a "dashboard" of graphical buttons and other functions in the form of icons and objects on the display screen. The user interacts with the system using a mouse to point-and-click. For example, press an icon button and the function is performed. Other GUI tools are more dynamic and involve things like moving an object on the screen, which invokes a function. For example, a slider bar is moved back and forth to determine a value associated with a parameter of a particular operation (e.g., setting the scale of a map).[2]

GRASS

See **geographical resource analysis support system**.

graticule

A grid of parallels and meridians on a map.[4]

grid data

(1) One of many data structures commonly used to represent geographic entities. A raster-based data structure composed of square cells of equal size arranged in columns and rows. The value of each cell, or group of cells, represents the entity value. (2) A set of regularly spaced reference lines on the earth's surface, a display screen, a map, or any other object. (3) A distribution system for electricity and telephones.

GUI

See **graphical user interface**.

hardware

Components of a computer system, such as the CPU, terminals, plotters, digitizers, or printers.

hierarchical

This type of data storage refers to data linked together in a treelike fashion, similar to the concept of family lines, where data relations can be traced through particular arms of the hierarchy. These data are dependent on the data structure.

hierarchy

Refers to information that has order and priority.

IGES

See **initial graphics exchange specification**.

image

A graphic representation or description of an object that is typically produced by an optical or electronic device. Common examples include remotely sensed data such as satellite data, scanned data, and photographs. An image is stored

as a raster data set of binary or integer values representing the intensity of reflected light, heat, or another range of values on the electromagnetic spectrum.
Remotely sensed images are digital representations of the earth.[2]

impedance
The amount of resistance (or cost) required to traverse through a portion of
a network such as a line, or through one cell in a grid system. Resistance may
be any number of factors defined by the user such as travel distance, time, speed
of travel times the length, slope, or cost.

index
A specialized lookup table or structure within a database which is used by
an RDBMS or GIS to speed searches for tabular or geographic data.

infrastructure
The fabric of human improvements to natural settings that permits a community, neighborhood, town, city, metropolis, region, or state to function.

initial graphics exchange specification (IGES)
An interim standard format for exchanging graphics data among computer
systems.[1]

integer
A number without a decimal. Integer values can be less than, equal to, or
greater than zero.

integrated terrain unit mapping (ITUM)
The process of adjusting terrain unit boundaries so there is increased coincidence between the boundaries of interdependent terrain variables such as hydrography, geology, physiography, soils, and vegetation units. Often, when this
is performed, one layer or unit of geographical/descriptive information contains
more than one central theme.

intelligent infrastructure
The result of automating infrastructure information management using modern computer image and graphics technology integrated with advanced database
management systems (DBMSs); used for spatially linked and networked facilities and land records systems. In addition, intelligent infrastructure systems manage work processes that deal with design, construction, operation, and maintenance of infrastructure elements.[5]

item
A field or column of information within an RDBMS.

ITUM
See **integrated terrain unit mapping**.

jaggies

A jargon term for curved lines that have a stepped or saw-tooth appearance on a display device.[1]

join

To connect two or more separate geographic data sets.

key

An item or column within an RDBMS that contains a unique value for each record in the database.

kriging

An interpolation technique based on the premise that spatial variation continues with the same pattern.

LAN

See **local area network**.

lat/long

See **latitude; longitude**

latitude

A method to measure the earth representing angles of a line extending from the center of the earth to the earth's surface. With the equator representing 0 degrees, angles are measured in degrees north or south until 90 degrees is obtained at the north and south poles. Lines of latitude are often called *parallels*.

layer

A logical set of thematic data, usually organized by subject matter.

library

A collection of repeatedly used items such as a symbol library; often-used graphics objects shown on a map or often-used program subroutines.[5]

line

(1) A set of ordered coordinates that represents the shape of a geographic entity too narrow to be displayed as an area (e.g., contours, street centerlines, and streams). A line begins and ends with a node. (2) A line on a map (e.g., a neatline).

local area network (LAN)

Computer data communications technology that connects computers at the same site. Computers and terminals on a LAN can freely share data and peripheral devices, such as printers and plotters. LANs are composed of cabling and special data communications hardware and software.[2]

longitude

A method to measure the earth representing angles of a line extending from the center of the earth to the earth's surface. With a line extending from the north to the south poles and passing through Greenwich, England, as 0 degrees, angles are measured in degrees east or west until 180 degrees is obtained at the opposite side of the earth from 0 degrees longitude. Lines of longitude are often called *meridians*.

macro

A set of instructions used by a computer program or programs. These are usually stored in a text file and invoked from a program that reads this text file as if the commands were typed interactively.

many-to-one relate

A relate in which many records can be related to a single record. A typical goal in relational database design is to use many-to-one relates to reduce data storage and redundancy.[2]

map projection

A mathematical model for converting locations on the earth's surface from spherical to planar coordinates, allowing flat maps to depict three-dimensional features. Some map projections preserve the integrity of shape; others preserve accuracy of area, distance, or direction.[2]

map units

The coordinate units in which the geographic data are stored, such as inches, feet, meters, or degrees, minutes, and seconds.

meridian

A line running vertically from the north pole to the south pole along which all locations have the same longitude. The prime meridian (0°) runs through Greenwich, England. Moving left or right of the prime meridian, measures of longitude are negative to the west and positive to the east up to 180° halfway around the globe.[2]

metropolitan statistical area (MSA)

A single county or group of contiguous counties that define a metropolitan region, usually with a central city with at least 50,000 inhabitants; in the past these have been called standard metropolitan statistical areas (SMSAs) and standard metropolitan areas (SMAs); the precise definitions and changes therein are established by the U.S. Office of Management and Budget.[4]

minimum bounding rectangle

The rectangle defined by the map extent of a geographic data set and specified by two coordinates: xmin, ymin and xmax, ymax.[2]

minor civil division (MSD)

The primary political or administrative subdivision of a county.[4]

model

(1) An abstraction of reality. Models can include a combination of logical expressions, mathematical equations, and criteria that are applied for the purpose of simulating a process, predicting an outcome, or characterizing a phenomenon. The terms *modeling* and *analysis* are often used interchangeably, although the former is more limited in scope. (2) Data representation of reality (e.g., spatial data models include the arc-node, georelational model, rasters or grids, and TINs).

neatline

A border line commonly drawn around the extent of a map to enclose the map, legend, scale, title, and other information, keeping all of the information pertaining to that map in one "neat" box.

network

(1) A system of interconnected elements through which resources can be passed or transmitted (e.g., a street network with cars as the resource or electric network with power as the resource.) (2) In computer operations, the means by which computers connect and communicate with each other or with peripherals.

network analysis

The technique utilized in calculating and determining relationships and locations arranged in networks, such as in transportation, water, and electrical distribution facilities.[5]

node

(1) The beginning or ending location of a line. (2) The location where lines connect. (3) In graph theory, the location at which three or more lines connect. (4) In computers, the point at which one computer attaches to a communication network.

northings

The *y*-coordinates in a plane-coordinate system. *See also* **eastings**.[4]

operating system (OS)

Computer software designed to allow communication between the computer and the user. For larger computers, it is usually supplied by the manufacturer. The operating system controls the flow of data, the interpretation of other programs, the organization and management of files, and the display of information. Commonly known OSs are VMS, VM/IS, UNIX, DOS, and OS/2.

OS

See **operating system**.

output

The results of processing data.

overshoot

That portion of a line digitized past its intersection with another line. Sometimes referred to as a dangling line.

pan

To move the spatial view of data to a different extent without changing the scale.

parallel

(1) A property of two or more lines that are separated at all points by the same distance. (2) A horizontal line encircling the earth at a constant latitude. The equator is a parallel whose latitude is 0°. For example, measures of latitude are positive up to 90° above the equator and negative below.[2]

pathname

The direction(s) to a file or directory location on a disk. Pathnames are always specific to the computer operating system. Computer operating systems use directories and files to organize data. Directories are organized in a tree structure; each branch on the tree represents a subdirectory or file. Pathnames indicate locations in this hierarchy.[2]

peripheral

A component such as a digitizer, plotter, or printer that is not part of the central computer but is attached through communication cables.

pixel

One picture element of a uniform raster or grid file. Often used synonymously with **cell**.

plane-coordinate system

A system for determining location in which two groups of straight lines intersect at right angles and have as a point of origin a selected perpendicular intersection.[4]

planimetric map

A large-scale map with all features projected perpendicularly onto a horizontal datum plane so that horizontal distances can be measured on the map with accuracy.[4]

PLSS

See **public land survey system**.

point

(1) A single x,y coordinate that represents a geographic feature too small to be displayed as a line or area (e.g., the location of a mountain peak or a building location on a small-scale map.)[2] (2) Some GIS systems also use a point to identify the interior of a polygon.

polygon

A vector representation of an enclosed region, described by a sequential list of vertices or mathematical functions.

precision

(1) If applied to paper maps or map databases, it means exactness and accuracy of definition and correctness of arrangement.[5] (2) If applied to data collection devices such as digitizers, it is the exactness of the determined value (i.e., the number 134.98988 is more precise than the number 134.9). (3) The number of significant digits used to store numbers.

primary key

The central item or column within an RDBMS that contains a unique value for each record in the database, such as the unique number assigned to each parcel within a county.

projection

See **map projection.**

public land survey system (PLSS)

A rectangular survey system that utilizes six sq. mile townships as its basic survey unit. The location of townships is controlled by baselines and meridians running parallel to latitude and longitude lines. Townships are defined by range lines running parallel (north to south) to meridians and township lines running parallel (east to west) to baselines. The PLSS was established in the United States by the Land Ordinance of 1785.[3]

quadrangle

A four-sided region, usually bounded by a pair of meridians and a pair of parallels.[4]

quadtree

A spatial index that breaks a spatial data set into homogeneous cells of regularly decreasing size. Each decrement in size is one-fourth the area of the previous cell. The quadtree segmentation process continues until the entire map is partitioned. Quadtrees are often used for storing raster data.

raster data

Machine-readable data that represent values usually stored for maps or images and organized sequentially by rows and columns. Each "cell" must be rectangular, but not necessarily square, as with grid data.

RDBMS
See **relational database management system.**

record
In an attribute table, a single "row" of thematic descriptors.

rectify
The process by which an image or grid is converted from image coordinates to real-world coordinates. Rectification typically involves rotation and scaling of grid cells and thus requires resampling of values.[2]

relate
An operation establishing a connection between corresponding records in two **relations** using an item common to both. Each record in one table is connected to one or more records in the other table that share the same value for a common item.

relation
A type of data storage involving tabular data where the storage structure conforms to the following six rules: (1) **columns** are uniquely named; (2) data in each **row** are unique; (3) there is no implied column order; (4) there is no row order; (5) each column of each row contains a single value; and (6) entries in each column are of the same kind.

relational database management system (RDBMS)
A database management system with the ability to access data organized in **relations** that may be related together by a common field (item). An RDBMS has the capability to recombine the data items from different files, thus providing powerful tools for data usage.[2]

relational join
The process of connecting two **relations** of descriptive data by relating them by a key item, then merging the corresponding data. The common key item is not duplicated in this process.

resolution
1. The accuracy at which the location and shape of map **features** can be depicted for a given map scale. For example, at a map scale of 1:63,360 (1 in. = 1 mile), it is difficult to represent areas smaller than 1/10-mile wide or 1/10-mile in length because they are only 1/10-in. wide or long on the map. In a larger scale map, there is less apparent reduction, so feature resolution more closely matches real-world features. As map scale decreases, resolution also diminishes because features must be smoothed, simplified, or not shown at all. (2) The size of the smallest feature that can be represented in a surface. (3) The number of points in x and y in a **grid** (e.g., the resolution of a USGS one-degree DEM is 1201×1201 mesh points).[2]

route

A process that establishes connections through a **network** or **grid** from a source to a destination. A network example would be to establish a route through a network of streets from a fire station to the fire. A grid example would be to move soil particles from a ridgetop to a stream based on equations developed by soil scientists. The determination of both these routes usually takes into consideration **impedances**.

row

(1) A record in an attribute table. (2) A horizontal group of cells in a grid or pixels in an image.

rubber-sheeting

A procedure to adjust the entities of a geographic data set in a nonuniform manner. From- and to-coordinates are used to define the adjustment.

scale

The relationship existing between a distance on a map and the corresponding distance on the earth. Often used in a 1:24,000 form, which means that 1 unit of measurement on the map equals 24,000 of the same units on the earth's surface.

scanning

Also referred to as *automated digitizing* or *scan digitizing*. A process by which information originally in hard copy format (paper print, mylar™ transparencies, microfilm aperture cards) can be rapidly converted to digital raster form (pixels) using optical readers.[5]

single-precision

A lower level of coordinate significance based on the possible number of digits that can be stored for each coordinate. Single-precision numbers can store up to seven significant digits for each coordinate and thus retain a precision of ±5 m in an extent of 1,000,000 m. **Double-precision** numbers can store up to 15 significant digits (typically 13 to 14 significant digits) and therefore retain the significance of much less than 1 m at a global extent.[2]

sliver polygon

A relatively narrow **feature** commonly occurring along the borders of polygons following the overlay of two or more geographic data sets. These sliver polygons also occur along map borders when two maps are overlayed. It is a result of inaccuracies of the coordinates of coincident lines in either or both maps.

smoothing

A process to generalize data and remove variation.

software
   A computer program that provides the instructions necessary for the hardware to operate correctly and to perform the desired functions. Some kinds of software are operating system, utility, and applications.[5]

soundex
   A phonetic spelling (up to six characters) of a street name, used for address matching. Each of the 26 letters in the English alphabet are replaced with a letter in the soundex equivalent:

   English: A B C D E F G H I J K L M N O P Q R S T U V W X Y Z
   Soundex: A B C D A B C H A C C L M M A B C R C D A B W C A C

Where possible, geocoding uses a soundex equivalent of street names for faster processing. During geocoding, initial candidate street names are found using soundex, then real names are compared and verified.[2]

spatial index
   A means of accelerating the drawing, spatial selection, and entity identification by generating geographic-based indexes based on an internal sequential numbering system.

spatial model
   Analytical procedures applied with a GIS. There are three categories of spatial modeling functions that can be applied to geographic data objects within a GIS: (1) geometric models (such as calculation of Euclidian distance between objects, buffer generation, and area and perimeter calculation); (2) coincidence models (such as polygon overlay); and (3) adjacency models (pathfinding, redistricting, and allocation). All three model categories support operations on geographic data objects such as **points**, **lines**, **polygons**, **TINs**, and **grids**. Functions are organized in a sequence of steps to derive the desired information for analysis.[2]

spike
   (1) An overshoot line created erroneously by a scanner and its **raster** software. (2) An anomalous data point that protrudes above or below an interpolated surface representing the distribution of the value of an attribute over an area.[2]

spline
   A method to mathematically smooth spatial variation by adding vertices along a line. *See also* **densify** for a slightly different method for adding vertices.

spot value
   See **z-value**.

spot elevation
   See **z-value**.

SQL
   See **structured query language**.

string
   See **line**.

structured query language (SQL)
   A syntax for defining and manipulating data from a relational database. Developed by IBM in the 1970s, it has since become an industry standard for query languages in most **RDBMSs**.[2]

surface
   A representation of geographic information as a set of continuous data in which the map features are not spatially discrete; that is, there is an infinite set of values between any two locations. There are no clear or well-defined breaks between possible values of the geographic feature. Surfaces can be represented by models built from regularly or irregularly spaced sample points on the surface.[2]

surface model
   Digital abstraction or approximation of a **surface**. Because a surface contains an infinite number of points, some subset of points must be used to represent the surface. Each model contains a formalized data structure, rules, and x, y, z point measurements that can be used to represent a surface.[2]

syntax
   A set of rules governing the way statements can be used in a computer language.[1]

table
   (1) Usually referred to as a **relation** or *relational table*. The data file in which the relational data reside. (2) A file that contains ASCII or other data.

template
   (1) A geographic data set containing boundaries, such as land-water boundaries, for use as a starting place in automating other geographic data sets. Templates save time and increase the precision of spatial overlays. (2) A map containing neatlines, north arrow, logos, and similar map elements for a common map series but lacking the central information that makes one map unique from another. (3) An empty tabular data file containing only item definitions.

thematic map
   A map that illustrates one subject or topic either quantitatively or qualitatively.[4]

theme

A collection of logically organized geographic objects defined by the user. Examples include streets, wells, soils, and streams.

Thiessen polygons

Polygons whose boundaries define the area that is closest to each point relative to all other points. Thiessen polygons are generated from a set of points. They are mathematically defined by the perpendicular bisectors of the lines between all points. A triangulated irregular network (TIN) structure is used to create Thiessen polygons.[2]

TIGER

See **topologically integrated geographic encoding and referencing data**.

tile

A part of the database in a GIS representing a discrete part of the earth's surface. By splitting a study area into tiles, considerable savings in access times and improvements in system performance can be achieved.[1]

TIN

See **triangulated irregular network**.

topographic map

A map of land-source features, including drainage lines, roads, landmarks, and usually relief or elevation.[4]

topologically integrated geographic encoding and referencing (TIGER) data

A format used by the U.S. Census Bureau for the 1990 census to support census programs and surveys. TIGER files contain street address ranges along lines and census tract/block boundaries. These descriptive data can be used to associate address information and census/demographic data to coverage features.[2]

topology

(1) The spatial relationships between connecting or adjacent **features**, such as **arcs, nodes, polygons**, and **points**. For example, the topology of an arc includes its from- and to-nodes, and its left and right polygons. Topological relationships are built from simple elements into complex elements: points (simplest elements); arcs (sets of connected points); areas (sets of connected arcs); and routes (sets of sections that are arcs or portions of arcs). Redundant data (coordinates) are eliminated because an arc may represent a linear feature, part of the boundary of an area feature, or both. Topology is useful in GIS because many spatial modeling operations do not require coordinates, only topological information. For example, to find an optimal path between two points requires a list of which arcs connect to each other and the cost of traversing along each arc in each direction. Coordinates are only necessary to draw the path after it is cal-

culated.[2] (2) The relationship between connecting componenets of an **enterprise GIS**: servers, PCs, workstations, terminals, and networks.

transformation

The process of converting data from one coordinate system to another through translation, rotation, and scaling.

triangulated irregular network (TIN)

A representation of a surface derived from irregularly spaced sample points and breakline features. The TIN data set includes topological relationships between points and their proximal triangles. Each sample point has an x,y coordinate and a surface or z value. These points are connected by edges to form a set of nonoverlapping triangles that can be used to represent the surface. TINs are also called *irregular triangular mesh* or *irregular triangular surface model.*[2]

tuple

Synonym of **record**.

unbundled

Refers to software sold separately from hardware. *See also* **bundled**.

undershoot

A **digitized** line that does not quite reach a line that it should intersect. As with an overshoot, this is also sometimes referred to as a **dangling arc**.

universal transverse Mercator (UTM)

A widely used planar coordinate system, extending from 84° north to 80° south **latitude** and based on a specialized application of the transverse Mercator projection. The extent of the coordinate system is broken into 60, 6-degree (**longitude**) zones. Within each zone, coordinates are usually expressed as meters north or south of the equator and east from a reference axis. For locations in the Northern Hemisphere, the origin is assigned a false **easting** of 500,000 and a false **northing** of 0. For locations in the Southern Hemisphere, the origin is assigned a false easting of 500,000 and a false northing of 10,000,000.

UTM

See **universal transverse Mercator**.

vector data

A coordinate-based data structure commonly used to represent map **features**. Each linear feature is represented as a list of ordered x,y coordinates. **Attributes** may be associated with the feature (as opposed to a raster data structure, which associates attributes with a grid cell). Traditional vector data structures include double-digitized polygons and **arc-node** models.[2]

vertex

One point along a line.

z-value

The elevation value of a surface at a particular x,y location. Also, often referred to as a **spot value** or **spot elevation**.[2]

zoom

To display a smaller or larger region instead the present spatial data set extent in order to show greater or lesser detail.

REFERENCES

1. Burrough, P. A., *Principles of Geographical Information Systems for Land Resources Assessment*, Butler & Tanner, Somerset, UK, 1990.
2. Environmental Systems Research Institute, Inc., *ARC/INFO®* Data Model, Concepts, and Key Terms*, ESRI, Inc., Redlands, CA, 1991.
3. Huxhold, W., An *Introduction to Urban Geographic Information Systems*, Oxford University Press, New York, 1991.
4. Monmonier, M. and Schnell, G. A., *Map Appreciation*, Prentice Hall, Englewood Cliffs, NJ, 1988.
5. Montgomery, G. and Juhl, G., *Intelligent Infrastructure Workbook*, A-E-C Automation Newsletter, Fountain Hills, AZ, 1990.

* Registered Trademark of Environmental Systems Research Institute, Redlands, CA.

Index

359